HYDRAULICS FOR FIREFIGHTING

Online Services

Delmar Online
For the latest information on Delmar's new series of Fire, Rescue and Emergency Response products, point your browser to:
http://www.firescience.com

Online Services

Delmar Online
To access a wide variety of Delmar products and services on the World Wide Web, point your browser to:
http://www.delmar.com
or email: info@delmar.com

A Division of Thomson Learning™

HYDRAULICS FOR FIREFIGHTING

William F. Crapo

DELMAR
THOMSON LEARNING™

Australia Canada Mexico Singapore Spain United Kingdom United States

Hydraulics for Firefighting
William F. Crapo

Business Unit Director:
Alar Elken

Acquisitions Editor:
Mark W. Huth

Development:
Dawn Daugherty

Executive Marketing Manager:
Maura Theriault

Channel Manager:
Mona Caron

Executive Production Manager:
Mary Ellen Black

Production Editor:
Barbara L. Diaz

Art/Design Coordinator:
Rachel Baker

Cover Image:
Clark O. Martin, Jr.

Printed in Canada
1 2 3 4 5 XXX 05 04 03 02 01 00

For more information contact Delmar,
3 Columbia Circle, PO Box 15015,
Albany, NY 12212-5015.

Or find us on the World Wide Web at
http://www.delmar.com

Library of Congress Cataloging-in-Publication Data

Crapo, William F.
Hydraulics for firefighting / William F. Crapo
p. cm
Includes bibliographical references and index.
ISBN 0-7668-1905-1 (alk. paper)
1. Fire extinction—Water supply. 2. Fire streams.
3. Hydraulics. I. Title.
TH9311.C73 2001
628.9'252—dc21

2001023109

NOTICE TO THE READER

Publisher does not warrant or guarantee any of the products described herein or perform any independent analysis in connection with any of the product information contained herein. Publisher does not assume, and expressly disclaims, any obligation to obtain and include information other than that provided to it by the manufacturer.

The reader is expressly warned to consider and adopt all safety precautions that might be indicated by the activities described herein and to avoid all potential hazards. By following the instructions contained herein, the reader willingly assumes all risks in connection with such instructions.

The Publisher makes no representation or warranties of any kind, including but not limited to, the warranties of fitness for particular purpose or merchantability, nor are any such representations implied with respect to the material set forth herein, and the publisher takes no responsibility with respect to such material. The publisher shall not be liable for any special, consequential, or exemplary damages resulting, in whole or part, from the readers' use of, or reliance upon, this material.

This book is dedicated to the memory of Brenna and Sunny.

Contents

Preface

It has been said that the person who knows *how* will always have a job, but the person who knows *why* will be his boss. This book is written for the person who aspires to be, or is already, an officer, or just wants to have a thorough understanding of hydraulics.

The goal of this book is to provide that in-depth knowledge of hydraulics. Hydraulics is more than just knowing how to find a pump pressure that works or knowing what an impeller is. Having a true understanding of hydraulics means that you understand the appropriate laws of physics and some chemistry. It also means that you be able to apply the correct formula to find an answer, regardless of what variables you are given. After all, anyone can find the answers to the common problems; only through an in-depth knowledge of the subject can you find the answers to unusual problems.

Although this book is intended for the community college fire science curriculum, it should be easily understood by anyone interested in learning hydraulics, from rookie firefighter to seasoned officer. In writing this book I have attempted to meet the requirements of both NFPA 1001, *Fire Fighter Professional Qualifications*, and NFPA 1002, *Fire Apparatus Driver/Operator Professional Qualifications.*

ORGANIZATION

This book has been organized in a sequence that builds on itself. The sequence of chapters is intended to be a logical progression that makes the learning process obvious. Throughout the book, various applicable laws of physics are mentioned to illustrate that hydraulics is part of a larger discipline called *physics*.

Chapter 1 defines a base for studying the chapters that follow. It sets down the facts and details about water. Without this information, you will never be able to understand the remainder of this book.

Chapters 2 through 6 concentrate on teaching the principles necessary to the study of hydraulics and provide most of the formulas. These formulas provide the basis of calculating engine pressure in chapters 10 and 11. More than just presenting the problems, chapters 2 through 6 also put the formulas in perspective. Where appropriate, the origin of the formula is illustrated and its relationship to fire service hydraulics proved.

Chapters 7 and 8 are dedicated to understanding the pump. They are intended to demystify the operating principles of the pump at a level few people, other than engineers, commonly study, but in terms any firefighter should easily grasp.

Chapter 9 discusses several principles and formulas not accounted for in the prior chapters. In a sense, this chapter ties up some loose ends before getting into calculating engine pressure.

Chapters 10 and 11 use the knowledge accumulated in the prior nine chapters to solve pump problems. New principles needed to solve problems are also introduced: principles such as calculating engine pressure for parallel lines, wyed lines and siamesed lines.

Chapter 12 introduces the student to both basic and advanced issues dealing with water supply. The basics of water supply are studied as well as a method for calculating how much water is needed to fight fires. Additionally, the section on how to get the most from the water supply puts real life bias on a subject not well understood.

Chapter 13 is intended to give the firefighting personnel insight into testing of sprinkler and standpipe systems. Here again, we generally accept what the engineers give us, but at some point we need to have the knowledge to verify that these systems really work. Finally, the subject of fireground formulas is raised with the intent of steering the serious student of hydraulics away from fancy guesstimations and toward more accurate calculations.

It may seem like a major chapter has been omitted, one that contains fireground engine pressure calculations. They are here, however, in chapters 10 and 11. The basic engine pressure formula used throughout both of these chapters is easily adapted to fireground use.

WHY THIS BOOK WAS WRITTEN

In my many years associated with the fire service I have not found a single source that contained information I considered appropriate to a complete understanding of hydraulics for the person riding the fire engines. Some books are too detailed, some too basic. An experienced fire officer has written this text with the intention of providing the necessary understanding of hydraulics in terms that can be understood by all. This text enables the newest firefighter to understand the laws, principles, and formulas involved, but without sacrificing the technical accuracy or detail demanded by the needs of today's fire service. It is intended to give the firefighter a thorough understanding of the laws of hydraulics, familiarity with the many formulas that are a part of knowing hydraulics, and a working knowledge of pumps, all at a level befitting the professional status of firefighting personnel, either volunteer or career.

Another consideration in writing this book is an update of many of the common formulas. In the past, many books have been written with the assumption that a cubic foot of water weighs 62.5 pounds. The correct weight of water was actually known, but it was easier to round it off to 62.5 when manually calculating formulas. In reality, a cubic foot of water is more correctly 62.4 pounds, the exact weight depending on what source you use. This issue may seem insignificant, but it has a ripple effect through many of the formulas necessary to hydraulics. While realistically any differences may be minor, it is important to be as accurate as possible, thus the formulas have been changed to reflect the weight of a cubic foot of water as 62.4 pounds. Another reason to use the more accurate weight in this book is that today we have the benefit of inexpensive scientific calculators that can easily handle the numbers.

MATH AND ROUNDING

In many respects this is a math book. Every chapter contains from one to more than a dozen formulas. In order to be able to work these formulas, it is necessary that the student of hydraulics be able to perform basic algebra and have a working knowledge of basic plain and solid geometry. Some of the formulas in this book may look intimidating, but they all are solvable with basic algebra and geometry. Even the intimidating looking Hazen-Williams formula in Chapter 12 is solvable with basic algebra. Anyone unsure about either basic algebra or geometry should take the time to review the subjects. Appendix A has been included to provide formulas for finding area and volume for the most common geometric shapes.

In most instances where calculations render more than two figures to the right of the decimal, these numbers are rounded to two significant figures. Some numbers have traditionally been used with three or four places to the right of the decimal; in instances where those figures are used, they will be shown in traditional format. In all instances, answers are rounded to just two decimal places. Where rounding to whole numbers is appropriate, it is noted in the appropriate section.

Many of the formulas used in this text are not used every day. They are, however, needed for a thorough understanding of the subject. Without being able to use the formulas included in this text, it is impossible to master hydraulics.

ACKNOWLEDGEMENTS

Writing a book is far from an individual task. Many people beyond those directly connected with reviewing and publishing this book are deserving of mention.

First, I would like to thank my wife, Karen, for her understanding during the long hours of researching and writing. She has been supportive throughout every endeavor I have undertaken since I met her. I would also like to thank my good friend, Bill Whelan, for his review and input. I asked Bill to perform an independent review due to my respect for his technical knowledge and his ability to uncover technical glitches. He also brings an unbiased perspective to the review process because he is not associated with the fire service.

Others deserving of recognition include Dr. Dave Thomas, senior fire protection engineer for the Fairfax County, Virginia, Department of Fire and Rescue. Dave made many suggestions and comments over the course of working on this book that have proved beneficial from both a technical and practical perspective. Thanks to Dr. Douglas Giacoli who supplied me with an updated description of how a radial flow pump imparts energy to the water. Special thanks go to my friend, Captain Robert Lynch, also of the Fairfax County, Virginia, Department of Fire and Rescue for his review of Chapter 13.

I would be remiss if I did not mention the input of the reviewers assembled by Delmar to review this book. Their input keeps works like this grounded in reality while meeting the needs of the service.

Delmar and the author would like to thank the following reviewers for their valuable suggestions and comments:

Gary Coley
Fox Valley Technical College
Neenah, WI

Timothy Flannery
John Jay College
New York City, NY

Mike McKenna
American River College
Antelope, CA

Tommy Abercrombie
Tarrant County College
Ft. Worth, TX

David Hauger
Westmoreland County Community College
New Stanton, PA

Dr. Terry Heyns
Lake Superior State University
Sault Ste. Marie, MI

Clark Custodio
Mission College
Santa Clara, CA

Last, but not least, I want to thank the people at Delmar who thought enough of the idea for this book to let me give it a try, and to their patience and dedication to achieving the best possible end result. Hopefully through the collective work of everyone involved, this book will become an integral part of fire service training well into the future.

Introduction to Hydraulics

Learning Objectives

Upon completion of this chapter, you should be able to:

- Define hydraulics and explain its origin as a science and its history within the fire service.
- Understand the chemical and physical properties of water.
- Understand the importance, as the mechanisms of extinguishment, of both the latent heat and specific heat properties of water.
- Calculate the volume of water needed to extinguish a fire by both BTUs absorbed and steam generated.
- Calculate the volume and weight of water in common-shaped containers.

INTRODUCTION

The study of the science of hydraulics as we use it in the fire service today began more than 250 years ago (Figure 1-1). The science began as a study of water flow in relation to water supply, irrigation, river control, and waterpower. Men such as Euler, Bernoulli, and Pascal, to name a few, stated the laws and principles we study today.

hydraulics
the science of water (or other fluids) at rest and in motion

Hydraulics can be defined as the science of water (or other fluids) at rest and in motion. Hydraulics is an applied science as opposed to a theoretical science in that it is the product of experimentation and observation. The laws, principles, and formulas found in this book are the end product of years of experimentation and observation. A theoretical science, conversely, is just that, only theory. For example, time travel is currently a theoretical science. Until someone actually travels backward or forward in time, it is only a theory.

The science of hydraulics originally began as a study of water alone. It was later that the need to study other liquids and gases borrowed principles and formulas from hydraulics. While the principles and formulas developed for hydraulics served well as a beginning, they were, for some purposes, too inexact. A new, more exact science was necessary. The new science, mechanics of fluids, includes hydraulics as a specialized phase and only one of its many subspecialties.

Figure 1-1 *Fire, the Destroyer: Without a thorough knowledge of hydraulics we can never defeat it.*

hydrostatics
the study of water at rest

hydrodynamics
the study of water in motion

The study of hydraulics is divided into two primary areas of study, **hydrostatics** and **hydrodynamics**. Hydrostatics is the study of water at rest. Chapter 2 is devoted entirely to hydrostatics. All the principles in Chapter 2 deal with water at rest, whether in an open container subject only to atmospheric pressure or in a closed, pressurized container. Hydrodynamics, however, is the study of water in motion. Hydrodynamics is more typically what we think of when we think of the fireground application of hydraulics, such as calculating friction loss and gallons-per-minute flow. Calculations associated with velocity and nozzle reactions are also dynamic calculations.

Some of the principles studied as hydrostatics also apply to hydrodynamics. With a little careful study and logic it is easy to identify which ones they are.

HISTORY OF HYDRAULICS IN THE FIRE SERVICE

Although the basic principles applying to hydraulics in the fire service are at least 250 years old, specific application of hydraulic principles, through engineering, to water supply needs began in earnest around 1889. It was then that engineers such as Shedd, Fanning, Freeman, and Knichling started doing serious work on calculating water supply needs for fire protection and associated water distribution systems. In 1910 the National Board of Fire Underwriters (NBFU), known today as the Insurance Services Office (ISO), published its first fire flow recommendations and requirements.

The person most recognized as the champion of the scientific application of hydraulics to the fire service is Fred Shepherd, considered the father of fire service hydraulics. He wrote his first book about hydraulics for the fire service in 1917. The book, *Practical Hydraulics for Firemen*, was written for the average firefighter, not engineers.

WATER

Firefighters all know how important water is to their job. Without it, firefighting as we know it would be impossible. Water has many characteristics and properties that make it an ideal fire suppression agent. It has been said that if water did not exist naturally, the fire service would have had to invent it.

■ Note
Water is a chemical compound of two parts hydrogen and one part oxygen.

To begin with, water is abundant, can be stored easily, and is relatively noncorrosive. Water itself is a chemical compound made up of two parts of elemental hydrogen and one part of elemental oxygen. The universally recognized compound symbol H_2O is derived from water's composition. Water is chemically stable because of its strong hydrogen-oxygen bond. This bond gives water a polar structure, making it a polar molecule. Due to its strong polar structure, water reacts favorably with many compounds, causing them to go into solution. In fact, water reacts adversely with so few chemicals that in inorganic chemistry it is a universal solvent.

Properties of Water

To understand hydraulics it is necessary to have a thorough understanding of the physical properties of water. The basic properties of water are summarized in Table 1-1. The following list explains the most important properties:

1. In addition to its liquid state, water has a solid state (ice) and a vapor state (steam and humidity).
2. Water, like all substances, gets denser as it approaches its freezing point. However, unlike most other substances, water actually gets less dense just before it freezes. In fact, water is densest at 38°F. Because the ice is less dense than the water, it floats, forming an ice layer that insulates the rest of the water. Due to the high specific heat of water, the remainder of the body of water will not freeze.

Note
Water gets less dense just before it freezes.

3. At its densest, a cubic foot of fresh water weighs 62.4 pounds. A cubic foot of saltwater weighs 64 pounds (see Figure 1-2).
4. A cubic foot of water occupies 1,728 cubic inches. One gallon of water occupies 231 cubic inches.

Example 1-1 If there are 1,728 cubic inches of water in a cubic foot and 231 cubic inches in a gallon, how many gallons are in a cubic foot?

ANSWER 1,728 cubic inches ÷ 231 cubic inches = 7.48 gallons in a cubic foot

5. A cubic foot of water contains 7.48 gallons (see Figure 1-3).

Example 1-2 If a cubic foot of water weighs 62.4 pounds, what does a gallon weigh?

ANSWER 62.4 pounds ÷ 7.48 gallons per cubic foot = 8.34 pounds per gallon

Table 1-1 *Basic properties of water.*

	Gallon	Cubic foot
Weight	8.34 pounds	62.4 pounds
Cubic inches	231	1,728
BTU absorbed	9,346	69,908[a]
Cubic feet of steam	213.9	1,600
Gallons	1	7.48
Specific heat = 1 BTU per pound of water		
Latent heat of vaporization = 970.3 BTU per pound of water		

[a]9,346 × 7.48 = 69,908 BTUs

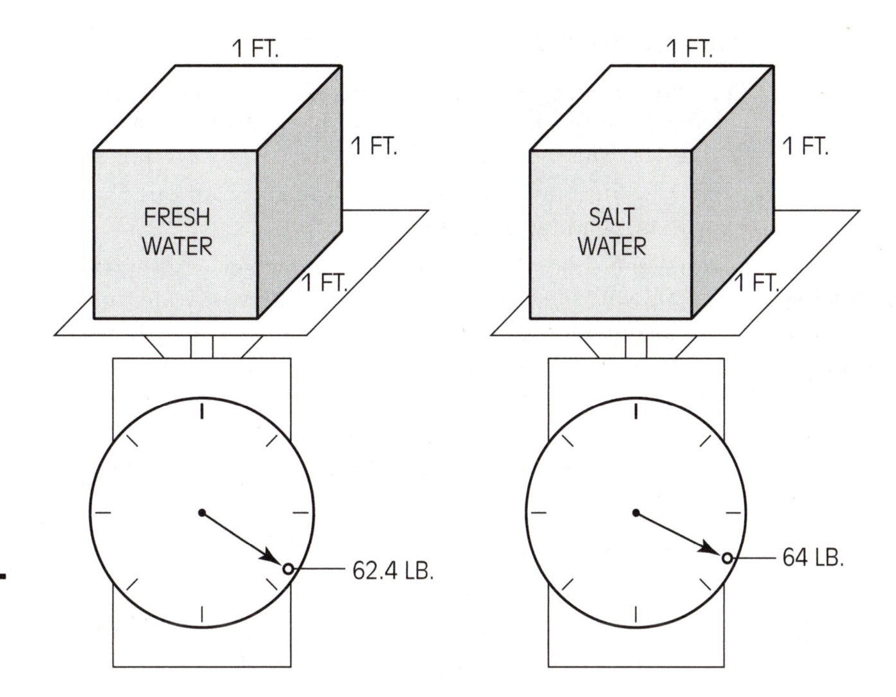

Figure 1-2 *Weight of 1 cubic foot of water.*

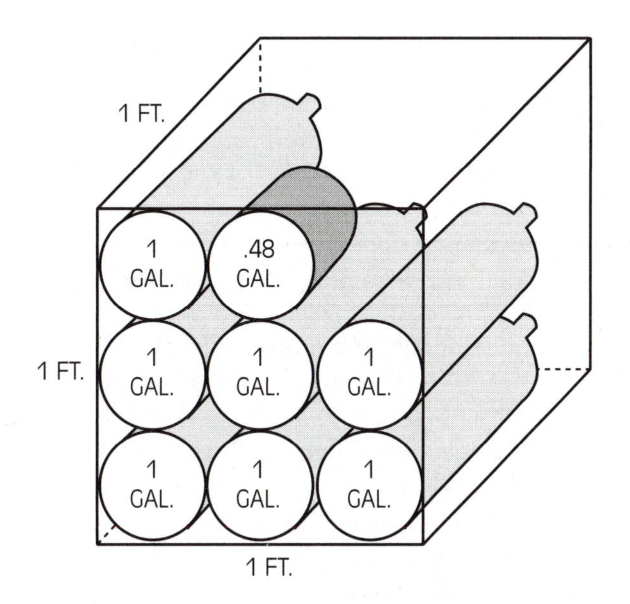

Figure 1-3 *There are 7.48 gallons in 1 cubic foot.*

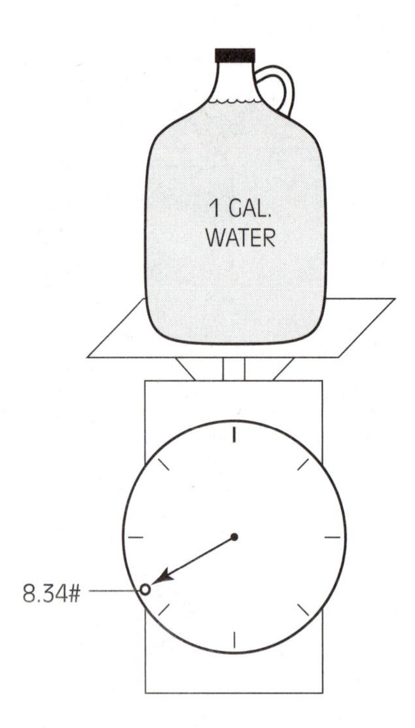

Figure 1-4 *Weight of 1 gallon of water.*

6. A gallon of water weighs 8.34 pounds (see Figure 1-4).
7. The boiling point of water is 212°F.
8. The freezing point of water is 32°F.
9. Triple point (a unique temperature and pressure where the substance exists in all three states at the same time) of water is 32°F at .08 pounds per square inch (psi) pressure.
10. Water is noncompressible.

BTU
the amount of heat needed to raise the temperature of 1 pound of water by 1°F at 60°F

In addition to these general properties of water, there are also some specific properties that are responsible for water's value as an extinguishing agent. In order to understand some of these figures it is necessary to understand what a British thermal unit (BTU) is. A **BTU** is defined as the amount of heat needed to raise the temperature of one pound of water by 1°F at 60°F.

■ Note
The latent heat of vaporization is 970.3 BTUs.

11. The latent heat (explained in the following section) of fusion is 143.4 BTUs per pound of water.
12. The latent heat of vaporization is 970.3 BTUs per pound of water.
13. Water has a specific heat of 1 BTU per pound of water.
14. Water expands at a ratio of 1:1,600 when converted to steam at 212°F (see Figure 1-5). At higher temperatures the expansion ratio is much higher.

■ Note
Water has a specific heat of 1 BTU per pound of water.

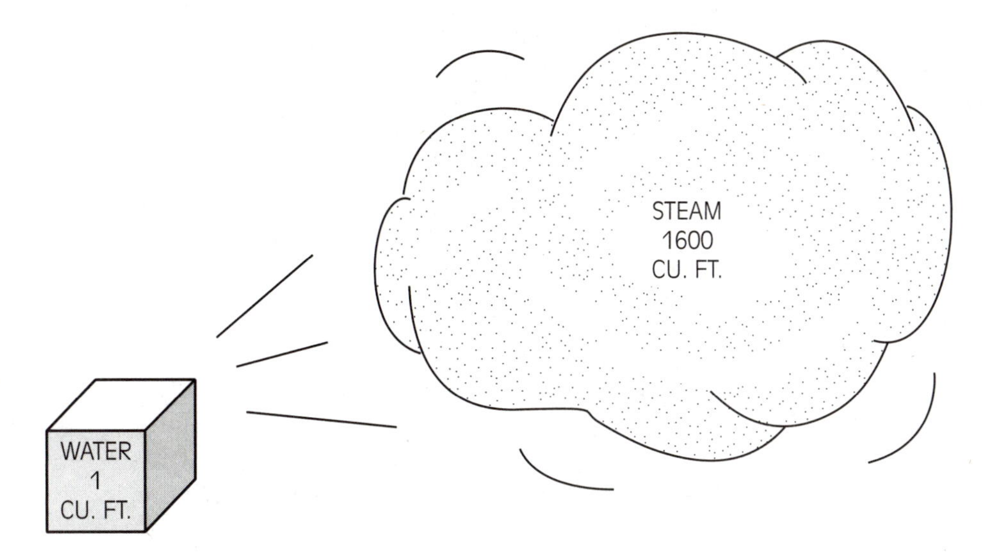

Figure 1-5 *Water expands 1,600 times when converted to steam at 212°F.*

Latent Heat and Specific Heat

specific heat
the amount of heat, in BTUs, needed to raise the temperature of 1 pound of substance 1°F

When water is converted to steam, an endothermic (absorbs heat) chemical reaction takes place. The heat absorbed by this reaction is the heat given off by the fire. To truly understand what properties of water make it such a valuable extinguishing agent, it is necessary to understand just how specific heat and latent heat properties apply. **Specific heat** is the amount of heat, in BTUs, needed to raise the temperature of 1 pound of substance 1°F. As mentioned in the foregoing list, the specific heat of water is 1 BTU per pound. For example, to raise 1 pound of water from 70°F to 75°F, the water must absorb 5 BTUs of heat. To put it another way, it takes 5 BTUs to raise 1 pound of water 5°F.

Suppose we are applying water to a fire. Let us assume that the water starts out at 62°F. When it is applied to the fire, it vaporizes. As we already know, water vaporizes at 212°F. How much heat (in BTUs) will 1 gallon of water absorb going from liquid at 62°F to steam at 212°F? To answer this question we need to first calculate the specific heat of a gallon of water (Example 1-3) and then the latent heat of vaporization of the same gallon (Example 1-4).

Example 1-3 How many BTUs will it take to raise 1 gallon of water to 212°F but not vaporize it?

ANSWER To raise 1 pound (weight) of water from 62°F to 212°F requires specific heat of 212°F – 62°F = 150 BTUs, but the problem calls for the specific heat (amount of BTUs) for a gallon (volume) of water. To find the specific heat for a gallon of water, we need to multiply the 150 BTUs for 1 pound of water by the weight of 1 gallon of water, which is 7.48 pounds.

Specific heat = *weight of water* × *change in temperature*

We can then use the formula Specific heat = weight of water × change in temperature, or specific heat = 7.48 × 150 = 1,251 BTUs. We now know that 1 gallon of water will absorb 1,251 BTUs going from ambient temperature (assumed to be 62°F) to 212°F.

latent heat the amount of heat absorbed by a substance when it changes state

Next, we need to understand latent heat properties. **Latent heat** is the amount of heat absorbed by a substance when it changes state. But first, a note of caution: When we talk about latent heat, we are talking about the heat absorbed as the substance changes state, but not temperature. At sea level, water that has been heated to 212°F remains at 212°F until all the water is vaporized to steam at 212°F. For our purposes we are concerned exclusively with the amount of heat absorbed when water vaporizes. For the sake of being thorough, we need to first mention the latent heat of fusion, which is the amount of BTUs absorbed when a pound of water goes from its solid state to its liquid state. That figure has been determined to be 143.4 BTUs. Since we do not put ice on a fire, it does not enter our calculations of BTUs absorbed.

Latent heat of vaporization, the amount of heat absorbed when water goes from its liquid state to its vapor state, is critically important in calculating the BTUs absorbed by water. When a pound of water goes from a liquid form at 212°F to steam at 212°F, it absorbs 970.3 BTUs just changing state. Again, just as with specific heat, it applies to each *pound* of water.

Example 1-4 How many BTUs will a gallon of water absorb going from a liquid state at 212°F to steam at 212°F?

ANSWER Because we already know the latent heat of vaporization, we only need to determine how many latent heat BTUs are needed for a gallon of water. Just like calculating the specific heat, we simply multiply the BTUs by the weight of water. The formula is Latent heat = weight of water × 970.3, or latent heat = 8.34 × 970.3 = 8,092 BTUs. We now know that the latent heat of vaporization of a gallon of water is 8,092 BTUs.

Latent heat = *weight of water* × 970.3

Now we know how much heat is absorbed by a gallon of water going from ambient temperature to 212°F. We also know how much heat is absorbed when water changes state. To truly appreciate the importance of specific heat and latent heat properties, we now put them together.

Example 1-5 How many BTUs are absorbed when a gallon of water goes from ambient temperature to steam?

ANSWER In Example 1-3 we determined that a gallon of water absorbs 1,251 BTUs going from 62°F to 212°F. In Example 1-4 we determined that a gallon of water absorbs 8,092 BTUs changing state from liquid to vapor. To find the answer to Example 1-5 we simply put the two together: 1,251 BTUs + 8,092 BTUs = 9,346 BTUs per gallon of water (see Figure 1-6).

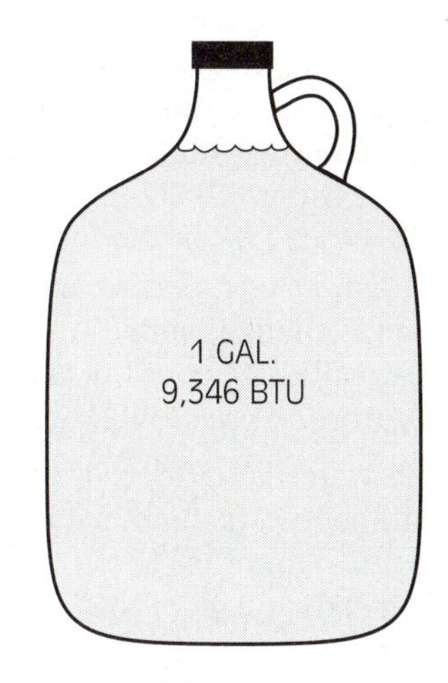

Figure 1-6 *One gallon of water absorbs 9,346 BTUs going from 62°F to steam.*

NFF
an estimation of the amount of water, in gpm, needed to extinguish a fire in a specific building

Gallons of water = *BTUs given off* ÷ 9,346

The practical application of Example 1-5 is that it tells us exactly how much heat a gallon of water will absorb, if completely vaporized, when applied to a fire, which is one method of determining needed fire flow (NFF). The **NFF** is an estimation of the amount of water, in gallons per minute (gpm), needed to extinguish a fire in a specific building. If we know the potential BTU output of the fuel burning, we can easily find the amount of water needed to extinguish it. Simply divide the potential BTU output by 9,346 to obtain the gallons of water necessary to absorb the BTUs being given off. This formula is Gallons of water = BTUs given off ÷ 9,346. There are a couple of catches, though: The water must be applied all at once and must be applied directly to the burning material so there is no runoff. Otherwise we fail to extinguish the fire, despite our most valiant efforts. This method of calculating how much water is needed to completely absorb the BTUs given off assumes the fuel is consumed all at once. Fires, however, burn over time; therefore, the BTU output is spread out over time. More practically, we only need to absorb the amount of BTUs being given off at any point in time to extinguish the fire.

fire load
the amount of combustibles per room or structure

$FLD = A \times fld$

When calculating **fire load**, the amount of combustibles per room or structure, it is given in pounds of combustibles per square foot. For instance, a bedroom is calculated at 4.3 pounds of combustibles per square foot. The fire load of the entire room is calculated by multiplying the fire load per square foot times the area of the room. We can make a formula for calculating the fire load: $FLD = A \times fld$. In this formula FLD is the fire load of the entire room, A is the area of the room, and fld is the fire load per square foot. A 225-square-foot bedroom would then have a fire load of $FLD = A \times fld$ or $FLD = 225 \times 4.3 = 967.5$. By multiplying the fire load by the BTUs given off per pound, the total BTU output can be determined.

Example 1-6 How many BTUs will be given off by 967.5 pounds of ordinary combustibles? How many gallons of water are needed to absorb the total BTUs given off?

ANSWER Ordinary combustibles generate about 8,000 BTUs per pound, so 967.5 pounds of combustibles would put off 967.5 pounds × 8,000 BTUs per pound = 7,740,000 BTUs. To find the gallons of water necessary to absorb the BTUs given off: Gallons of water = BTUs given off ÷ 9,346 or gallons of water = 7,740,000 ÷ 9,346 = 8,323.8 rounded to 8,324 gallons of water.

Conversion of Water to Steam

The method just given for calculating the amount of water needed to absorb the BTUs given off by a fire assumes the water is applied directly to the fuel. It also assumes that the steam generated will be lost. In such instances it is impossible to take advantage of the ability of steam to smother the fire. However, when and where possible we should always take advantage of the smothering properties of steam to extinguish the fire. The most basic way to determine how much water is needed for extinguishment is to determine the volume of the room (in cubic feet), then divide that volume by 1,600 (the expansion ratio of water to steam). However, this figure only tells us how many cubic feet of water are needed to extinguish the fire. We must then convert the cubic feet of water to gallons of water.

Example 1-7 How much water is required to completely fill a bedroom on fire with steam if the room is 15 feet × 15 feet?

Cubic feet of water = $V \div 1{,}600$

Gallons = $V \times 7.48$

ANSWER First determine the volume of the room, and, unless otherwise known, assume the height of the room is 8 feet. The volume of the room, then, is 15 feet × 15 feet × 8 feet = 1,800 cubic feet. Now to determine the gallons of water needed, we first divide the volume of the room by 1,600, which is the amount of steam generated by one cubic foot of water. Using the formula Cubic feet of water = $V \div 1{,}600$, where V is the volume of the room in cubic feet, the answer is $V \div 1{,}600 = 1.125$ cubic feet of water. Finally, to determine how many gallons of water this figure represents, use the formula Gallons = $V \times 7.48$, where V is the cubic feet of water. This gives us an answer of $1.125 \times 7.48 = 8.4$ gallons of water.

This amount of water does not seem like much considering the amount usually put on most fires. For it to take just 8.4 gallons of water would require an ideal situation. The room would have to be unvented, and we would have to have perfect conversion of water to steam. Neither one of these conditions occurs at a real fire, but this example still illustrates that a simple room-and-contents fire can be extinguished with a minimum of water.

■ Note

Assume a practical limit of 100 cubic feet per gallon for converting water into steam.

This method of calculating the volume of steam necessary to extinguish a fire is the basis of some of the more popular formulas for calculating NFF. However the popular formulas for determining the NFF assume a practical limit of 100 cubic feet per gallon for converting water into steam. This assumption is necessary for two reasons: (1) there is rarely a perfect conversion of water to steam and (2) steam is always lost. One popular formula, the Iowa State formula for determining needed fire flow, requires that the volume of the room or building be divided by 100, giving the amount of water needed in 30 seconds to extinguish the fire. The Iowa State formula would require 18 gallons of water in 30 seconds for the problem in Example 1-7 ($1,800 \div 100 = 18$), or just about twice the water calculated in Example 1-7. To bring the foregoing method in line with more conservative formulas (in firefighting, conservatism is good), simply double the results. The answer to Example 1-7 then becomes 16.8 or 17 gallons of water, which is still not much water! Regardless of the method of calculating NFF, they all only represent a beginning in determining what is needed.

Example 1-8 If the room in Example 1-7 can be filled with steam with just 17 gallons of water, how many BTUs would be absorbed?

BTUs absorbed = $9{,}346 \times gal$

ANSWER In Example 1-5 we determined a gallon of water would absorb 9,346 BTUs going from ambient temperature to steam. Simply multiply 9,346 by the number of gallons needed, found in the answer to Example 1-7, using the formula BTUs absorbed = $9{,}346 \times$ gal. Thus we have $9{,}346 \times 17$ gallons = 158,882 BTUs will be absorbed.

Calculating Volume and Weight of Water

At some point, it inevitably becomes necessary to calculate the volume of a container of water, and from there we must calculate the gallons and weight of the water. Appendix A contains several geometric shapes and formulas for calculating their area and volume. The shapes that firefighters should be most concerned with are cylinders (hose, water towers and tankers), spheres (water towers), and rectangles (reservoirs and pools). Most of these are pretty straightforward, but at times a little ingenuity is called for in calculating volume.

$V = L \times W \times H$

Most problems call for simple calculations, such as the volume of a water tank on a pumper. These involve a simple formula, $V = L \times W \times H$, where V is the volume of the container in cubic feet, L is the length of the container in feet, W is the width of the container in feet, and H is the height of the container in feet.

Once the volume in cubic feet is found, regardless of the shape of the container, finding the volume in gallons is easy. Multiply the cubic feet of the container by the number of gallons in a cubic foot. The formula for finding how many gallons of water are contained in a given size container, once we know the volume in cubic feet, is Gallons = $V \times 7.48$.

Example 1-9 What is the volume, in cubic feet and gallons, of a tank 10 feet long, 5 feet wide, and 1.33 feet (16 inches) deep? (These measurements represent the approximate shape of a water tank on some fire apparatus.)

ANSWER Begin by finding the volume in cubic feet by multiplying length by width by height: 10 feet $\times$ 5 feet $\times$ 1.33 feet = 66.5 cubic feet. To convert this figure to gallons, Gallons = $V \times 7.48$, or gallons = $66.5 \times 7.48 = 497.42$ gallons.

Gallons = $V \div 231$

If the need arises to find the volume of a small container, it can sometimes be easier to find the volume in cubic inches and then convert it to gallons. The formula is the same, $V = L \times W \times H$, only in this instance all dimensions must be in inches and the volume will be given in cubic inches. Then, to find the volume in gallons we must divide by 231, the number of cubic inches in a gallon. The formula for finding gallons when cubic inches are known is: Gallons = $V \div 231$, where Gallons is the number of gallons in the container, V is the volume of the container in cubic inches, and 231 is the number of cubic inches in a gallon.

Example 1-10 What is the volume in cubic inches, and how many gallons of water are in a container 18 inches × 18 inches × 24 inches?

ANSWER 18 inches × 18 inches × 24 inches = 7,776 cubic inches. Divide the volume of 7,776 cubic inches by 231 to get 33.66 gallons of water.

One of the more useful shapes you may need to find the volume for is an in-ground swimming pool. In rural areas pools can be useful as supplementary water sources. They are also not the easiest shapes for which to calculate volumes because of their multiple depths. They usually have a shallow end that extends for several feet at one end and a deep end that extends for several feet at the other end. In between, the depth drops off at an angle. These sections can be figured as three separate problems, with the shallow end and deep end separately as rectangles and the sloped area in between as a trapezoid (see Figure 1-7A). An easier way is to

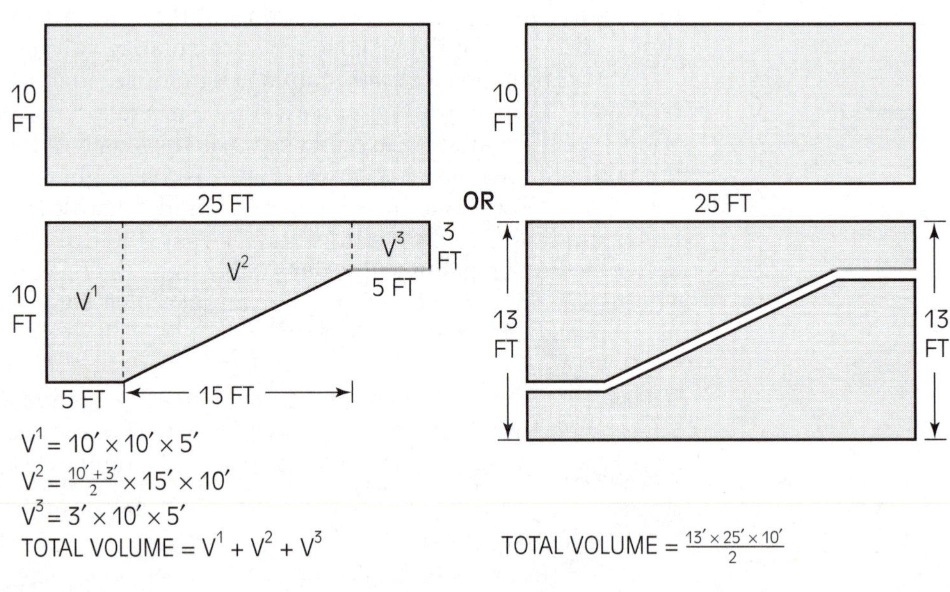

Figure 1-7 *Calculating the volume of a pool.*

double the pool back-to-back on itself (see Figure 1-7B). You now have a rectangle that is easier to work with. Find the volume of the doubled pool and take half of it. It will not be exact, but unless you calculate for all the rounded corners and curved surfaces, any method is only a reasonable approximation. If the pool is composed of multiple curved or irregularly shaped areas, contact the owner or manufacturer for the volume.

One statistic that always seems to be asked at some point is the volume and weight of charged hose. It actually has a useful purpose in illustrating to new recruits and interested observers just how much work is really involved in firefighting. The following method can be used to find the volume, in cubic feet and gallons, of any cylinder.

$V = .7854 \times D^2 \times H$ (or L)

The formula for finding the volume of a cylinder is $V = .7854 \times D^2 \times H$ (or L). In this formula V is volume in either cubic feet or cubic inches, .7854 is a constant used to find the area of a circle when using its diameter to solve for area, D is the diameter of the cylinder, and H (or L) is the height (or length) of the cylinder. The units in the answer depend on the units of D and H. If D and H are in inches, the answer will be in cubic inches. If D and H are in feet, the answer will be in cubic feet. There is one stipulation on D and H: They must be in the same units. If one dimension is in feet and the other one in inches, convert the inches to feet by dividing by 12.

$V = 7.48 \times .7854 \times D^2 \times H$ (or L)

$V = 5.87 \times D^2 \times H$ (or L)

If the volume is in cubic feet, this same formula can be used to find the volume in gallons. Simply multiply the answer by the number of gallons in a cubic foot. The formula then becomes $V = 7.48 \times .7854 \times D^2 \times H$ (or L). Because multiplication can be done in any sequence, we can multiply the 7.48 by .7854 to get a constant of 5.87. By using this constant we eliminate a step, creating a shortcut, and the formula becomes $V = 5.87 \times D^2 \times H$ (or L).

Example 1-11 How much water is in a 50-foot section of charged 2½-inch hose?

ANSWER $V = 5.87 \times D^2 \times L$

$= 5.87 \times (2.5/12)^2 \times 50$ feet

$= 5.87 \times (.208)^2 \times 50$ feet

$= 5.87 \times .043 \times 50$ feet

$= 12.6$ gallons of water

$W = V \times 8.34$

If we need to know the weight of a volume of water, we only need to multiply the number of gallons by the weight of a gallon of water. This requires a formula of $W = V \times 8.34$, where W is the weight of the water and V is the volume of water in gallons.

Example 1-12 What is the weight of 12.6 gallons of water?

ANSWER $W = V \times 8.34$

$= 12.6 \times 8.34$

$= 105$ pounds

If we do not need to know the volume in gallons but only need to know the weight of water in the cylinder, we can create another shortcut. If we were to write an entire formula to find volume in gallons and then find the weight, it could be written as a single formula: Weight = $8.34 \times 7.49 \times .7854 \times D^2 \times H$. Again, since multiplication can be done in any sequence, we can create a constant of 49.06, or 49, simply by multiplying $8.34 \times 7.49 \times .7854$. The new formula for finding the weight of the water in a cylinder becomes $W = 49 \times D^2 \times H$ (or L).

$W = 49 \times D^2 \times H$ (or L)

Example 1-13 What is the weight of water in a 50-foot section of 2½-inch hose?

ANSWER $W = 49 \times D^2 \times H$

$= 49 \times (2.5/12)^2 \times 50$ feet

$= 49 \times (.208)^2 \times 50$ feet

$= 49 \times .043 \times 50$ feet

$= 105.35$ pounds or, rounded down to 105 pounds.

Both formulas, while shortcuts, are perfectly legitimate. However, before using the shortcuts it is important to fully understand how the formulas work. Also, where and when we round off can affect the answer. In most cases the difference due to rounding will not be significant.

A NOTE OF CAUTION

A note of caution concerns some of the figures given as standards in this book. In some instances, these figures will not agree with other references, but with good reason. The figures in this book represent a recent calculation performed for the purpose of checking the validity of the older numbers before including them in this book. In many instances the older figures have been replaced with the results of the recent calculations. An example is the weight of a column of water 1inch × 1inch × 1foot tall. Older books give it a weight of .434 pounds, but it is calculated in Chapter 2 to be .433 pounds. This difference is because the older figure was calculated using the weight of a cubic foot of water as 62.5 pounds, not 62.4 pounds. The 62.4 pound figure is the more accurate figure. The 62.5 pound figure, from which .434 was derived, was rounded off for ease of calculation. In each instance where this book uses a different figure than has been used in the past, the calculations from which the figures were derived are included.

APPLICATION ACTIVITIES

You are now ready to put your knowledge of water and its properties to work.

Problem 1 How many gallons of water are in a reservoir that measures 100 feet by 50 feet and is 10 feet deep?

ANSWER $V = L \times W \times H$

$= 100 \text{ feet} \times 50 \text{ feet} \times 10 \text{ feet}$

$= 50{,}000 \text{ cubic feet}$

To convert to gallons:

$\text{Gallons} = 50{,}000 \times 7.48$

$= 374{,}000$

Problem 2 A 10,000-square-foot warehouse has a fire loading of 5 pounds per square foot of common combustibles. If a pound of common combustibles generates 8,000 BTUs per pound when it burns, what is the minimum gallons of water needed to absorb the total BTUs of a fire if the warehouse is fully involved?

ANSWER First calculate the total fire load using the fire load formula:

$FL = A \times fl$

$= 10{,}000 \times 5$

$= 50{,}000 \text{ pounds}$

Next, calculate the BTU output of the fire load:

$\text{BTU} = \text{BTUs per pound of combustible} \times \text{pounds of combustibles}$

$= 8{,}000 \times 50{,}000$

$= 400{,}000{,}000$

In Example 1-5 it was determined that each gallon of water applied to a fire will absorb 9,346 BTUs. Therefore:

$\text{Gallons} = 400{,}000{,}000 \div 9{,}346$

$= 42{,}799$ gallons of water are needed to absorb the total BTUs

Problem 3 An office is 10,000 cubic feet. How many gallons of water will it take to completely fill the space with steam?

ANSWER We previously solved a similar problem by dividing the volume of the room by 1,600 to determine the cubic feet of water needed to fill the room with steam. Then we multiplied the cubic feet of water by 7.48 to get the gallons of water. If we know that 7.48 gallons of water generate 1,600 cubic feet of steam, we can easily find the amount of steam generated by one gallon by dividing 1,600 by 7.48: $1600 \div 7.48 = 213.9$, or 214 cubic feet of steam are generated per gallon of water. Now, to determine how many gallons of water are needed to fill a given space with steam we only need to divide the volume of the room by 214. (Remember, earlier it was said that a conservative figure of 100 cubic feet of steam per gallon of water was used by the Iowa State Formula. The calculation in Example

1-7 was then shown to be off by a factor of two; now we know why. One gallon of water *actually* expands into 214 cubic feet of steam.) A new formula for calculating how many gallons of water are needed to fill a room with steam becomes Gallons = $V \div 214$. Thus,

Gallons = $V \div 214$

Gallons = $V \div 214$
= $10{,}000 \div 214$
= 46.72 gallons of water are needed to completely fill the office with steam.

Multiply this answer by 2 to arrive at the more conservative figure of 93.44, or 94, gallons of water to extinguish the fire.

Problem 4 An Air Crane helicopter, with a special apparatus for dropping water on forest fires, has a capacity of 2,000 gallons of water. How much does the water add to the weight of the helicopter?

ANSWER Weight of water = $2{,}000 \times 8.34$ pounds per gallon of water
= 16,680 pounds

Problem 5 How many BTUs will be absorbed by 100 gallons of water if the water is completely vaporized?

ANSWER This problem is easy. Multiply 100 by the number of BTUs absorbed per gallon of water:

BTUs = $100 \times 9{,}346$
= 934,600

Summary

Hydraulics, as a science, began more than 250 years ago. At first it was intended only to answer questions about water. In the late 1800s engineers began making their first serious calculations of water supplies for fire protection purposes.

Water is an ideal fire suppression agent, but to use water to its maximum potential it is necessary to understand both its physical and chemical properties. Understanding these properties also plays a role in our understanding of the laws, principles, and formulas known as the science of hydraulics.

The most basic hydraulic calculations involve finding the volume of water storage facilities and weight of water. Further, in order to be able to estimate the water requirements for a particular fire or potential fire, it is necessary to be able to calculate the amount of water needed to extinguish it. Modern calculations meant to estimate NFF are based on supplying the amount of water needed to either generate enough steam to smother the fire or absorb the BTUs being given off.

Review Questions

1. A salesperson is selling a tanker that weighs 19,182 pounds more when full than when empty.
   A. How many gallons of water does it hold?
   B. What is the volume of the tank in cubic feet?
2. A water tower with a 20-foot sphere at the top to store the water will hold how many gallons?
3. What is the weight of a tank of water if the tank is 10 feet long and 4½ feet in diameter?
4. What is the weight of the tank in Question 3 if the tank is full of saltwater?
5. Two hose streams are operating into the second floor of a small warehouse for one half-hour. If one line is flowing at 210 gpm and the other one is flowing at 265 gpm, how much weight will be added to the weakened floor structure if only half of the water is converted to steam? (Water not converted to steam becomes part of the live load of the building.)
6. How many BTUs will be absorbed if 2 cubic feet of water at 62°F are heated to 200°F?
7. The average living room is assumed to have 3.9 pounds of combustibles per square foot. In a fire, how many gallons of water will be needed to absorb the BTUs from each square foot if each pound of combustibles is putting off 8,000 BTUs?
8. A hose line with a 125-gpm tip can be expected to extinguish a room of what volume if the line operates for 30 seconds? Assume 100% conversion of water to steam.
9. What is the weight of one gallon of saltwater?

10. What is the volume of the swimming pool in Figure 1-7 in

    A. Cubic feet?

    B. Gallons?

11. How many BTUs will be given off from a fire in a library if the building is 75 feet by 100 feet and is one story in height? The fire load is 23.6 pounds per square foot.

12. How many gallons of water will it take to absorb all the BTUs given off in Question 6?

List of Formulas

Finding specific heat:

Specific heat = *weight of water* × *change in temperature*

Calculating latent heat of a given weight of water:

$$\text{Latent heat} = \textit{weight of water} \times 970.3$$

Finding the amount of water needed to absorb the BTUs being given off:

$$\text{Gallons of water} = \textit{BTUs given off} \div 9{,}346$$

Fire load formula:

$$FLD = A \times fld$$

Finding how much water is needed to fill a room with steam:

$$\text{Cubic feet of water} = V \div 1{,}600$$

Calculating how much water is in a container:

$$\text{Gallons} = V \times 7.48$$

Finding how many BTUs will be absorbed by the number of gallons of water:

$$\text{BTUs absorbed} = 9{,}346 \times \textit{gallons}$$

Volume of all rectangular boxes:

$$V = L \times W \times H$$

Finding gallons when volume is in cubic inches:

$$\text{Gallons} = V \div 231$$

Volume of a cylinder in cubic inches or cubic feet:

$$V = .7854 \times D^2 \times H \text{ (or } L\text{)}$$

Volume of a cylinder in gallons:

$$V = 7.48 \times .7854 \times D^2 \times H \text{ (or } L\text{)}$$

Direct calculation of gallons of water in a cylinder:

$$V = 5.87 \times D^2 \times H \text{ (or } L\text{)}$$

Weight of water in a container:

$$W = V \times 8.34$$

Direct calculation of weight of water in a cylinder:

$$W = 49 \times D^2 \times H \text{ (or } L\text{)}$$

Gallons of water needed to fill a room with steam:

$$\text{Gallons} = V \div 214$$

Force and Pressure

Learning Objectives

Upon completion of this chapter, you should be able to:

- Understand the difference between force and pressure.
- Calculate pressure given either elevation or force and area.
- Calculate force given pressure and area.
- Understand the six principles of pressure.
- Understand the difference between absolute pressure and relative pressure.
- Calculate absolute pressure from relative pressure.
- Differentiate between static pressure, residual pressure, and flow pressure.

INTRODUCTION

The study of hydraulics must begin with a thorough understanding of both force and pressure. Studying force and pressure together is logical, because their definitions are interrelated. Further, it is necessary to understand the six principles of fluid pressure as a basis for our study of hydraulics. Together, these definitions and six principles of fluid pressure lay the foundation for an accurate understanding of hydraulics.

PRESSURE

pressure
force per unit of area

$P = \frac{F}{A}$

Pressure is defined as the *force per unit of area*, or in the form of an algebraic expression, $P = F/A$. For now let us accept the definition of force, in this equation, simply as the weight of the fluid, in pounds, and area can be in any commonly accepted units. Units of area can be in square yards or square feet, but it is almost universally given in square inches. Therefore, force in pounds, exerted over an area measured in square inches, gives us a reading of pressure in pounds per square inch (psi).

Recall from Chapter 1 that the weight of a cubic foot of water is 62.4 pounds. This figure is also the force, in pounds, of the water on the inside bottom surface of the container. So we can say that the force of 1 cubic foot of water over an area of 12 inches by 12 inches is 62.4 pounds. Notice force is assigned units of pounds. If the container is 12 inches × 12 inches × 12 inches, it can correctly be stated that it represents both a force of 62.4 pounds and a pressure of 62.4 pounds per square foot if we were using square feet as our unit of pressure. But we want our units of pressure to be in psi. Since our example is a true cubic foot of water, then the surface where the force is acting is 12 inches by 12 inches or 144 square inches. By using the formula $P = F/A$, we can calculate pressure from force.

Example 2-1 What is the pressure in psi of a cubic foot of water if it is exerted over 1 square foot.

ANSWER The force of a cubic foot of water, 62.4 pounds, divided by 144 square inches

$$\begin{aligned} P &= F/A \\ &= 62.4 \div 144 \\ &= .433 \text{ psi} \end{aligned}$$

■ Note
A column of water 1 inch by 1 inch and 1 foot tall exerts a pressure of .433 psi.

We now know that a column of water 1 inch by 1 inch and 1 foot tall exerts a pressure of .433 psi (see Figure 2-1). We have just calculated pressure from force.

In general, in situations where firefighters need to calculate pressure, the force is not known. Therefore, we need another way to calculate pressure. Because we now know a column of water is capable of exerting a pressure of

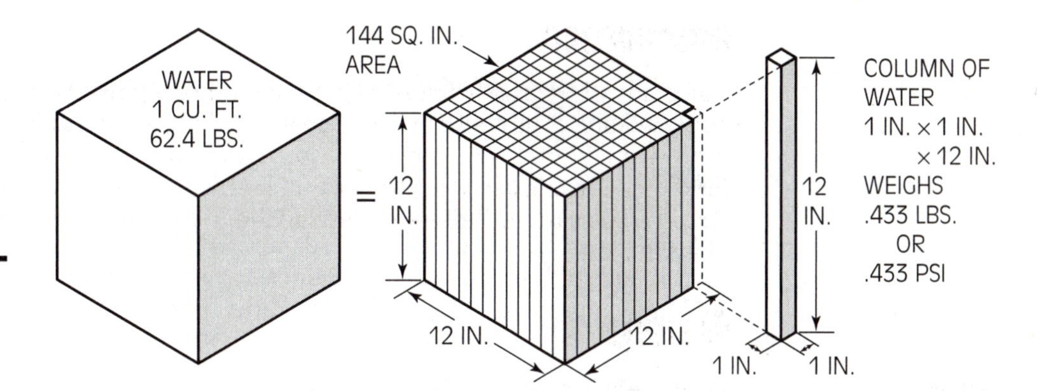

Figure 2-1 *Pressure exerted by water 1 inch × 1 inch × 1 foot tall.*

$P = .433 \times H$

.433 psi for every foot of elevation, we can calculate pressure by using the equation, $P = .433 \times H$. In this formula H stands for the elevation or height of water, in feet, over the point where the measurement is taken. It is often referred to as the elevation head.

Example 2-2 What is the pressure at the base of a column of water 10 feet tall?

ANSWER Multiply .433 by the elevation head

$$P = .433 \times H$$
$$= .433 \times 10$$
$$= 4.33 \text{ psi}$$

Calculating Elevation from Pressure

$H = 2.31 \times P$

Suppose we know the pressure at a given point and want to find the elevation of the surface of the water. We can actually use this same formula by rearranging it to read $H = P/.433$. Or we can create a new formula that allows us to multiply to find the answer. The new formula is $H = 2.31 \times P$, where H is the elevation head of water, 2.31 (the reciprocal of .433) is 2.31 feet and is calculated by dividing 1 by .433, and P is the pressure. In this formula the constant 2.31 tells us that every pound of pressure represents 2.31 feet of elevation.

Example 2-3 If there is a pressure of 65 psi at the base of a water tank, how high is the water inside?

ANSWER Use the new formula, $H = 2.31 \times P$

$$= 2.31 \times 65$$
$$= 150.15 \text{ feet}$$

Six Principles of Fluid Pressure

Now that we know what pressure is, it is important to study the six principles of fluid pressure. These principles explain the behavior of pressure in fluids.

Principle 1 *Pressure is exerted perpendicular to any surface on which it acts.* Simply stated, pressure must act at right angles to any surface with which it is in contact (see Figure 2-2). If pressure were imposed at other than right angles, the fluid would tend to flow out of the container or bunch up in the center. But this does not happen, proving that pressure is exerted at right angles to its container.

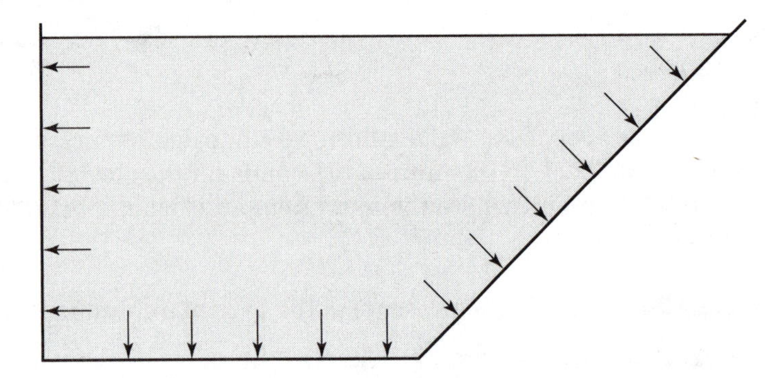

Figure 2-2 *Pressure is exerted perpendicular to any surface on which it acts.*

Principle 2 *Pressure in a fluid acts equally in all directions.* To understand this principle requires a bit of imagination. Visualize a bucket full of water. Then, at a point in the center of the volume of water visualize a single drop of water. From this drop pressure is exerted equally in all directions, sideways, up, down at 45-degree angles, and so forth with equal force (see Figure 2-3A).

A more familiar example is fire hose (see Figure 2-3B). Visualize a point of pressure, in the exact middle of the stream of water in the hose. If that pressure is truly equal in all directions, the hose, which has no intrinsic shape of its own, should be round when charged, and it is.

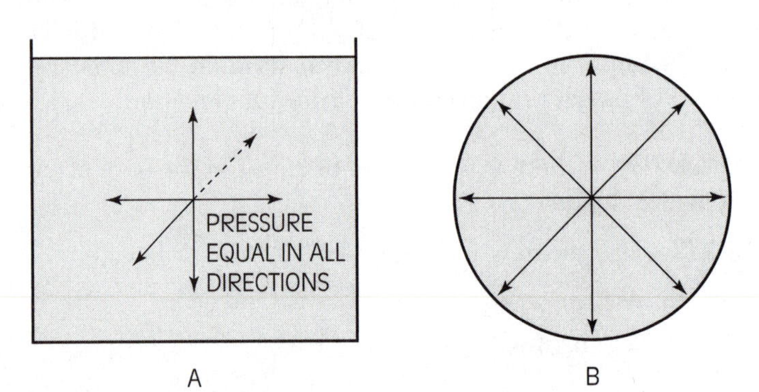

Figure 2-3 *Pressure in a fluid acts equally in all directions.*

Principle 3 *If pressure is applied to a confined liquid, that pressure is transmitted to every point within the liquid without reduction in intensity.* This principle, also known as Pascal's principle, is named for Blaise Pascal (1623–1662), a French philosopher and scientist who first postulated it.

Imagine a tank of water 2 feet in diameter and 10 feet long. On the top of this tank is a cylinder with an area of exactly 10 square inches (see Figure 2-4). With no external pressure applied, there would be a pressure of .866 psi at the bottom of the tank (.433 × 2 feet of water), .433 psi at midpoint, and 0 at the top. If 100 psi of pressure is applied to the tank by the cylinder, there would now be a total pressure of 100.866 psi at the bottom, 100.433 psi at midpoint and 100 psi at the top.

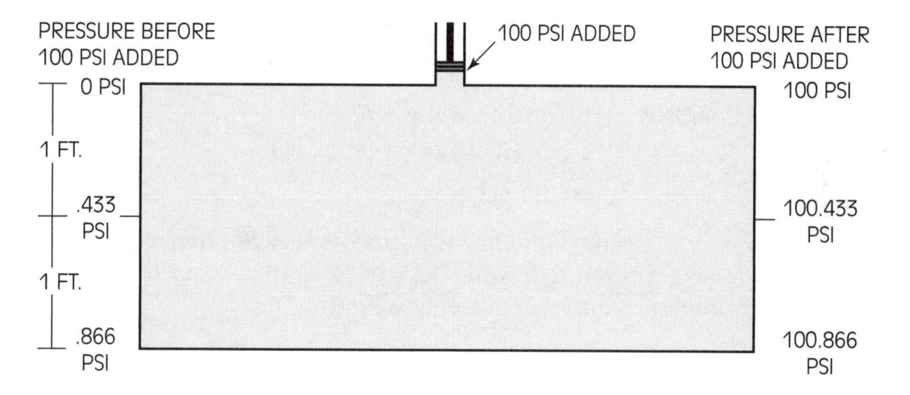

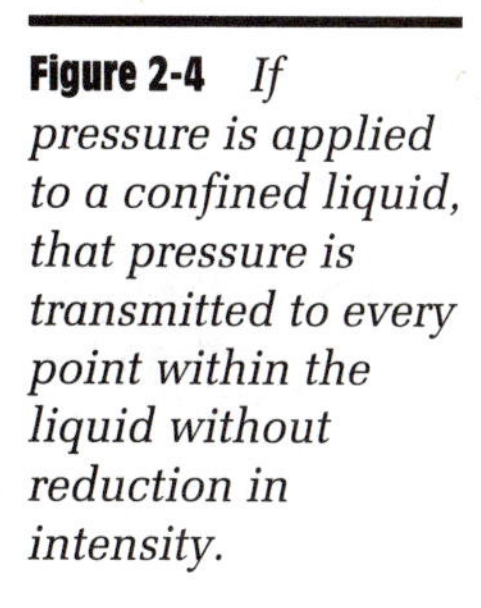

Figure 2-4 *If pressure is applied to a confined liquid, that pressure is transmitted to every point within the liquid without reduction in intensity.*

Principle 4 *The pressure of a liquid in an open vessel is proportional to the depth of the liquid.* We use this principle to determine the pressure at the 1 foot and 2 foot level in our illustration of Principle 3. Each segment of water 1inch × 1inch × 12 inches tall exerts .433 psi. Therefore every foot of water above the point where we measure the pressure adds another .433 psi (see Figure 2-5).

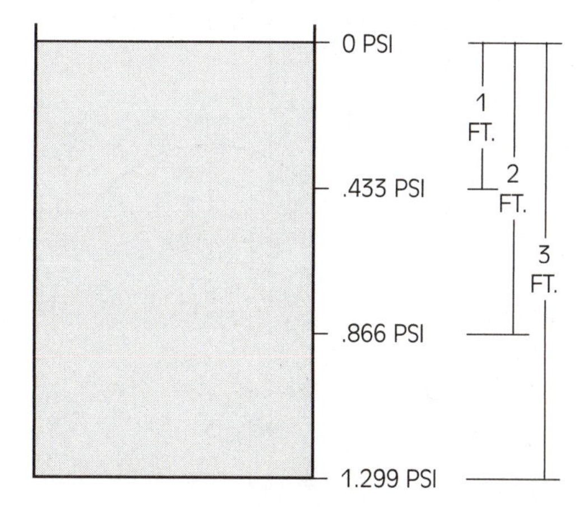

Figure 2-5 *The pressure of a liquid in an open container is proportional to the depth of the liquid.*

Principle 5 *The pressure of a liquid in an open container is proportional to the density of the liquid.* This principle is the twin of Principle 4, only in this principle reference is made not to the height of liquid in the container, but to the density of the liquid. Because different liquids have different densities, the pressure they exert for a given elevation is different. The only liquid, other than water, we would be concerned with in pump operations, is mercury. Mercury is used to measure the amount of work a pump does at draft. Let us use mercury to illustrate this principle.

By assigning a density of 1 to water, mercury on the same scale has a relative density of 13.6, or in other words, mercury weighs 13.6 times as much as water. This means that if a cubic foot of water exerts a force of 62.4 pounds, a cubic foot of mercury exerts a force of 848.64 pounds.

Example 2-4 If a cubic foot of mercury exerts a force of 848.64 pounds, how much pressure will a column of mercury 1 inch × 1 inch × 12 inches tall exert?

ANSWER Use the formula $P = F/A$

$$P = 846.64 \div 144$$
$$= 5.89 \text{ psi}$$

There is another way to compare the density of water to mercury. It would take a column of water 13.6 inches tall to exert the same pressure as a column of mercury only 1inch tall (see Figure 2-6).

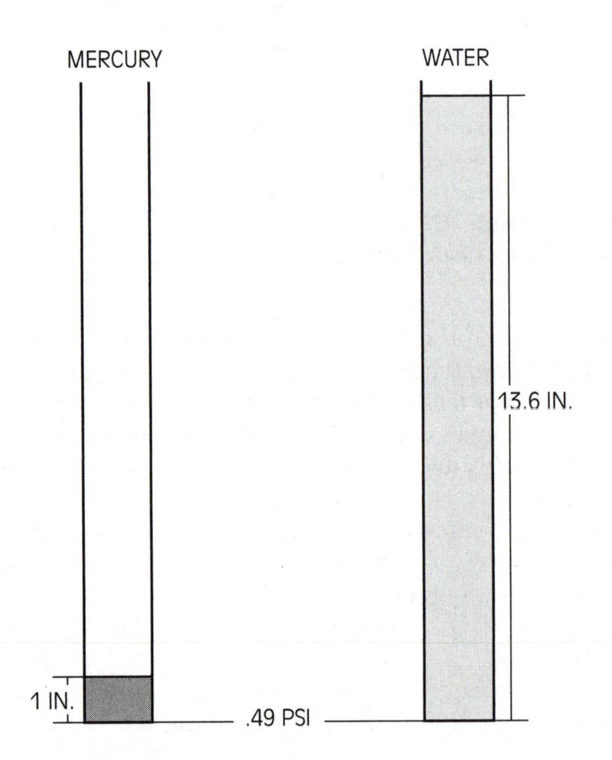

Figure 2-6 *The pressure of a liquid in an open container is proportional to the density of the liquid.*

Principle 6 *Liquid pressure on the bottom of a container is unaffected by the size and shape of the container.* This principle is similar to Principle 4 and Principle 5. But Principle 6 goes further by saying that for a given fluid, the shape of the container does not affect the pressure at the base, or at any point for that matter. The pressure of any size or shape of container is always the same at any point along a horizontal plane. Only the distance between the top and bottom (or other points of measurement) of the liquid, not the shape or volume of the container, has any affect on the pressure (see Figure 2-7).

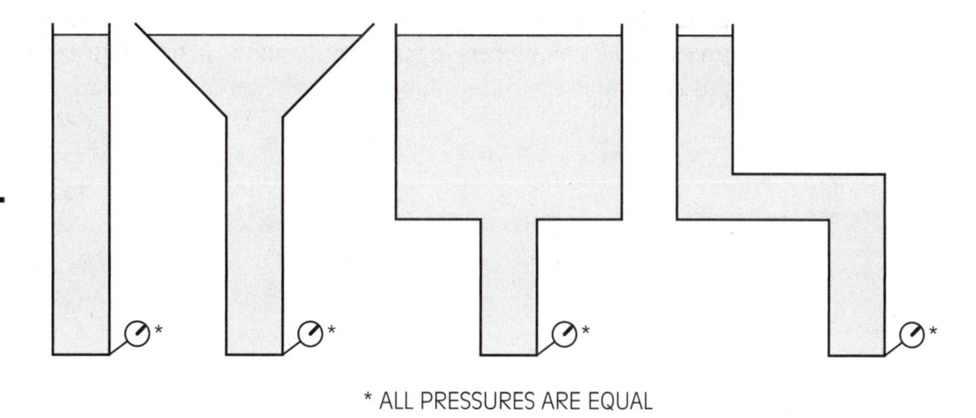

Figure 2-7 *Liquid pressure on the bottom of a container is unaffected by the size and shape of the container.*

Relative Pressure versus Absolute Pressure

There are two methods for measuring pressure. Just as the weight of a column of water 1inch × 1inch is called pressure, and for every 1foot of water we have .433 pounds of pressure, the atmosphere (air) around us also has weight. After all, air is a fluid with mass and volume that obeys the laws of pressure we have just outlined, but with some variation due to the fact that gasses are compressible while liquids are not. This means that a column of air (gas) is actually denser at the bottom than at the top. The pressure of air at sea level is 14.7 psi. This measure is referred to as one atmosphere and is the standard used for atmospheric pressure. When we expose a gauge to the atmosphere at sea level it should naturally read 14.7 psi, but most gauges do not.

relative pressure pressure indicated with 0 psi = atmospheric pressure

absolute pressure pressure indicated when 14.7 psi = atmospheric pressure

Why gauges do not brings us to the concept of relative pressure (sometimes referred to as gauge pressure) versus absolute pressure. **Relative pressure** can be defined as the pressure indicated with 0 psi = atmospheric pressure. **Absolute pressure** can be defined as the pressure indicated when 14.7 psi = atmospheric pressure. Technically all gauges read atmospheric pressure even when at rest. Some read it in absolute terms and indicate 14.7 (or whatever the atmospheric pressure is) while most read 0 or relative pressure, which means that whatever pressure is registered on a relative gauge, the absolute pressure is 14.7 psi higher.

Example 2-5 What formula converts relative pressure to absolute pressure?

Absolute pressure = relative pressure + atmospheric pressure

ANSWER Absolute pressure = Relative pressure + Atmospheric pressure

Distinguishing between absolute and relative pressure might seem like a trivial distinction, because the pressure is only off by the value of atmospheric pressure, but it is actually a critical distinction. Without the distinction between the two methods of reading pressure, there is no negative pressure, or vacuum. Figure 2-8 illustrates the difference between absolute and relative pressure.

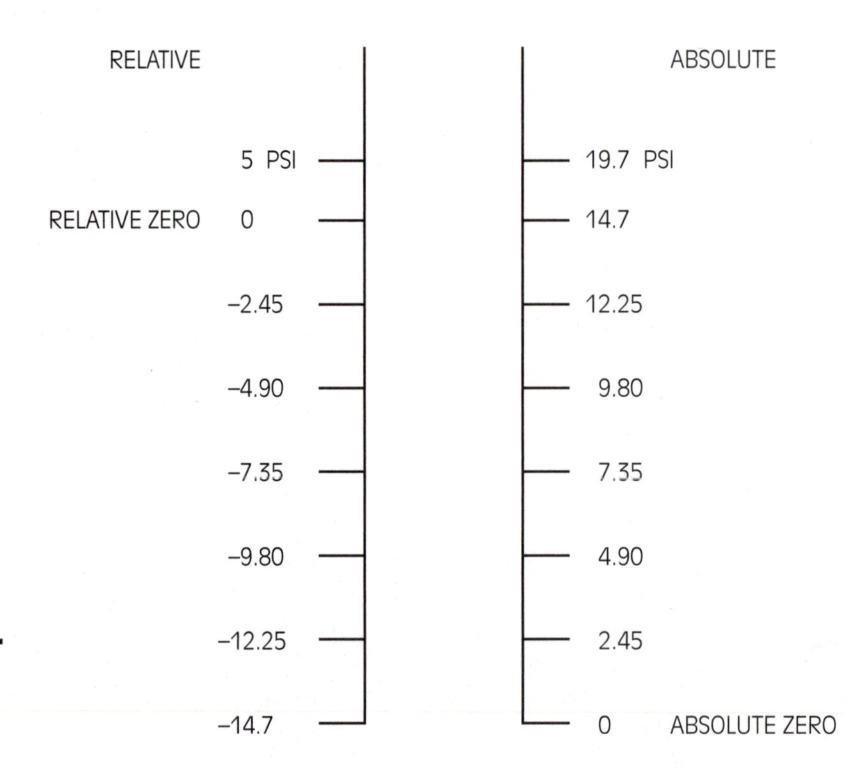

Figure 2-8 *Relative versus absolute pressure scales.*

Einstein has been quoted as saying, "Nature abhors a vacuum." In fact, vacuums, in terms of negative pressure, technically do not exist. What we refer to in the fire service as a **vacuum** is just a relative pressure below the surrounding atmospheric pressure. We usually identify this vacuum in reference to inches of mercury. (This concept is fully explained in Chapter 8.) On an absolute scale, pressure below atmospheric is still a positive pressure (see Figure 2-8). This distinction is important to understand and its relevance will become more apparent later when we discuss pump theory and drafting. In absolute terms, a vacuum is the absence of any pressure.

vacuum
relative pressure below the surrounding atmospheric pressure

Types of Pressure

Before leaving our discussion of pressure, we must define some types of pressure. These definitions are used throughout the remainder of this text.

They are:

Static pressure

Residual pressure

Flow pressure

static pressure
pressure of a fluid at rest

Static pressure is the pressure of a fluid at rest. For example, when a pumper is first hooked up to a hydrant and the hydrant is charged (or opened) but no water is flowing, the intake gauge will read static pressure.

residual pressure
pressure remaining when water is flowing

Residual pressure is the pressure remaining when water is flowing. With a pumper hooked up to a hydrant and the hydrant charged, after a discharge is opened and water is flowing the pressure on the intake gauge will go down from its static reading. The new pressure is residual pressure.

flow pressure
pressure measured as the water leaves the nozzle or other opening

Flow pressure is the pressure measured by means of a pitot gauge as the water leaves the nozzle or other opening. By using a pitot gauge in the stream of water as it exits a nozzle the velocity of the water is converted into psi to give a flow pressure.

FORCE

force
pressure times the area it is exerted against

Force is best defined as pressure times the area it is exerted against. To best understand this concept, and to get a better understanding of the difference between force and pressure (see Figure 2-9). In Figure 2-9, we have two containers, each with 3 cubic feet of water. The containers measure 1 foot × 1 foot × 3 feet. In Figure 2-9A the container is standing on end on a table. Because it is standing on end, it has an area of 144 square inches in contact with the table. The pressure would be .433 × 3 or 1.299 psi because the container is 3 feet high. Force then becomes 144 square inches times 1.299 psi, or 187.06 pounds.

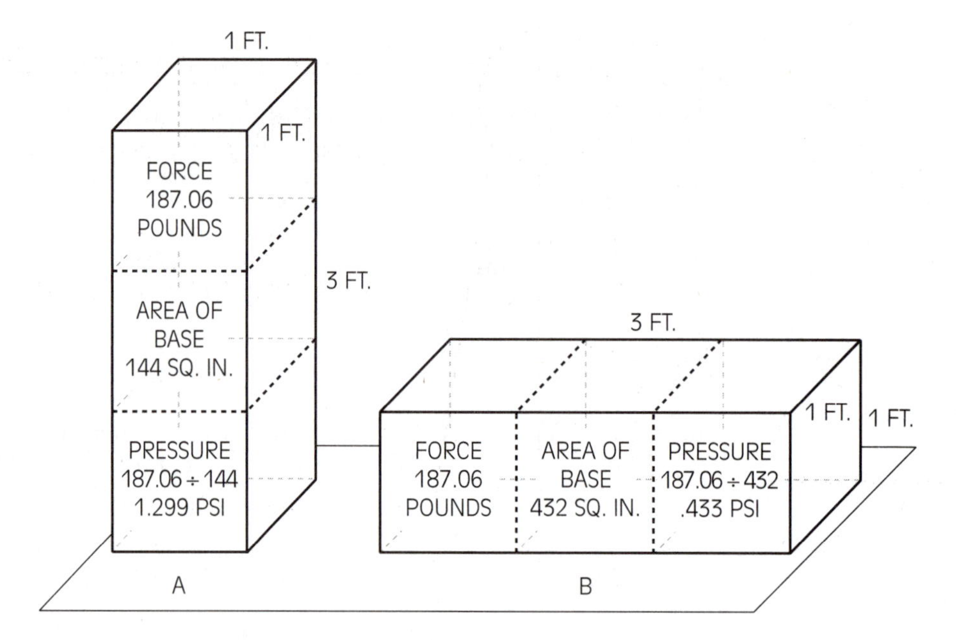

Figure 2-9 *The force of both containers is equal, but the pressure is different.*

In Figure 2-9B, the second container has been placed on its side. The force is exerted over an area of 432 square inches, and the pressure in this example is .433 psi because the elevation is just 1 foot. Force in this example is 432 square inches times .433 psi, which is 187.06 pounds, which is the same as in Figure 2-9A. We can verify the force in Figure 2-9 is correct because in both cases the force is equal to the weight of water in the container. Therefore, the force should be 3 cubic feet times 62.4 pounds per cubic foot, or 187.2 pounds. Note that in both of these examples force stays the same because the weight of the water is the same, but the pressure varies depending on the orientation of the container.

Earlier we used a definition of force as the weight of the fluid. However, this definition is not always true. For example, look at Figure 2-10. Because Principle 4 and Principle 6 tell us that the pressure at the bottom of the water tank is proportional to the depth of the liquid regardless of shape, then the pressure at the base of the water tower in Figure 2-10 is 43.3 psi. If the area of the pipe leading from the tank is 100 square inches, there should be a force of 4,330 pounds (43.3 psi × 100 square inches). But if we use the weight of the fluid (water) to calculate the force, then there should be the force of 31,741.84 gallons (total volume of sphere and supply pipe) × 8.34 pounds per gallon or 264,726.94 pounds. How do we account for the difference?

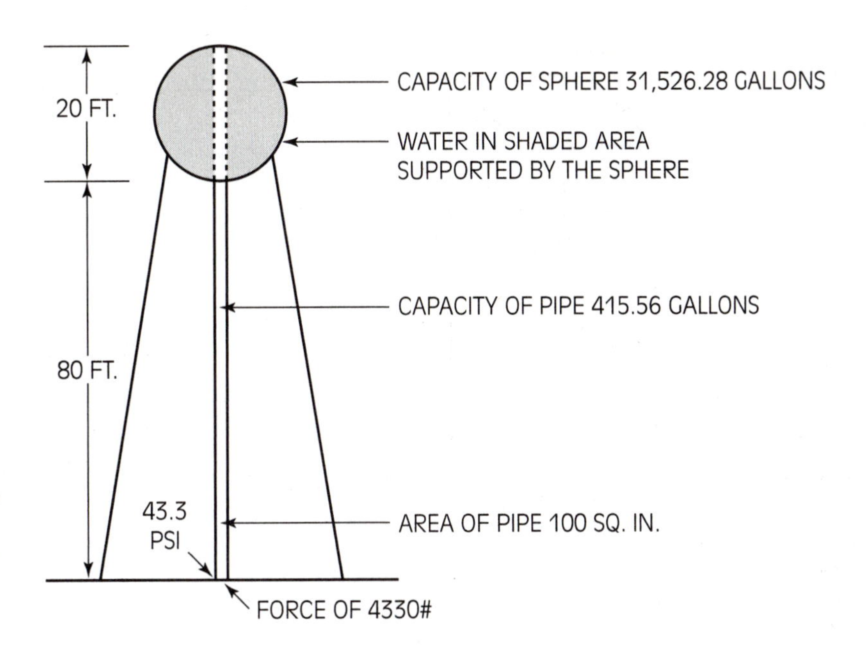

Figure 2-10 *When force is not equal to the weight of the water.*

In this example, the force we are measuring is only the weight exerted by a column of water 100 square inches by 100 feet tall (4,332 pounds). It is the area represented by the unshaded portion of the tank in Figure 2-10 and the pipe below it. The force (weight) of the water in the shaded area of the tank in Figure 2-10 is actually bearing against the tank itself, not on the ground. Remember that the pressure is exerted perpendicular to the surface on which it works. Consequently the tank supports the force of the water in the shaded area of Figure 2-10.

Another example of the distinction is shown in Figure 2-11. In Figure 2-11A, the container is holding 12 cubic feet of water, while the container in Figure 2-11B is holding 30 cubic feet of water. In both examples the pressure at the base is determined by the elevation of the water. The pressure is 10 feet × .433 psi or 4.33 psi. Each container has an area at the base of 432 square inches (12 inches × 36 inches = 432 square inches), making the force in both examples 1,870.5 pounds. (Accept this as fact for now; later you'll be shown how to calculate it.) In Figure 2-11B the weight of 30 cubic feet of water is 1,872 pounds. This difference of only 1.5 pounds is easily explained by how and where formulas are rounded off. In Figure 2-11A however, we only have 12 cubic feet of water or 748.8 pounds or 1,121.2 pounds less than the 1,870 pounds of force exerted at the bottom. Yet in both examples the force and pressure indicated are correct. How is this possible?

net force
a partial force

The answer to this apparent contradiction is net force. **Net force** is a partial force. In other words, several net forces may have to be added together to obtain

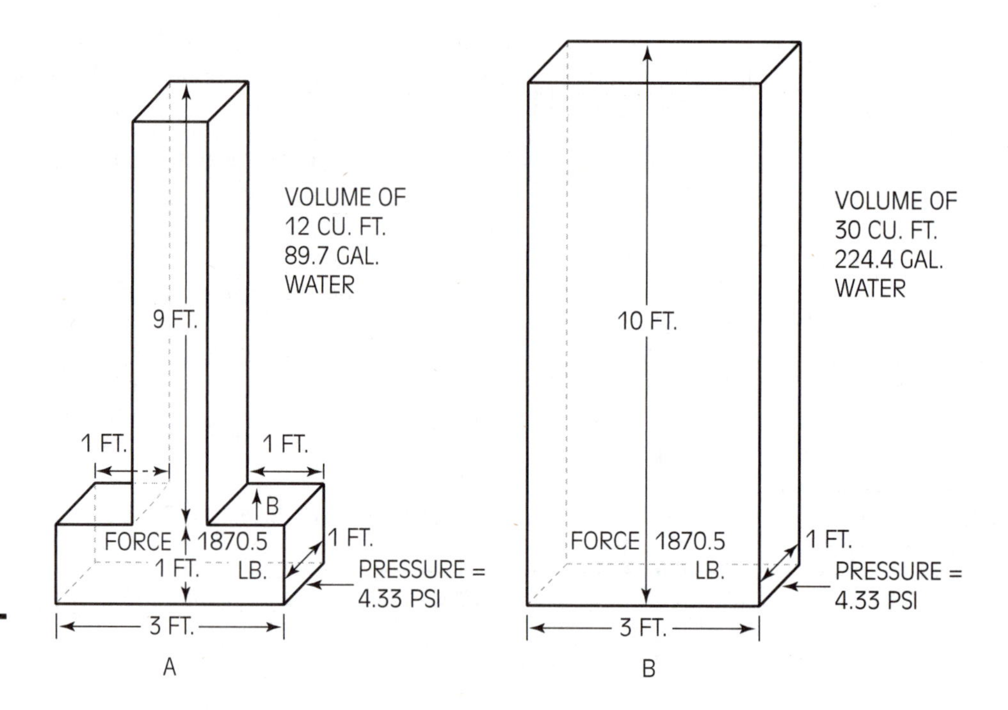

Figure 2-11 *Net force.*

the total force. In the example in Figure 2-11A, there is a force directed up at point B. With 9 feet of water (elevation head) determining the pressure at this point (remember Principle 4 and Principle 2), there is a pressure of 3.897 psi directed up. This pressure is acting on a total of 288 square inches, with a resulting net force of (288 square inches × 3.897 psi) 1,122.3 pounds. Add this net force to the weight of 12 gallons of water (748.8 pounds), which in this example is also a net force, and we have a total force of 1,122.3 pounds + 748.8 pounds = 1,871.1 total pounds of force. Figure 2-11A and Figure 2-11B, are in agreement. (Again this small difference is due to rounding.)

The purpose of this exercise in net force is to clarify the following explanation of force. Force on a surface is a function of gravity, density of the fluid, the height (or piezometric plane) of the fluid, and the area it is exerted against. To simplify this, we can say that the influence of gravity and density gives us a constant for water of .433 psi per foot of elevation. This figure was previously shown to be the pressure exerted by a 1-foot column of water. To calculate the force at the bottom of any container regardless of shape or size, simply multiply .433 × height (or piezometric head) × the area. The piezometric plane referred to is simply the equivalent elevation head (H) of water at the point the force is measured (see Figure 2-12). This calculation gives the following formula for calculating force when height (or elevation) is known: $F = .433 \times H \times A$.

$F = .433 \times H \times A$

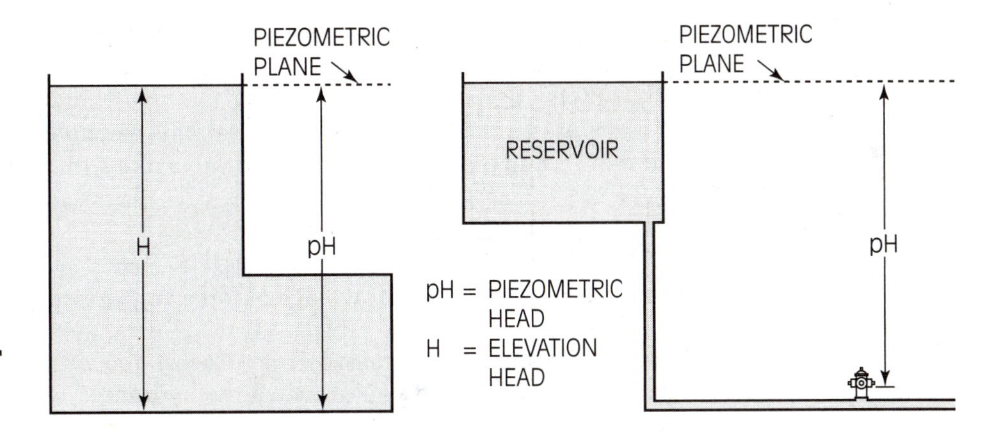

Figure 2-12
Piezometric plane.

$F = P \times A$

How do we calculate force when pressure is known? Recall that $.433 \times H$ is pressure. If we substitute P for $.433 \times H$ in the formula $F = .433 \times H \times A$, we arrive at the formula, $F = P \times A$, where F is force, P is pressure, and A is area. This formula is just a rearrangement of $P = F/A$.

Example 2-6 Using the new formula for force, prove 1,870.5 pounds is the correct force in Figure 2-11a and Figure 2-11b.

ANSWER In both cases,

$P = .433 \times H$
$= .433 \times 10$ feet
$= 4.33$ Using this figure in

or

$F = .433 \times H \times A$
$= .433 \times 10 \times 432$
$= 1{,}870.56$ pounds

$F = P \times A$ gives
$= 4.33$ psi $\times$ 432 square inches
$= 1{,}870.56$ pounds

This example of net force illustrates the point that the total force at the bottom of a container may not be equivalent to the weight of the liquid. Therefore, whenever calculating force, always use the formula $F = P \times A$. It is accurate 100 % of the time.

While on the subject of force, it should be noted at this point that force, not pressure, is the primary cause of most problems we normally associate with pressure. For example, we always say that too much pressure on a hose line makes it difficult to open the nozzle. However, if you calculate the area over which the pressure is working on the ball valve and multiply it by the pressure, you will find the force can be quite formidable. It is technically force on the nozzle preventing it from opening, not pressure.

By knowing this we can also use force to our advantage. A primary example is how a low pressure of 15 to 20 psi on the airside can hold back a water pressure of 60 or more psi in a differential dry pipe valve of a sprinkler system.

APPLICATION ACTIVITIES

Now you can put your new knowledge of force and pressure to work.

Problem 1 If the surface of a reservoir is 123 feet above the only fire hydrant in town, what pressure would you expect at the hydrant?

ANSWER $P = .433 \times H$

$= .433 \times 123$

$= .433 \times 123 = 53.25$ or 53 psi

Problem 2 If the pressure at the only hydrant in town is 63 psi, what is the elevation head of the water supply?

ANSWER $H = 2.31 \times P$

$= 2.31 \times 63$

$= 145.53$ feet above the hydrant

Problem 3 If a charged fire hydrant has a pressure of only 10 psi on the 4½-inch blind cap, what is the force on the cap?

ANSWER $F = P \times A$

$A = .7854 \times D^2$

The formula for the area of a circle is $A = .7854 \times D^2$, where A is the area and D is the diameter. (This formula is more convenient than the conventional $A = \pi \times r^2$ because we universally use diameter in the fire service. Either one will work, so use what you feel comfortable with.)

$A = .7854 \times D^2 = .7854 \times 4.5^2 = .7854 \times 20.25 = 15.9$ square inches

$F = 10$ psi $\times 15.9$

$= 159$ pounds of force on the cap

This amount of force is why it is so difficult to operate gates (valves) or remove blind caps under pressure.

Problem 4 What is the pressure at the bottom of a container of water if the force on the bottom of the container is 25 pounds and it measures 5 inches × 5 inches?

ANSWER $P = F/A$

$A = 5 \times 5 = 25$ square inches

$P = 25$ pounds of force/25 square inches $= 1$ psi

Problem 5 If atmospheric pressure at sea level is the only pressure in consideration, how high will it raise a column of water?

ANSWER $H = 2.31 \times P$

$= 2.31 \times 14.7$

$= 33.96$ feet

We now know that atmospheric pressure at sea level will raise a column of water 33.957 feet. This fact is critical in Chapter 8 when we discuss drafting.

Summary

The most basic concepts of hydraulics require understanding of force and pressure. Force and pressure are mutually dependent definitions because each is contained in the other's definition.

Fluids at rest must obey the six laws of fluid pressure. These six laws cover any possible pressure situation. Pressure itself can be measured in absolute or relative terms. And pressure can be measured as static, residual, or flow pressures.

Force is no more than the cumulative amount of pressure working over a given area. It is measured in pounds and is the true culprit in causing gates and other valves to work hard. The cumulative effect of even a small pressure over a large area can be catastrophic under the right circumstances.

Review Questions

1. What is the difference between force and pressure?
2. An imaginary line on the same plane as the level of the surface of an open container of water, used as a reference for determining elevation head is called __________.
3. The pressure reading from water at rest is called __________.
4. If a water tank is 60 feet high, what pressure will it have at its base?
5. What is the force in Question 4 if the base has an area of 2 square feet?
6. How much pressure is needed to lift water 35 feet?
7. How high will a pressure of 70 psi lift water?
8. A water tank is 10 feet in diameter and 40 feet high.
   A. How much force is exerted on the bottom of the tank if the tank is half full?
   B. How much pressure is at the base when it is full?
   C. If the tank is three-quarters full and 100 psi of air pressure is maintained in the air space, what is the force on the bottom of the tank?
9. How much force is there on a 4½-inch blind cap of a hydrant if it has a pressure on it of 65 psi?
10. How high is the piezometric plane that represents the water elevation of the reservoir that supplies the pressure to the hydrant in Question 9?
11. If you were to attach a pressure gauge to one of the 2½-inch outlets of a hydrant and then open the hydrant, allowing water to flow from one of the remaining outlets, what type of pressure would the gauge read?

List of Formulas

Finding pressure when force and area are known:

$$P = \frac{F}{A}$$

Finding pressure when the height (elevation) of water is known:

$$P = .433 \times H$$

Finding the height of water when the pressure it creates is known:

$$H = 2.31 \times P$$

Finding force when the height of the water and the area it acts on are known:

$$F = .433 \times H \times A$$

Finding force when the pressure and area are known:

$$F = P \times A$$

Finding area from diameter:

$$A = .7854 \times D^2$$

Finding absolute pressure:

Absolute pressure =
relative pressure + atmospheric pressure

Bernoulli's Theorem

Learning Objectives

Upon completion of this chapter, you should be able to:

- Understand the basic principles of Bernoulli's theorem.
- Understand the concept of conservation of energy.
- Distinguish between potential and kinetic energy.
- Understand the application of Bernoulli's theorem as it applies to hydraulics.
- Apply Bernoulli's equation to solve problems.
- Apply Torricelli's equation to calculate velocity.

INTRODUCTION

What might be considered the single most important theorem in understanding hydraulics was developed by the Swiss mathematician and physicist Daniel Bernoulli (1700–1782) and first published in 1738. I say most important because it applies to such a wide variety of situations. In fact, even beyond our study of hydraulics, everyday life is full of examples of Bernoulli's theorem.

Bernoulli's theorem is responsible for some everyday phenomena that we normally take for granted. For example, Bernoulli's theorem is responsible for giving lift to wings on aircraft. Or, have you ever wondered why the canvas top of a convertible automobile bulges up as it goes down the highway? Or how is it that a sailboat can actually sail into the wind? The answer to each question is Bernoulli's theorem.

But since this is a book about hydraulics, what examples apply to hydraulics? The first one that comes to mind is listed on the pump panel of all pumping apparatus. Note the plate that lists specifications for testing the pumps. As the required test pressure goes up, the rated capacity goes down. Bernoulli's theorem is also responsible for explaining how a foam eductor works and how a nozzle converts pressure to velocity.

The study of Bernoulli's theorem requires understanding of what appear to be complicated algebraic equations. In reality, the equation is not that complicated. In fact, the equation known as Bernoulli's equation can be, and is, broken down into very simple terms. Remember that the purpose of this chapter is to explain the theory in terms that can be understood.

CONSERVATION OF ENERGY

■ Note
Conservation of energy means that energy can neither be increased nor decreased in any system. It can only change from one form to another or be transferred from one body to another.

Daniel Bernoulli first postulated his theory to explain conservation of energy as it pertained to hydraulics. Conservation of energy means that energy can neither be increased nor decreased in any system. It can only change from one form to another or be transferred from one body to another.

To better understand the concept of conservation of energy, let us consider how energy can change from one form to another. A match has energy in the form of a chemical. When it is struck, the chemical energy is converted to heat and light energy. How much heat and light the match will give off is in proportion to the amount of chemical on the head of the match. To alter the amount of heat and light, the amount of chemical on the head of the match must be changed.

Next, we need to understand how energy can be transferred from one object to another. To understand this transfer of energy, consider a batter as he swings at a pitch. If he hits the ball squarely, the energy he has put into the bat will be transferred to the ball. If the batter put enough energy into his swing, the ball will sail over the outfield fence. In short, how far the ball will go depends on how much energy was in the bat and transferred to the ball.

BERNOULLI'S THEORY

Bernoulli's theory
where the velocity of a fluid is high, the pressure is low, and where the velocity is low, the pressure is high

Bernoulli's theory states, Where the velocity of a fluid is high, the pressure is low, and where the velocity is low, the pressure is high. In this theory, Bernoulli is saying that we can have energy, as it pertains to hydraulics, either in the form of pressure or velocity, but not both at the same time. One will be dominant. In short, energy can be transferred between pressure and velocity, but if one increases, the other must decrease. Before we continue with Bernoulli's theorem, we need to review the concept of energy.

Energy

energy
the ability to do work

potential energy
energy of position or stored energy

Energy is the ability to do work. This ability is in the form of either potential energy or kinetic energy. **Potential energy** can be defined as energy of position or stored energy. There are different kinds of potential energy. For example, a bucket of water held in the air (see Figure 3-1) has gravitational potential energy due to its position in relationship to the floor. A spring has elastic potential energy.

If the bucket in Figure 3-1 is released, it will fall to the floor and cause work to be done. The amount of work (dent it puts in the floor) will be in relation to the weight of water in the bucket and the distance (position) the bucket was from the ground. This principle can actually be put in the form of an equation; $PE_{gravity} = mgy$. In this equation $PE_{gravity}$ is the gravitational potential energy, m is mass of the object, g is the acceleration due to gravity, and y is the vertical distance. For our purposes, it is sufficient to use this equation simply to confirm our definition of potential energy as energy of position.

kinetic energy
energy of motion

Kinetic energy is defined as energy of motion. In the example with the bucket of water, if the bucket is released and allowed to fall (see Figure 3-2), it will eventually hit the floor, but it will take time. The distance that an object moves in a measured amount of time is called *velocity*. Velocity introduces motion to our definition of kinetic energy. Catch the bucket at any point on the way down and you will feel the energy.

Kinetic energy can also be defined in terms of an equation. Technically this equation refers to translational (straight line) kinetic energy, as opposed to rotational kinetic energy. The equation is $KE = \frac{1}{2}mv^2$, where KE is kinetic energy, m is the mass of the object, and v is the velocity of the object. Velocity in the equation for kinetic energy confirms kinetic energy as energy of motion.

Total PE = Total KE

The next logical step is to put both of these formulas together. In a system where energy is conserved, we can say that Total PE = Total KE, as illustrated in Figure 3-3. The bucket at position A represents total PE. The bucket at position B represents total KE. The energy at each position is the same.

What happens if we were to release the bucket and measure the energy at the halfway point? At point C in Figure 3-3 we have one-half PE (half the PE has been converted to KE) and one-half KE (half the PE still exists). Together it equals the total energy in the system. Or $\frac{1}{2}PE + \frac{1}{2}KE$ = Total KE = Total PE. In short,

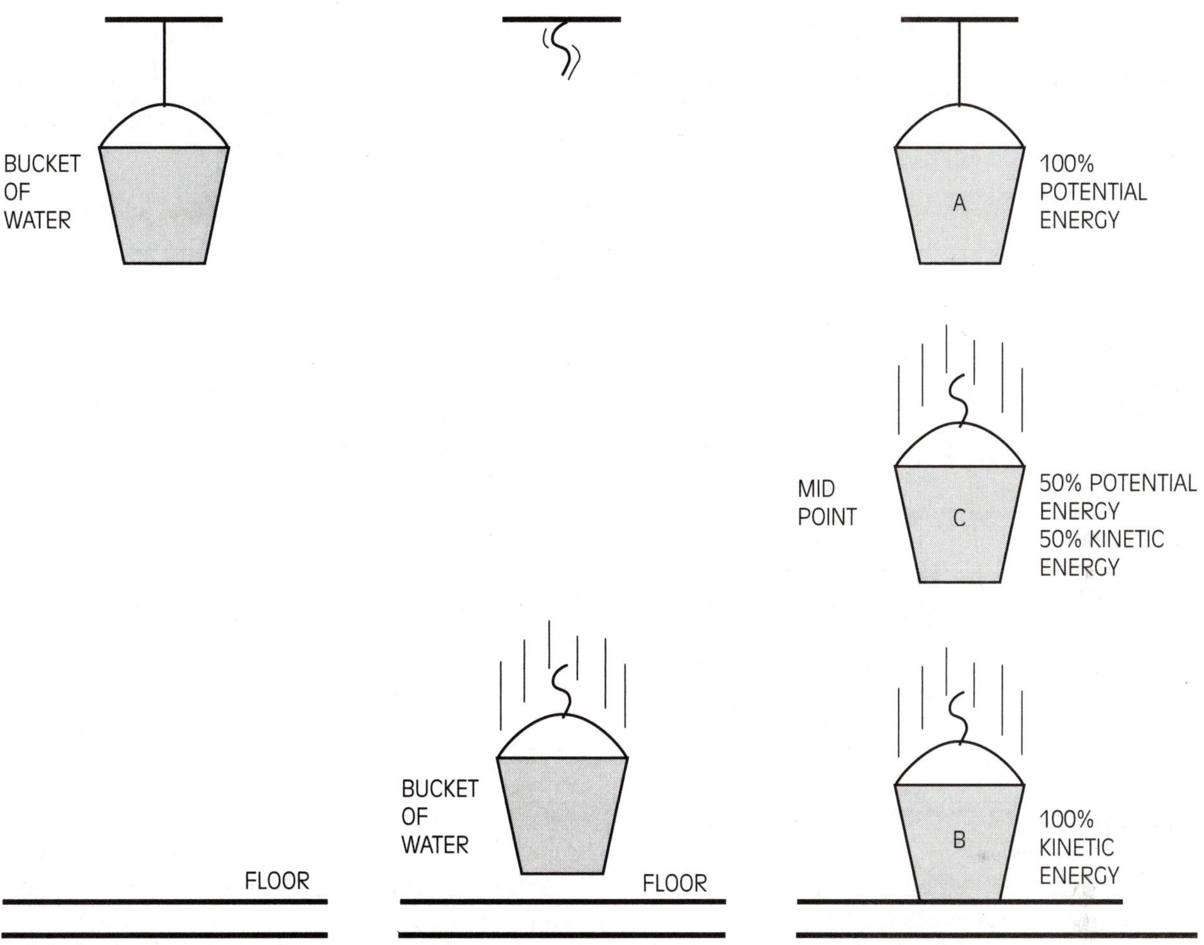

Figure 3-1 *Example of gravitational potential energy.*

Figure 3-2 *The bucket has kinetic energy.*

Figure 3-3 *Total potential energy = total kinetic energy.*

■ Note
If potential energy and kinetic energy exist at the same time, they must be proportional and add up to the total energy in the system.

energy can exist in the form of either potential energy or kinetic energy, but we cannot have total potential energy and total kinetic energy at the same time. If potential energy and kinetic energy exist at the same time, they must be proportional and add up to the total energy in the system. Energy can be transferred from potential to kinetic and back, but it can not be increased or deceased.

BERNOULLI'S EQUATION

No good scientific theory is worth its weight without a corresponding mathematical formula to prove it, and Bernoulli's theorem is no different. The purpose of the equation is to prove that energy in (*PE*) is equal to energy out (*KE*). Therefore,

Bernoulli's equation is $P_1 + \frac{1}{2}V_1^2 + gh_1d = P_2 + \frac{1}{2}V_2^2d + gh_2d$. The total energy on each side of the equals sign must be the same because each side represents total energy of different points in the same system. Each side of the equals sign has potential for a mixture of both potential and kinetic energy. For our purpose, we can redefine the terms of the equation to a simpler form as follows:

P = Pressure head = PH

$\frac{1}{2}V^2d$ = Velocity head = VH

ghd = Elevation head = EH

$PH_1 + EH_1 + VH_1 = PH_2 + EH_2 + VH_2$

The term head in each of the definitions above refers to an equivalent in feet. For example, 20 psi is equivalent to a pressure head of 46.18 feet. This allows us to restate Bernoulli's theory in the form of an equation as $PH_1 + VH_1 + EH_1 = PH_2 + VH_2 + EH_2$. **Bernoulli's equation** allows us to make comparisons of the total energy between any two points within a system. Our answer is expressed in feet, instead of some technical scientific jargon, but it is valid just the same. The following two examples demonstrate how this equation is used:

Example 3-1 A water tank with water to the 20-foot level has a pressure gauge at the base of the tank that reads a pressure of 8.66 psi. Using Bernoulli's equation, compare the total head at point 1 and point 2 (see Figure 3-4).

ANSWER

$PH_1 = 0$ $PH_2 = 8.66$ psi

$EH_1 = 20$ feet $EH_2 = 0$

$VH_1 = 0$ $VH_2 = 0$

The 8.66 psi must be converted into feet.

The equation is written:

$PH_1 + EH_1 + VH_1 = PH_2 + EH_2 + VH_2$

0 + 20 feet + 0 = 8.66 psi + 0 + 0

0 + 20 feet + 0 = (8.66 psi × 2.31) + 0 + 0

0 + 20 feet + 0 = 20 feet + 0 + 0

20 feet = 20 feet

Both sides of the equation are in agreement.

Example 3-2 This example is the same as Example 3-1, only it is a closed system under 50 psi pressure (see Figure 3-5).

ANSWER

$PH_1 = 50$ psi $PH_2 = 58.66$ psi

$EH_1 = 20$ feet $EH_2 = 0$

$VH_1 = 0$ $VH_2 = 0$

Because of Pascal's principle, the pressure at point 2 would be 58.66 psi.

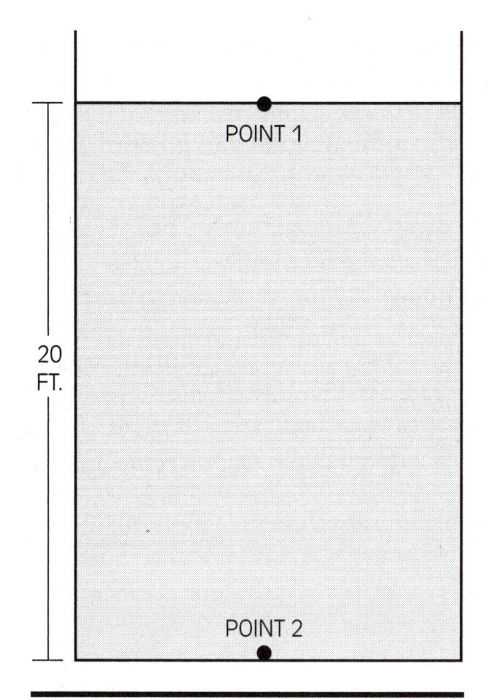

Figure 3-4 *Are point 1 and point 2 in agreement?*

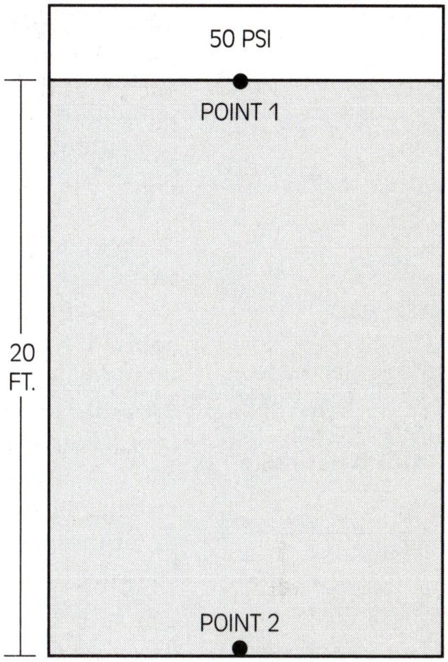

Figure 3-5 *Is total energy at point 1 equal to total energy at point 2?*

The equation is written:

$$PH_1 + EH_1 + VH_1 = PH_2 + EH_2 + VH_2$$

$$50 \text{ psi} + 20 \text{ feet} + 0 = (50 \text{ psi} + 8.66 \text{ psi}) + 0 + 0$$

$$(50 \text{ psi} \times 2.31) + 20 \text{ feet} + 0 = (58.66 \times 2.31) + 0 + 0$$

$$115.\ 5 \text{ feet} + 20 \text{ feet} + 0 = 135.\ 5 \text{ feet} + 0 + 0$$

$$135.45 \text{ feet} = 135.45 \text{ feet}$$

Both sides of the equation are in agreement.

In each of the two examples the equations have balanced, which proves that the energy in each system has remained constant, or in other words, has been conserved.

■ Note

Bernoulli's theorem assumes a system has no friction loss.

A word of caution is in order here. Bernoulli's theorem assumes a system has no friction loss. In real life, friction loss is as certain as death and taxes and needs to be taken into account. When applying the concepts in Bernoulli's theorem to real life, keep the friction loss in mind.

APPLICATION OF BERNOULLI'S THEOREM

Bernoulli's theorem has many everyday applications in hydraulics, most notably in the pump itself. As the capacity of the pump increases, the maximum discharge pressure decreases. For example, in order to flow the capacity of a pump, the pump is limited to a maximum of 150 psi discharge pressure. If we limit our discharge to 70% of the pump's capacity, we can increase the maximum discharge pressure to 200 psi. If we are content with only 50% of the capacity of the pump, we can pump at a maximum of 250 psi discharge pressure.

venturi tube
a restriction in a conduit intended to increase velocity with a corresponding reduction in pressure

A **venturi tube** is another excellent example of how Bernoulli's theorem is applied in the real world of hydraulics. A venturi tube is a restriction in a conduit intended to increase velocity with a corresponding reduction in pressure (see Figure 3-6). Notice that the cross-sectional areas at points A and C are the same. However, the cross-sectional area at point B is smaller. Water flowing through the conduit at points A and C will have the same velocity. Because the area in the restricted portion of the conduit is less than at points A or C, the water has to flow faster through point B. Another way to say this is that the velocity at point B will be faster than at A or C.

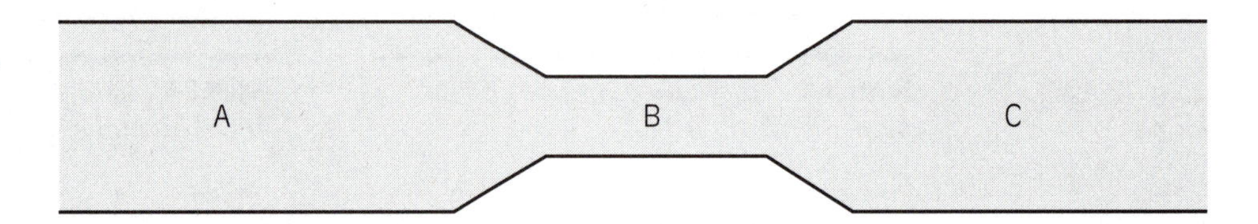

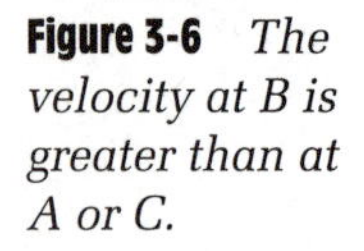

Figure 3-6 *The velocity at B is greater than at A or C.*

Now recall exactly what Bernoulli's theorem says: Where the velocity of a fluid is high, the pressure is low, and where the velocity is low, the pressure is high. If we apply this theorem to the venturi tube in Figure 3-6 we realize that the pressure at point B is less than the pressure at either point A or point C.

If our venturi tube is configured with a pickup tube attached at point B (see Figure 3-7), and the velocity of water is high enough, the pressure at point B can be reduced below atmospheric pressure. When this happens, the pickup tube can be used to draft a fluid, such as foam concentrate, out of a container. This effect is referred to as a venturi. A venturi tube designed to pick up foam concentrate is called a *foam eductor.*

Torricelli's theorem
the velocity of water escaping from an opening below the surface of a container of water will have the same velocity, minus exit losses, as if it were to fall the same distance

Torricelli's Theorem

A special application of Bernoulli's theorem is called Torricelli's theorem, even though the Italian physicist and mathematician Evangelista Torricelli (1608–1647) first proved it about one hundred years before Bernoulli was even born. **Torricelli's theorem** says that the velocity of water escaping from an opening below the

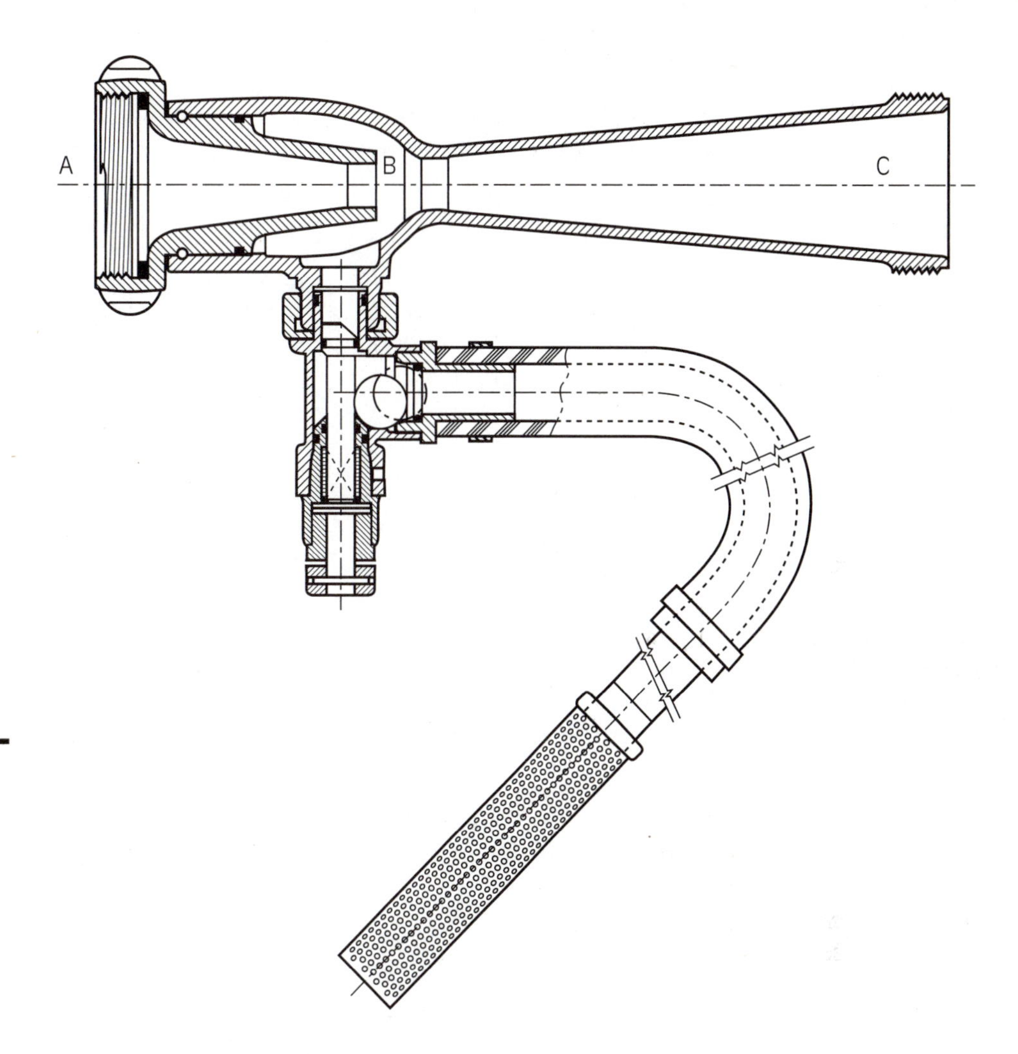

Figure 3-7 *Foam eductor. Points A, B, and C are the same as in Figure 3-6. Courtesy of Elkhart Brass Manufacturing Company, Inc.*

$Ve = \sqrt{2g(y_2 - y_1)}$

surface of a container of water will have the same velocity, minus exit losses, as if it were to fall the same distance. This means that the velocity of water at a discharge point a given distance below the surface of a body of water will be accelerated by the force of gravity exactly as if the water were to free-fall the same distance. The only thing we need to do to get the actual velocity is factor in the friction loss at the discharge (see Figure 3-8). This leads to the formula for **Torricelli's equation**, $Ve = \sqrt{2g\,(y_2 - y_1)}$, where Ve is velocity in feet per second (fps), g is the gravitational acceleration constant of 32 feet per second per second, and $y_2 - y_1$ is the elevation difference between the top of the container and the point where the water is being discharged.

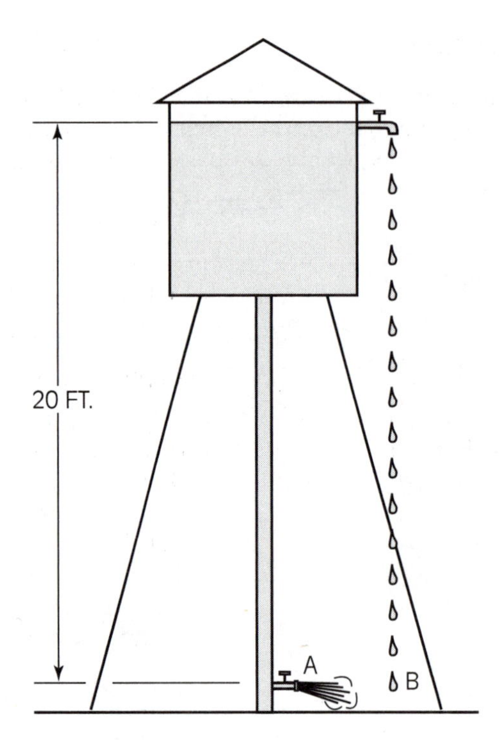

Figure 3-8 *Velocity at A is the same as at B.*

If we rewrite the equation, substituting H for $y_2 - y_1$, and simplify it as far as possible, the equation becomes:

$$Ve = \sqrt{2g\,H}$$
$$= \sqrt{2 \times 32 \times H}$$
$$= \sqrt{64 \times H}$$
$$= 8\sqrt{H}$$

$Ve = 8\sqrt{H}$

The formula for determining velocity from a discharge, when the elevation head is known is now $Ve = 8\sqrt{H}$, where H is the elevation or elevation head.

Example 3-3 What is the velocity of a discharge 20 feet below the surface of an open container?

ANSWER To solve this problem we use the formula $Ve = 8\sqrt{H}$

$$Ve = 8\sqrt{H}$$
$$= 8\sqrt{20}$$
$$= 8 \times 4.47$$
$$= 35.76 \text{ fps}$$

Revised Bernoulli's Equation

Torricelli's equation can be rewritten as, $H = Ve^2/2g$. If we insert this equation directly into Bernoulli's equation, we convert velocity directly to feet (velocity head). If we also insert an expression for pressure so as to get a direct equivalent in feet, we get a revised Bernoulli's equation. Following are the revised equivalents for *PH*, *EH*, and *VH*:

$PH = P/W$	where P = pressure and W = weight of water 1 inch × 1 inch × 1 foot tall
$EH = Z$	H is replaced with Z to indicate elevation
$VH = Ve^2/2g$	As explained previously.

The revised form of Bernoulli's equation becomes $P_1/W + Z_1 + Ve_1^2/2g = P_2/W + Z_2 + Ve_2^2/2g$. We can also insert the constants we already know, that is W = .433 and 2g = 64, and get $P_1/.433 + Z_1 + Ve_1^2/64 = P_2/.433 + Z_2 + Ve_2^2/64$

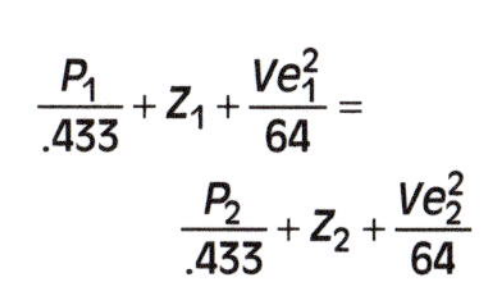

This revised form of Bernoulli's equation saves a couple of steps. There is no need for separate calculations, which then have to be put into the equation, to convert pressure into equivalent feet or velocity into equivalent feet. Just plug in the appropriate figures for pressure and/or velocity and do the calculations. The answer will be in feet.

Example 3-5 A water tower has a water level of 20 feet and a pressure of 50 psi. The discharge has been only partially opened, allowing a discharge velocity of 35.77 fps with a residual pressure of 50 psi. Use the revised Bernoulli's equation to compare point 1 and point 2 (see Figure 3-9).

ANSWER Simply insert the appropriate numbers in the equation.

$$P_1/.433 + Z_1 + Ve_1^2/64 = P_2/.433 + Z_2 + Ve_2^2/64$$

$$50 \text{ psi}/.433 + 20 \text{ feet} + 0 = 50 \text{ psi}/.433 + 0 + 35.77^2/64$$

$$115.47 \text{ feet} + 20 \text{ feet} + 0 = 115.47 + 0 + 20 \text{ feet}$$

$$135.47 \text{ feet} = 135.47 \text{ feet}$$

The revised formula gives us the answer without having to do separate calculations to find the *PH* or *VH*. This form of the equation is also the one found in the eighteenth edition of the *Fire Protection Handbook* (National Fire Protection Association, Quincy, Massachusetts, 1997).

APPLICATION ACTIVITIES

You are now ready to put your knowledge of Bernoulli's theorem to work.

Problem 1 A pumper is hooked up to a hydrant and pumping 100 psi to a line that extends down a hill with an elevation difference of 25 feet. Using Bernoulli's equation, calculate the total head for both points 1 and 2 (see Figure 3-10).

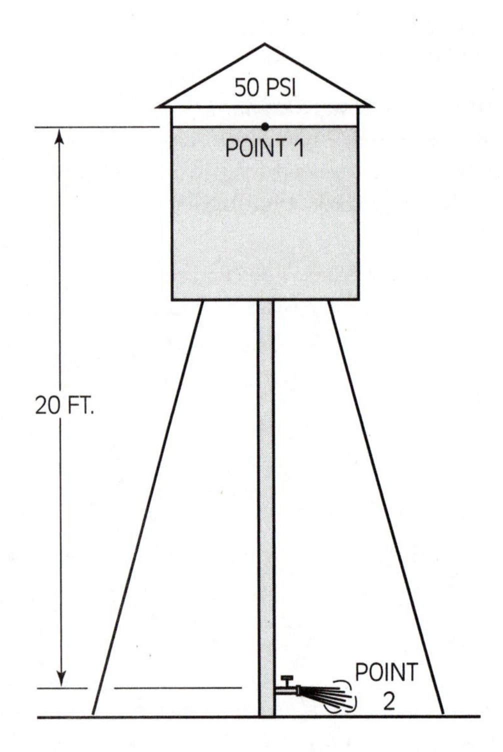

Figure 3-9 *Compare point 1 and point 2.*

ANSWER $PH_1 + EH_1 + VH_1 = PH_2 + EH_2 + VH_2$

100 psi + 25 feet + 0 = (100 psi + 10.83 psi) + 0 + 0

(100×2.31) + 25 feet + 0 = (110.83×2.31) + 0 + 0

231 feet + 25 feet + 0 = 256 + 0 + 0

256 feet = 256 feet

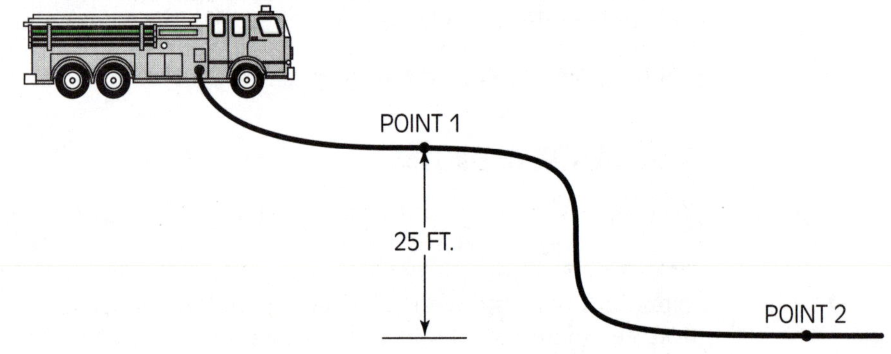

Figure 3-10 *What is the total head at point 1 and point 2?*

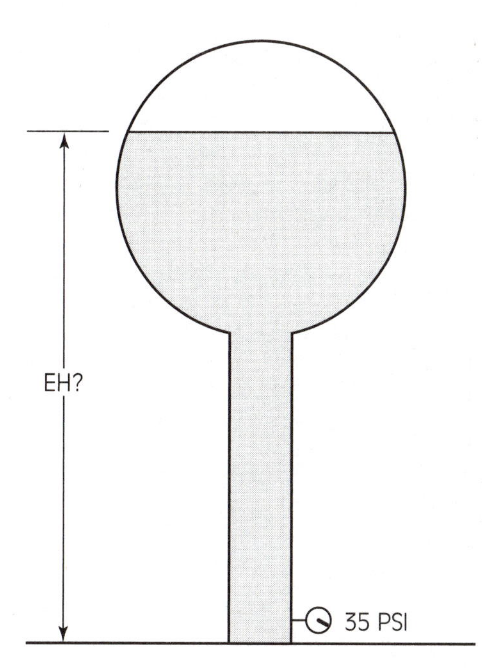

Figure 3-11 *What is the elevation head?*

Problem 2 Figure 3-11 shows a water tank with a pressure at the base of 35 psi. Using Bernoulli's equation, calculate EH_1 for point 1.

ANSWER

$$PH_1 + EH_1 + VH_1 = PH_2 + EH_2 + VH_2$$
$$0 + EH_1 + 0 = 35 \text{ psi} + 0 + 0$$
$$0 + EH_1 + 0 = (35 \times 2.31) + 0 + 0$$
$$0 + EH_1 + 0 = 80.85 \text{ feet} + 0 + 0$$
$$EH_1 = 80.85 \text{ feet}$$

Problem 3 Using Torricelli's equation, calculate the velocity at the discharge where the elevation head is 50 feet.

ANSWER Use the abbreviated form of Torricelli's equation, $Ve = 8\sqrt{H}$.

$$Ve = 8\sqrt{H}$$
$$= 8\sqrt{50} \text{ feet}$$
$$= 8 \times 7.07$$
$$= 56.56 \text{ fps}$$

Problem 4 Using Torricelli's equation, calculate the elevation needed to give a velocity of 113 fps.

ANSWER Rearrange Torricelli's equation to read $H = (Ve/8)^2$

$$H = (Ve/8)^2$$
$$= (113/8)^2$$
$$= 14.125^2$$
$$= 199.5 \text{ feet}$$

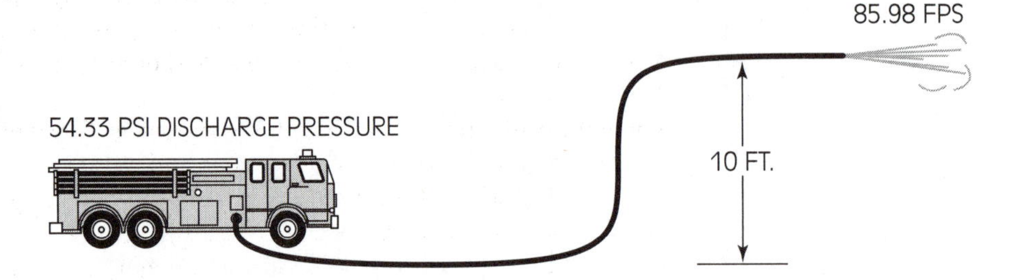

Figure 3-12 *Compare the pressure at the wagon with the pressure at the nozzle.*

Problem 5 A pumper is pumping at 54.33 psi to a smooth bore nozzle 10 feet above the pumper. The nozzle has a discharge velocity of 85.98 fps. Use the revised version of Bernoulli's equation to compare pressure at the wagon with pressure at the nozzle. Note: The engine pressure does not include any friction loss because Bernoulli's equation does not take it into consideration (see Figure 3-12).

ANSWER Using the revised form of Bernoulli's equation we get:

$$P_1/.433 + Z_1 + Ve_1^2/64 = P_2/.433 + Z_2 + Ve_2^2/64$$
$$54.33 \text{ psi}/.433 + (-10 \text{ feet}) + 0 = 0 + 0 + 85.98 \text{ fps}^2/64$$
$$125.5 + (-10) + 0 = 0 + 0 + 115.5$$
$$115.5 \text{ feet} = 115.5 \text{ feet}$$

Note: In this example, because the pumper is pumping uphill, the elevation represents a loss of pressure. Not a loss in the sense that the energy is lost, but a loss in a sense that the pump must compensate for this downward pressure or energy. By this we mean that the pressure created by the elevation, 4.33 psi, works against the pressure being created by the pumper, and by the time the pressure reaches the nozzle it is only 50 psi. In this example the pumper has had to pump at 54.33 psi in order to compensate for this downward pressure and still have a nozzle pressure of 50 psi. This pressure is called back pressure. We can more precisely define **back pressure** as the pressure exerted by a column of water against the discharge of a pump.

back pressure
the pressure exerted by a column of water against the discharge of a pump

Summary

Energy is fixed in any given system. It can change forms or be transferred to another object, but it cannot be increased or decreased. Bernoulli's theorem applies this theory to our study of hydraulics. Whether we are talking about the capacity of a pump or how a foam eductor works, Bernoulli's theorem is responsible.

We are able to test Bernoulli's theorem with his appropriately named equation, Bernoulli's equation, which allows us to compare both kinetic and potential energy at any two points in a system. The total energy at these two points must be equal.

A special application of Bernoulli's theorem is Torricelli's theorem, which addresses the relationship between elevation and velocity.

For the record, why does the roof of a convertible bulge out as the car goes down the road? The air traveling across the top of the car has velocity, but the air inside does not, even though it is traveling just as fast as the car. As Bernoulli's theorem states, where the velocity is high, the pressure is low. Thus the higher-pressure air inside the car pushes out on the only part of the car that will give, the canvas top!

Review Questions

1. In what year did Bernoulli first publish his theorem on conservation of energy, as it pertains to hydraulics?
2. How are pressure and velocity, in a hose, related to each other?
3. The ability to do work is called __________.
4. Give an example of potential energy not already given in this chapter.
5. If the contents of a bucket of water were to be dropped from an elevation of 10 feet, what would the energy measurement be at the halfway point? (Hint: Define in terms of potential or kinetic energy.)
6. What is the pressure head equivalent to 35 psi?
7. Are there any situations when the total energy into a system is not equal to the total energy out?
8. Torricelli's theorem correlates velocity to what factor or "head"?
9. In its original form, what does *g* stand for in Torricelli's equation?
10. What is the advantage of using the revised form of Bernoulli's equation, $P_1/.433 + Z_1 + Ve_1^2/64 = P_2/.433 + Z_2 + Ve_2^2/64$, instead of the $PH_1 + EH_1 + VH_1 = PH_2 + EH_2 + VH_2$ version?
11. A gauge reading 15 psi is attached to a tank outlet with water discharging at 30 fps. What is the elevation of water in the tank above the level of the outlet?

List of Formulas

Conservation of energy:

$$\text{Total } PE = \text{Total } KE$$

Bernoulli's equation:

$$PH_1 + EH_1 + VH_1 = PH_2 + EH_2 + VH_2$$

Finding velocity when the height is known.
Torricelli's equation simplified:

$$Ve = 8\sqrt{H}$$

Revised Bernoulli's equation:

$$\frac{P_1}{.433} + Z_1 + \frac{Ve_1^2}{64} = \frac{P_2}{.433} + Z_2 + \frac{Ve_2^2}{64}$$

Torricelli's equation:

$$Ve = \sqrt{2g(y_2 - y_1)}$$

Reference

Arthur E. Cote, P.E., Editor-in-chief, *Fire Protection Handbook,* 18th ed. Quincy, MA: National Fire Protection Association, 1997.

Velocity and Flow

Learning Objectives

Upon completion of this chapter, you should be able to:

- Calculate velocity, given either height or pressure.
- Calculate either height or pressure, given velocity.
- Calculate the change of velocity corresponding to a change of area.
- Understand the relationship between velocity and flow.
- Understand the relation between velocity and area.
- Calculate flow, given area of discharge and velocity.

INTRODUCTION

This chapter actually began in Chapter 3 with Torricelli's theorem, which introduces velocity to our study of hydraulics. After all, any time water leaves an opening (e.g., through a nozzle), it has velocity. Being able to calculate the velocity of water flowing from an opening is the first step in calculating how much water is flowing. There is a direct relationship between the velocity of water through an opening, the area of the opening, and the amount of water being discharged. Thus, just by using a simple formula of area times velocity we are able to calculate how much water is flowing.

In this chapter, the quantity of water being discharged is calculated in cubic feet. It is important that close attention be given to the relationship between the area of an opening and the velocity of the water (or other fluid) being discharged in calculating quantity of water being discharged. In Chapter 5 we take this process one step further by developing a formula that calculates discharge in gallons per minute.

Given a fixed flow of water, as the size of the hose changes so does velocity. This change is directly proportional to the area of the hose. This chapter shows how to calculate the change in velocity due to change in the size of hose and how to calculate the change in hose size when the change in velocity is known.

VELOCITY

Because velocity is critical in calculating flow, it is worth taking the time to review how velocity is calculated when elevation is known. Then, we add a new formula that is used to calculate velocity when pressure is known.

In Chapter 3 the concept of velocity was introduced in the form of Torricelli's theorem. Torricelli said that water exiting an opening will have the same velocity as if it were to free-fall the same distance. This property allows us to use the formula $Ve = \sqrt{2gH}$ to calculate velocity. A physicist would use this same formula to calculate the velocity of an object in free fall.

The fact that the gravitational acceleration constant is part of this formula should not be taken for granted. Remember that all objects on planet Earth are affected by gravity. When those objects are "falling," they are accelerating at a constant rate determined by the force of gravity, that is, the gravitational acceleration constant. In the study of hydraulics the object "falling" is water. In Chapter 3 we simplified the formula to $Ve = 8\sqrt{H}$.

Example 4-1 Water is flowing from an opening at the base of a water tank. If the water inside is 49 feet above the point of discharge, what is the velocity? (See Figure 4-1.)

ANSWER

$$\begin{aligned} Ve &= 8\sqrt{H} \\ &= 8\sqrt{49} \\ &= 8 \times 7 \\ &= 56 \text{ fps} \end{aligned}$$

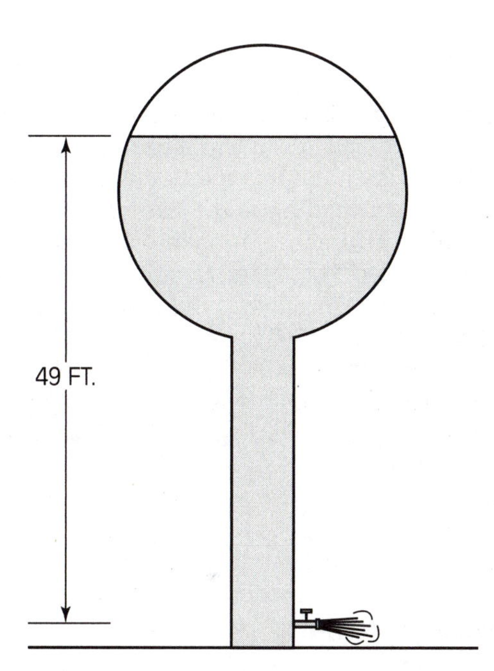

Figure 4-1 *Find the velocity at the discharge.*

Velocity from Pressure

For our purpose, however, this formula does have limits. We in the fire service usually need to be able to calculate velocity from pressure, not elevation. The good news is we can use the same basic formula, $Ve = \sqrt{2gH}$, but in order to use it we must replace H in the basic formula with an appropriate expression for pressure. Fortunately, we learned just such an expression in Chapter 2. That expression (formula) is $H = 2.31 \times P$. The formula for finding velocity from pressure then is $Ve = \sqrt{2 \times g \times 2.31 \times P}$. Recall that g is 32 feet per second per second.

$Ve = \sqrt{2 \times g \times 2.31 \times P}$

Just as we have sought to simplify other formulas, this one is no exception. We can simplify it as follows:

$$Ve = \sqrt{(2 \times 32 \times 2.31) \times P}$$

$$= \sqrt{147.84 \times P}$$

$$= 12.16\sqrt{P}$$

$Ve = 12.16\sqrt{P}$

By multiplying $2 \times 32 \times 2.31$ and then finding the square root of the product, we create a constant of 12.16, which is multiplied by the square of the pressure to get velocity. It will work in all instances where we need to calculate velocity from pressure.

Example 4-2 What is the discharge velocity of a smooth-bore nozzle with a 50 psi nozzle pressure?

ANSWER $Ve = 12.16\sqrt{P}$

$= 12.16\sqrt{50}$

$= 12.16 \times 7.07$

$= 89.97$ fps

Example 4-3 What is the discharge velocity on the 2½-inch discharge of a fire hydrant if it has a pressure of 36 psi?

ANSWER $Ve = 12.16\sqrt{P}$

$= 12.16\sqrt{36}$

$= 12.16 \times 6$

$= 72.96$ fps

Finding Height When Velocity Is Known

$$H = \left(\frac{Ve}{8}\right)^2$$

Just as it is necessary at times to be able to calculate the elevation of a water source when we know the pressure, a thorough understanding of hydraulics requires that we be able to find elevation when we know the velocity. To do this, first start with the formula for finding velocity when the elevation is known, $Ve = 8\sqrt{H}$. By rearranging the elements of the formula, in keeping with the appropriate laws of mathematics, we end up with the formula $H = (Ve/8)^2$. This formula is the same one we used in Chapter 3 to calculate the velocity head in Bernoulli's formula.

Example 4-4 Find the elevation that creates a velocity of 50 fps (see Figure 4-2).

ANSWER Use the formula $H = (Ve/8)^2$

$H = (Ve/8)^2$

$= (50/8)^2$

$= (6.25)^2$

$= 39.06$ feet

Finding Pressure When Velocity Is Known

$$P = \left(\frac{Ve}{12.16}\right)^2$$

Finding pressure from velocity is very similar to finding elevation from velocity. We start with the basic formula for finding velocity from pressure, $Ve = 12.16\sqrt{P}$, and convert it to find pressure. By following the appropriate laws of mathematics, the formula is rearranged to read $P = (Ve/12.16)^2$. This formula will calculate the pressure whenever we are given velocity at a point of discharge.

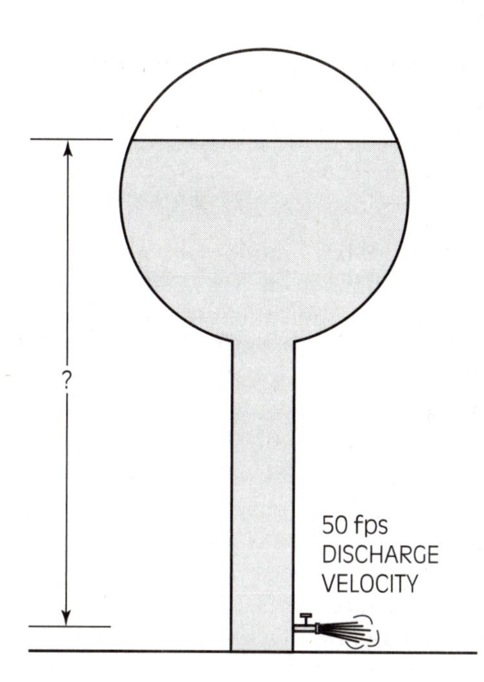

Figure 4-2 *Find the elevation of the water.*

Example 4-5 Find the pressure where the discharge velocity is 50 fps.

ANSWER Use the formula $P = (Ve/12.16)^2$

$$
\begin{aligned}
P &= (Ve/12.16)^2 \\
&= (50/12.16)^2 \\
&= (4.11)^2 \\
&= 16.89 \text{ psi}
\end{aligned}
$$

Notice that in both Examples 4-4 and 4-5 a velocity of 50 fps was used. In Example 4-4 we found that 50 fps is equivalent to an elevation of 39.06 feet. And in Example 4-5 it was found that a velocity of 50 fps is equivalent to a pressure of 16.89 psi. If the formulas in examples 4-4 and 4-5 are correct, 39.06 feet of elevation should be equal to 16. 89 psi.

Example 4-6 How much pressure is exerted by an elevation of 39.06 feet?

ANSWER From Chapter 2 we use the formula $P = .433 \times H$, to find pressure from elevation.

$$
\begin{aligned}
P &= .433 \times H \\
&= .433 \times 39.06 \\
&= 16.91 \text{ psi}
\end{aligned}
$$

The difference between the 16.91 psi and 16.89 psi is due to rounding. The formulas for finding height and pressure from velocity work. The formula, $H = 2.31 \times P$, can also be used to verify the answers to Examples 4-4 and 4-5.

AREA AND VELOCITY

$A_1 \times Ve_1 = A_2 \times Ve_2$

The velocity of water in a hose is a function of both the quantity of water flowing and the cross-sectional area of the hose. Just as in a venturi tube, if we reduce the size of the hose (see Figure 4-3), the velocity will increase. If we know the velocity in either size hose, we can calculate the velocity in the other. To do this we need to introduce the formula: Area_1 x Velocity_1 = Area_2 x Velocity_2 or $A_1 \times Ve_1 = A_2 \times Ve_2$. In this formula, velocity is given in fps. Using the formula $A_1 \times Ve_1 = A_2 \times Ve_2$, as long as we are given any three factors we can find the fourth.

This formula tells us that velocity of water in a hose, for a given quantity of flow, is inversely proportional to the change in area of the hose. More simply put, for a given flow, if the area of the hose is doubled, the velocity will be reduced by one-half, and if the area of the hose is reduced by one-half, the velocity will double.

A unique feature of this formula is that while velocity is always given in feet per second, area can be in any unit of area. What A_1 and A_2 do, in this formula, is establish a ratio of their respective areas. That ratio is the same whether the area is given in square inches, square feet, or even if given in square yards.

Example 4-7 If the velocity of water in the 2½-inch hose in Figure 4-3 is 10 fps, what is the velocity in the 1½-hose?

Answer Use the formula $A_1 \times Ve_1 = A_2 \times Ve_2$

where A_1 is the area of 2½-inch hose
Ve_1 is 10 fps
A_2 is the area of 1½-inch hose
Ve_2 is what we are determining

$$A_1 \times Ve_1 = A_2 \times Ve_2$$
$$(.7854 \times 2.5^2) \times 10 = (.7854 \times 1.5^2) \times Ve^2$$
$$4.9 \times 10 = 1.77 \times Ve_2$$
$$Ve_2 = 49/1.77$$
$$Ve_2 = 27.68 \text{ fps}$$

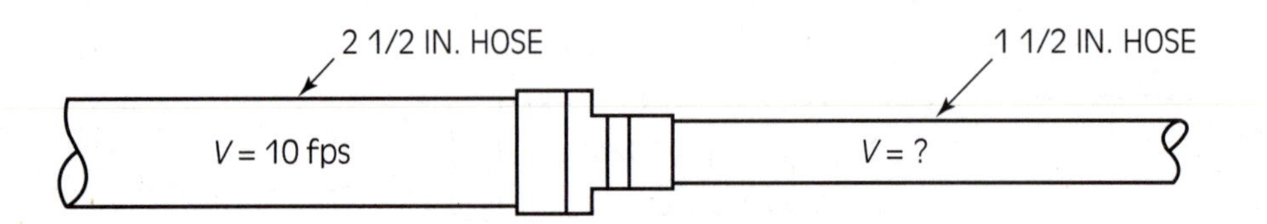

Figure 4-3 *Find the velocity in the 1½-inch hose.*

In Example 4-7, the area of the 2½-inch hose is 2.768 times larger than the area of the 1½-inch hose (4.9/1.77 = 2.768). And, the velocity in the 1½-inch hose is 2.768 times greater than in the 2½-inch hose (27.68/10 = 2.768). These figures verify that for a given flow, the velocity of the water is inversely proportional to the change in area of the hose.

At times it may be necessary to find the area of a second hose size when the velocity in both hose sizes is known. As already pointed out, we still use the formula $A_1 \times Ve_1 = A_2 \times Ve_2$, only in this case we are solving for A_2 instead of Ve_2.

Example 4-8 The velocity of water in a 3-inch hose is 10 fps. Personnel then change to a smaller hose with a velocity of 29.29 fps. What is the area, and size, of the smaller hose? (See Figure 4-4.)

ANSWER

$$A_1 \times Ve_1 = A_2 \times Ve_2$$
$$(.7854 \times D^2) \times 10 = A_2 \times 29.29$$
$$(.7854 \times 3^2) \times 10 = A_2 \times 29.29$$
$$7.07 \times 10 = A_2 \times 29.29$$
$$A_2 = 70.7/29.29$$
$$= 2.41$$

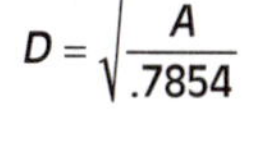

Now that we have the area of the hose, we need to find its diameter. To do that we rearrange the formula $A = .7854 \times D^2$ to read $D = \sqrt{A/.7854}$, which gives us the diameter of any circular opening when we know the area.

$$D = \sqrt{A/.7854}$$
$$= \sqrt{2.41/.7854}$$
$$= \sqrt{3.06}$$
$$= 1.75\text{-inch hose}$$

The velocity in the 1¾-inch hose is 2.9 (29.29/10 = 2.9) times the velocity in the 3-inch hose. The area of the 3-inch hose is 2.9 (7.06/2.41 = 2.9) times the area of the 1¾-inch hose.

In both Examples 4-7 and 4-8, we calculated the actual area of the hose involved. When a quick assessment of the area of two different size hose is needed, it is sufficient to simply use D^2 to make the comparison. In other words, to make

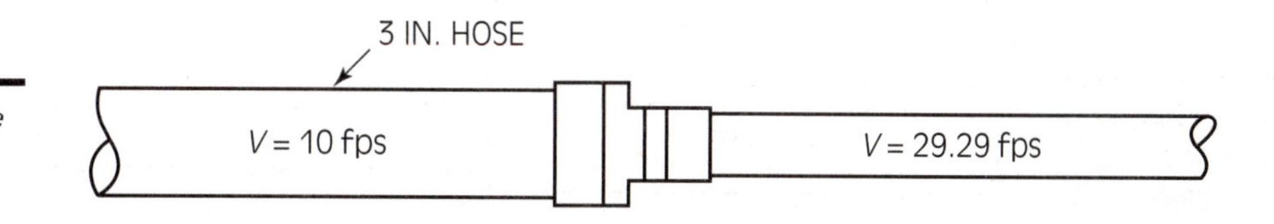

Figure 4-4 *Find the diameter of the smaller hose.*

■ Note
To make a quick assessment of the area of two different size hoses, "square and compare the diameter of the hose."

a quick assessment, simply "square and compare the diameter of the hose." It will give you the same relationship as if you took the extra time to actually find the area of the hose involved.

CALCULATING FLOW

Because we now know how to calculate velocity and we are aware of the relationship of area to velocity, we can now put these two functions together to find the quantity of water flowing. When calculating flow, we use velocity of the water as it exits an opening, such as a nozzle. The velocity of the water is a linear dimension that gives us the distance the water will flow in a second. When we multiply the velocity by the area of the opening, or nozzle size, the discharge becomes a volume.

This concept is easy to visualize if we think of the water being discharged as a cylinder. The area of the opening is the area of the cylinder, and the velocity, or distance the water will flow in 1 second, is the length. We already know that the volume of a cylinder is area times length, or height. If we substitute quantity for volume and velocity per second for length, we have the formula, $Q = A \times Ve$. Remember, since the quantity of water flowing in 1 second is a volume, the answer must be in cubic feet per second (cfs).

$Q = A \times Ve$

Example 4-9 How much water will flow through an opening that is 1 square foot in area, if it is traveling at a velocity of 89 fps? (See Figure 4-5.)

ANSWER

$$\begin{aligned} Q &= A \times Ve \\ &= 1 \times 89 \\ &= 89 \text{ cfs} \end{aligned}$$

In the fire service we are rarely, if ever, given the luxury of being given the area in units of square feet. When given the size of openings we are always given the diameter, never the area. This requires us to insert the formula for calculating area from diameter into the formula for finding quantity flow. The formula then becomes $Q = .7854 \times D^2 \times Ve$.

$Q = .7854 \times D^2 \times Ve$

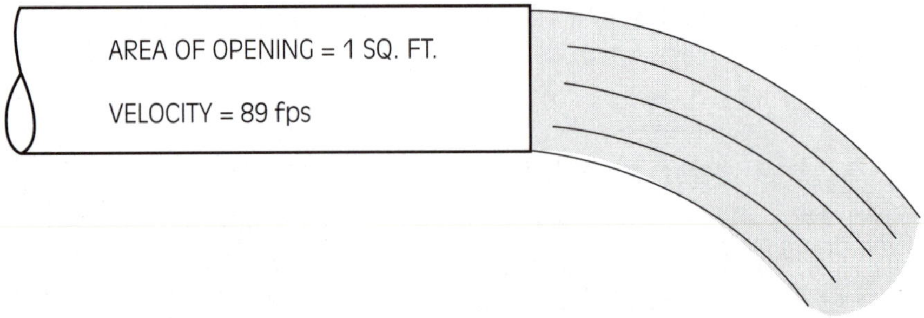

AREA OF OPENING = 1 SQ. FT.
VELOCITY = 89 fps

Figure 4-5 *Calculate the flow in 1 second.*

Example 4-10 How much water will an opening discharge if it is 1.128 feet in diameter and has a velocity of 10 fps?

ANSWER In this example, the velocity will discharge a cylinder of water 10 feet long in 1 second. Use the formula $Q = .7854 \times D^2 \times Ve$ to solve for volume.

$$Q = .7854 \times D^2 \times Ve$$
$$= .7854 \times 1.128^2 \times 10$$
$$= .7854 \times 1.272 \times 10$$
$$= .999 \times 10$$
$$= 9.99 \text{ or } 10 \text{ cubic feet}$$

It should be noted here that in the formula $Q = A \times Ve$, area and velocity must both be in the same units. Remember that we have already assigned units of feet to velocity, and we usually deal with area of hose and nozzles in units of square inches. To use this formula, when we are given opening sizes in inches, it will be necessary to convert the area of any opening into square feet. By multiplying the area in square inches by 1/144, square inches are converted to square feet. The 144 represents the number of square inches in a square foot.

In application in the fire service, $Q = A \times Ve$ is used to find flow through a circular opening of only a couple inches (at most). If we substitute $.7854 \times D^2 \times 1/144$ for A, we will automatically be finding the area of any small circular opening in square feet. The formula then becomes $Q = .7854 \times D^2 \times 1/144 \times Ve$. All we need to do is plug in the diameter and velocity and we have the amount of water that will flow in 1 second.

$$Q = .7854 \times D^2 \times \frac{1}{144} \times Ve$$

Example 4-11 How much water will flow in 1 second through a 1¼-inch opening, if it has a velocity of 86 fps? (See Figure 4-6.)

ANSWER

$$Q = .7854 \times D^2 \times 1/144 \times Ve$$
$$= .7854 \times 1.25^2 \times 1/144 \times 86$$
$$= .0085 \times 86$$
$$= .731 \text{ cfs}$$

This formula can be used any time we need to find the flow from any circular opening.

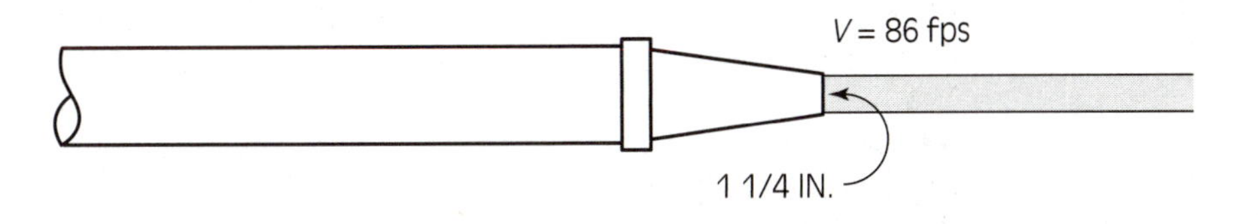

Figure 4-6 *How much water is flowing?*

Finding the Opening Diameter When Volume and Velocity Are Known

$A = \frac{Q}{Ve}$

Calculating the quantity of water discharged in 1 second is a straightforward formula. But how would you find the area of an opening if you knew the quantity of water being discharged and the velocity? We can easily solve for area using the same formula we use to calculate the quantity of water flowing. Just fill in Q with the cubic feet of water flowing each second, and then insert the velocity. Then solve for area by dividing Q by Ve. The formula to find the area then becomes: $A = Q/Ve$.

Example 4-12 What is the diameter of an opening that discharges water at the rate of 1.05 cubic feet per second with a velocity of 86 fps?

ANSWER Use the formula $A = Q/Ve$ to find the area of the opening.

$$A = Q/Ve$$
$$= 1.05/86$$
$$= .0122 \text{ square feet}$$

$D = \sqrt{\frac{A}{.7854} \times 144}$

Now that the area of the circle is known, we need to find the diameter of the circle. The formula to find the diameter of a circle is $D = \sqrt{A/.7854}$, where the diameter and area are in the same units. When the area is given in square feet and we need a diameter in inches we use the formula $D = \sqrt{A/.7854}$ and multiply $A/.7854$ by 144. The formula then becomes $D = \sqrt{(A/.7854) \times 144}$.

Example 4-13 Find the diameter of the circle in Example 4-12.

ANSWER Find the diameter of the opening using the formula $D = \sqrt{(A/.7854) \times 144}$.

$$D = \sqrt{(A/.7854) \times 144}$$
$$= \sqrt{(.0122/.7854) \times 144}$$
$$= \sqrt{.0155 \times 144}$$
$$= \sqrt{2.237}$$
$$= 1.496 \text{ inches}$$

$D = \sqrt{\frac{\frac{Q}{Ve}}{.7854} \times 144}$

The formulas $A = Q/Ve$ and $D = \sqrt{(A/.7854) \times 144}$ each perform a separate function. One formula finds area and the other formula converts that area from square feet to a diameter in inches. There may be times when these functions need to be performed separately, so it is useful to know how to use both of these formulas. In many instances it may be possible to directly calculate the diameter when given quantity and velocity. Doing so requires us to merge the two formulas, which gives the new formula, $D = \sqrt{([Q/Ve]/.7854) \times 144}$.

Example 4-14 What size opening will discharge .593 cubic feet of water per second at 86 fps velocity?

ANSWER In the formula, $D = \sqrt{([Q/Ve]/.7854) \times 144}$, replace Q with .593 cubic feet and Ve with 86 fps.

$$D = \sqrt{([Q/Ve]/.7854) \times 144}$$
$$= \sqrt{([.593/86]/.7854) \times 144}$$
$$= \sqrt{(.0069/.7854) \times 144}$$
$$= \sqrt{.0088 \times 144}$$
$$= 1.125\ (1\tfrac{1}{8})\ \text{inches}$$

Finding Velocity When Quantity and Area or Diameter Are Known

$$Ve = \frac{Q}{A}$$

When quantity and area are known, we can find velocity in a manner similar to how we found area when we were given the quantity and velocity. Rearrange the formula $Q = A \times Ve$ to read $Ve = Q/A$. Remember, quantity must be in cfs and area must be in square feet to give us a velocity in fps.

Example 4-15 What is the velocity of the flow from an opening of 1 square foot if it is discharging 50 cfs?

ANSWER $Ve = Q/A$

$$= 50/1$$
$$= 50 \text{ fps}$$

$$Ve = \frac{Q}{\left(.7854 \times D^2 \times \frac{1}{144}\right)}$$

Typically, in the fire service we need to find the velocity from openings of only inches. This is done by substituting $.7854 \times D^2 \times 1/144$ for A in the formula $Ve = Q/A$. The formula then becomes $Ve = Q/(.7854 \times D^2 \times 1/144)$.

Example 4-16 What is the velocity of discharge if the quantity discharged is .731 cfs and the opening is 1¼-inches?

ANSWER $Ve = Q/(.7854 \times D^2 \times 1/144)$

$$= .731/(.7854 \times 1.25^2 \times 1/144)$$
$$= .731/.7854 \times 1.56 \times .0069$$
$$= .731/.0085$$
$$= 86 \text{ fps}$$

CALCULATING FLOW IN ONE MINUTE

The next step in calculating flow is to determine how much water is flowing in units of time traditional to the fire service. We do not normally calculate flow from a nozzle in seconds: we calculate the flow in minutes. To get minutes from the units we have already been working with we simply multiply the answer by 60, the number of seconds in a minute. The answer will be in cubic feet per minute (cfm).

Example 4-17 In Example 4-11 it was determined that a 1¼-inch nozzle would have a discharge of .731 cubic feet per second if the water was flowing at 86 fps. What would the flow be in 1 minute?

ANSWER To find the flow in 1 minute when we already know a 1¼-inch nozzle will flow .731 cfs, multiply the .731 by 60.

$$Q = .731 \times 60$$
$$= 43.86 \text{ cfm}$$

$Q = .7854 \times D^2 \times \frac{1}{144} \times Ve \times 60$

What would really be helpful is a way to directly calculate quantity flow per minute in a single formula. To do this we start with the formula we already have for calculating quantity flow, $Q = .7854 \times D^2 \times 1/144 \times Ve$ and multiply by 60. Our new formula for directly calculating quantity flow in cfm becomes $Q = .7854 \times D^2 \times 1/144 \times \text{Ve} \times 60$. By simply inserting the diameter of the opening in inches and the velocity in fps, we get a Q that is cfm.

Example 4-18 What is the quantity flow, in cfm, from a 1¼-inch opening with a velocity of 86 fps?

ANSWER $Q = .7854 \times D^2 \times 1/144 \times Ve \times 60$

$$= .7854 \times 1.25^2 \times 1/144 \times 86 \times 60$$
$$= .7854 \times 1.56 \times .0069 \times 86 \times 60$$
$$= 43.62 \text{ cfm}$$

We know the answer of 43.62 cfm is correct because between Example 4-11 and Example 4-17 we already calculated the flow to be 43.86 cfm. The .24 difference is due to rounding.

Example 4-19 What is the flow for 1 minute from a 1-inch tip if it has a velocity of 86 fps?

ANSWER $Q = .7854 \times D^2 \times 1/144 \times Ve \times 60$

$$= .7854 \times 1^2 \times 1/144 \times 86 \times 60$$
$$= .7854 \times 1 \times .0069 \times 86 \times 60$$
$$= 27.96 \text{ cfm}$$

APPLICATION ACTIVITIES

Put your knowledge of velocity and flow to use by solving the following problems.

Problem 1 What is the velocity of discharge at an opening at the base of a tank if the water in the tank is 125 feet above the point of discharge and the tank holds 100,000 gallons of water? (See Figure 4-7.)

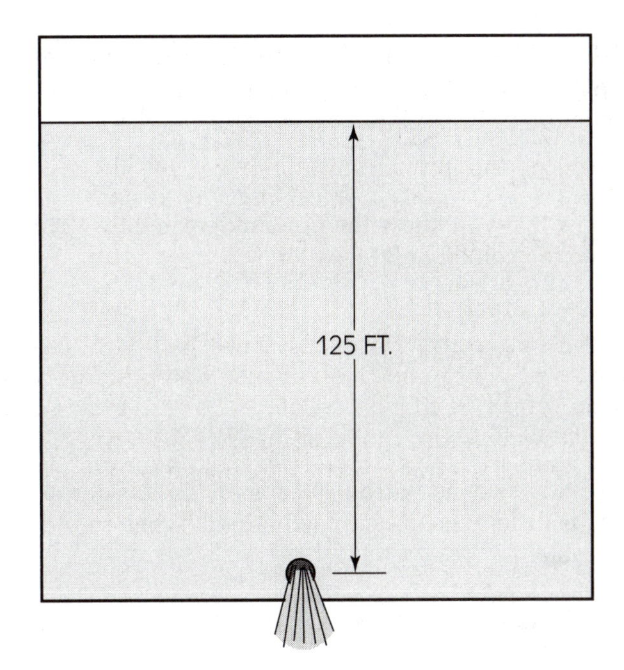

Figure 4-7 *What is the velocity of the water at the discharge?*

ANSWER We learned in Chapter 2 that the shape and volume of a tank are irrelevant in determining pressure at the base. When calculating the velocity from an opening at the base of a tank these principles hold true. The shape and volume have no effect on velocity, only the water level does. In order to calculate the velocity of water at the point of discharge, the only fact you need to know is the elevation of the water above the discharge. In this problem it is 125 feet.

$$Ve = 8\sqrt{H}$$

$$= 8 \times \sqrt{125}$$

$$= 8 \times 11.18$$

$$= 89.44 \text{ fps}$$

Problem 2 In the fire service we normally calculate discharge using pressure and not elevation of the water, because we can impart pressure to water by means of pumps, even where there is no elevation. In Problem 1, what would be the pressure at the point of discharge, and using the formula for calculating the velocity from pressure, what is the velocity of discharge?

ANSWER First calculate the pressure created by an elevation of 125 feet.

$$P = H \times .433$$
$$= 125 \text{ feet} \times .433$$
$$= 54.125 \text{ psi}$$

Now that you know the pressure created by 125 feet of elevation, calculate the velocity of the water.

$$Ve = 12.16 \times \sqrt{P}$$
$$= 12.16 \times \sqrt{54.125}$$
$$= 12.16 \times 7.357$$
$$= 89.46 \text{ fps}$$

This answer is within .02 fps of the answer arrived at in Problem 1. This difference is easily explained by the rounding of numbers in both problems.

Problem 3 What is the velocity of water in a 2½-inch hose if it is flowing water from a 1⅛-inch tip at 50 psi?

ANSWER First find the velocity for the 1⅛-inch tip at 50 psi.

$$Ve = 12.16 \times \sqrt{P}$$
$$= 12.16 \times \sqrt{50}$$
$$= 12.16 \times 7.071$$
$$= 85.98 \text{ fps}$$

Now calculate the velocity of water in the 2½-inch hose using the formula $A_1 \times Ve_1 = A_2 \times Ve_2$,

where A_1 = area of the 1⅛-inch tip
Ve_1 = velocity from the 1⅛-inch tip
A_2 = area of the 2½-inch hose
Ve_2 = velocity of water in the 2½-inch hose

$$A_1 \times Ve_1 = A_2 \times Ve_2.$$
$$.7854 \times 1.125^2 \times 85.98 = .7854 \times 2.5^2 \times Ve_2$$
$$.994 \times 85.98 = 4.9 \times Ve_2$$
$$85.46 = 4.9 \times Ve_2$$
$$Ve_2 = 85.46/4.9$$
$$Ve_2 = 17.44 \text{ fps}$$

Problem 4 Is the answer to Problem 3 correct?

ANSWER To verify the answer to Problem 3, we only need to compare the area of the 2½-inch hose to the area of the 1⅛-inch nozzle and the velocity of water in the hose to the velocity of water leaving the nozzle.

The area of the 2½-inch hose is 4.9/.994 or 4.93 times the area of the 1⅛-inch tip. The velocity of the 1⅛-tip is 85.98/17.44 or 4.93 times the velocity of the 2½-inch hose. The fact that both the ratio of the areas and the ratio of the velocities is exactly the same verifies the answer in Problem 3 is correct. Just remember that the relationship of area to velocity is an inverse one.

Problem 5 What is the quantity of water flowing if the opening is 1.5 square feet and the water is flowing at 3 fps?

ANSWER To find how much water is flowing, use the formula $Q = A \times Ve$.

$$
\begin{aligned}
Q &= A \times Ve \\
&= 1.5 \times 3 \\
&= 4.5 \text{ cfs}
\end{aligned}
$$

Problem 6 What is the quantity of water flowing if the opening is 2 inches in diameter and the water has a velocity of 108.76 fps?

ANSWER When finding the area of the 2-inch opening, remember to convert it to square feet.

$$
\begin{aligned}
Q &= A \times Ve \\
&= .7854 \times D^2 \times 1/144 \times Ve \\
&= .7854 \times 2^2 \times 1/144 \times 108.76 \\
&= .7854 \times 4 \times .0069 \times 108.76 \\
&= 2.36 \text{ cfs}
\end{aligned}
$$

Problem 7 How much water will flow in 1 minute from an opening at the base of a tank if the water level is 85 feet above the opening and the opening is 2½-inches in diameter?

ANSWER First find the velocity of the water.

$$
\begin{aligned}
Ve &= 8 \times \sqrt{H} \\
&= 8 \times \sqrt{85} \\
&= 8 \times 9.22 \\
&= 73.76 \text{ fps}
\end{aligned}
$$

Now calculate the amount of water flowing in 1 minute.

$$Q = .7854 \times D^2 \times 1/144 \times Ve \times 60$$
$$= .7854 \times 2.5^2 \times 1/144 \times 73.76 \text{ fps} \times 60$$
$$= .7854 \times 6.25 \times .0069 \times 73.76 \times 60$$
$$= 149.9 \text{ cfm}$$

Note: As the tank empties, the elevation of the water changes in relation to the discharge, which changes the velocity and flow of water. Problems like this one are meant to be approximate, because in real life you would have to recalculate the velocity and flow every few seconds to account for the elevation change as the tank emptied.

Summary

In order to determine just how much water is flowing, we need to understand the relationship of velocity to area. We must be able to calculate the velocity of water as it is discharging from an opening. Velocity can be calculated either from the elevation head of water over the point of discharge or from the pressure at the point of discharge.

Calculating the quantity of water flowing is like calculating the volume of a cylinder. The velocity is the length of the cylinder, and the dimensions of the opening give us the area of the cylinder (length times area equals volume). Knowing these facts allows us to calculate the volume of water being discharged in one minute.

When we know the velocity through one size hose, we can calculate the velocity through another size hose, given constant flow. We can use the formula $A_1 \times Ve_1 = A_2 \times Ve_2$ to find any variable, as long as we know the other three. Using the appropriate formula for calculating velocity of discharge from a nozzle, we can calculate the velocity of water in the hose supplying that nozzle.

This chapter has laid the groundwork for Chapter 5. In this chapter we ended with a formula for determining how much water was flowing in 1 minute. In Chapter 5 we expand that to calculating gallons per minute (gpm) and develop a simplified formula for calculating gallons per minute.

Review Questions

1. What is the gravitational acceleration constant?
2. Why is the gravitational acceleration constant important in determining velocity?
3. How did we arrive at the constant 8 in the formula $Ve = 8\sqrt{H}$
4. Calculate the velocity of water discharging from an open pipe if it is 37 feet below the water level.
5. What is the pressure where the discharge velocity is 89 fps?
6. How high is water in a tank if it creates a discharge velocity of 10 fps?
7. If the area of hose 1 is 1.5 times greater than the area of hose 2, what will the velocity of water be in hose 1 as compared to hose 2?
8. Find the area of hose 2 if hose 1 is $2\frac{1}{2}$ inches with a velocity of 10 fps, and the velocity of water in hose 2 is 20.42 fps?
9. How much water will flow in 1 second from a $1\frac{3}{4}$-inch opening if it has a velocity of 109 fps?
10. How many cubic feet of water will flow from a $1\frac{5}{8}$-inch nozzle, at 80 psi nozzle pressure in 1 minute?
11. How much water will flow from an opening 3 inches in diameter if the water has a velocity of 35 fps?
12. What size opening is needed to discharge 1.88 cubic feet of water per second if the velocity is 86 fps? Give the answer in square feet and square inches. What is the diameter of the opening?

List of Formulas

Finding velocity when pressure is known:

$$Ve = \sqrt{2 \times g \times 2.31 \times P}$$

Finding velocity when pressure is known, simplified:

$$Ve = 12.16\sqrt{P}$$

Finding height when velocity is known:

$$H = \left(\frac{Ve}{8}\right)^2$$

Finding pressure when velocity is known:

$$P = \left(\frac{Ve}{12.16}\right)^2$$

Comparison of area and velocity in two sizes of hose for same flow:

$$A_1 \times Ve_1 = A_2 \times Ve_2$$

Finding the diameter of a circle when area is known:

$$D = \sqrt{\frac{A}{.7854}}$$

Basic quantity flow formula:

$$Q = A \times Ve$$

Finding flow from circular opening:

$$Q = .7854 \times D^2 \times Ve$$

Finding flow from circular opening when diameter is in inches:

$$Q = .7854 \times D^2 \times \frac{1}{144} \times Ve$$

Finding area of opening when quantity of flow and velocity are known:

$$A = \frac{Q}{Ve}$$

Finding diameter in inches when area is in square feet:

$$D = \sqrt{\frac{A}{.7854} \times 144}$$

Finding diameter in inches when quantity and velocity are known:

$$D = \sqrt{\frac{\frac{Q}{Ve}}{.7854} \times 144}$$

Finding velocity when quantity and area are known:

$$Ve = \frac{Q}{A}$$

Finding velocity when quantity and diameter in inches are known:

$$Ve = \frac{Q}{.7854 \times D^2 \times \frac{1}{144}}$$

Calculating flow in 1 minute:

$$Q = .7854 \times D^2 \times \frac{1}{144} \times Ve \times 60$$

GPM

Learning Objectives

Upon completion of this chapter, you should be able to:

- Understand the origins of the formula used to calculate gallons per minute.
- Calculate gallons per minute.
- Calculate the nozzle diameter, where gpm and nozzle pressure are known.
- Calculate the nozzle pressure, where gpm and nozzle diameter are known.
- Calculate velocity, where gpm and nozzle diameter are known.
- Calculate the flow from sprinkler heads.

INTRODUCTION

In Chapter 4 we began to develop the formula that will ultimately give us the flow from a nozzle in gallons per minute (gpm). The origin of the formula has already been explained, from the basic formula $Q = A \times Ve$ to $Q = .7854 \times D^2 \times 1/144 \times Ve \times 60$, which gives us the quantity flow in cfm. The next logical step is to find gpm and then develop a simplified formula for calculating it.

This chapter also discusses the use of velocity in the formula for calculating gpm. It is far more convenient for us to calculate gpm from pressure, because it is a standard fire service unit of measurement. But the fact remains that the formula for calculating gpm requires an expression of velocity, so we find a way to introduce pressure into the formula

After the gpm formula has been developed, we use a variation of the formula to find the diameter of the nozzle when gpm and nozzle pressure are known and to calculate the nozzle pressure when gpm and nozzle diameter are known. The formula for finding nozzle pressure and nozzle diameter can be handy when testing new equipment or finding equivalent diameters when the flow from a fog nozzle is known.

As a bonus, we learn how to calculate the velocity of water being discharged when gpm and nozzle diameter are known. And finally, we learn how to calculate the flow from sprinkler heads.

CALCULATING GALLONS PER MINUTE

In Chapter 4, we developed the formula $Q = .7854 \times D^2 \times 1/144 \times Ve \times 60$ to make a direct calculation of the quantity of water flowing in cfm. The fire service, however, uses quantity flow in units of gallons per minute. Fortunately, because there are 7.48 gallons in a cubic foot, the conversion of cubic feet per minute to gallons per minute is just a matter of multiplying the cubic feet per minute by 7.48.

Example 5-1 If a 1-inch nozzle operating at 86 fps flows 27.96 cfm, how many gallons per minute is it flowing?

ANSWER Multiply the cubic feet of water that flows for 1 minute by 7.48.

$$\text{gpm} = 27.96 \times 7.48$$
$$= 209.14$$

This entire process can be combined into one formula, eliminating the need to do independent calculation. Additionally, by putting all these factors into one formula, we will eventually develop a gpm formula.

The new formula for directly calculating gpm where velocity and nozzle size are known is gpm = $.7854 \times D^2 \times 1/144 \times Ve \times 60 \times 7.48$. All that is needed is for

$gpm = 2.448 \times D^2 \times Ve$

the diameter and velocity to be inserted into the formula and the answer will be in gpm. But before we do that, since this formula finally has all the factors needed to calculate gpm, we can simplify it to a single constant and two variables, which makes using the formula much easier. To do that, we simply multiply $.7854 \times 1/144 \times 60 \times 7.48$ to arrive at a variable of 2.448. The new formula for calculating gpm when nozzle diameter and velocity are known is $gpm = 2.448 \times D^2 \times Ve$.

Example 5-2 What is the flow, in gpm, for a 1¼-inch tip, if the velocity is 86 fps?

ANSWER Insert 1¼-inch for D and 86 for Ve.

$$\begin{aligned} gpm &= 2.448 \times D^2 \times Ve \\ &= 2.448 \times 1.25^2 \times 86 \\ &= 2.448 \times 1.56 \times 86 \\ &= 328.95 \end{aligned}$$

Calculating gpm When Pressure Is Known

The formula given for finding gpm encompasses all the factors we need to calculate the flow from a nozzle. The one disadvantage of the formula is that we need to know the velocity of the water exiting the discharge. The velocity is nothing more than a measurement of the energy in the water as it leaves the nozzle. In the fire service we universally measure the energy in water in terms of pressure. Before we can use the new formula we just developed, we must convert the nozzle pressure to velocity. Fortunately, we learned a formula in Chapter 4 that converts pressure to velocity, $Ve = 12.16\sqrt{P}$. With this formula, we can convert pressure into velocity, and then insert the velocity into the formula for gpm. Or, we could substitute $12.16\sqrt{P}$ in the formula for gpm in place of Ve, eliminating a separate calculation. The formula for calculating gpm when the pressure is given then becomes $gpm = 2.448 \times D^2 \times 12.16 \times \sqrt{P}$.

$gpm = 2.448 \times D^2 \times 12.16 \times \sqrt{P}$

Example 5-3 Calculate the gpm from a 1¼-inch tip with a 50 psi nozzle pressure (see Figure 5-1).

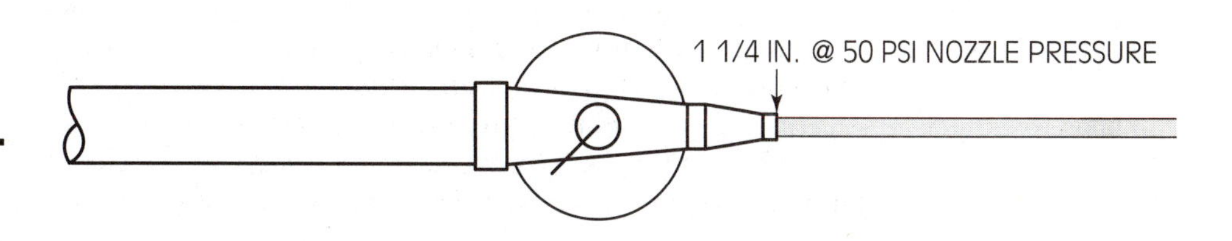

Figure 5-1
Calculate gpm.

ANSWER Substitute 1¼-inches for D and 50 for P.

$$\begin{aligned} \text{gpm} &= 2.448 \times D^2 \times 12.16 \times \sqrt{P} \\ &= 2.448 \times 1.25^2 \times 12.16 \times \sqrt{50} \\ &= 2.448 \times 1.56 \times 12.16 \times 7.07 \\ &= 328.31 \text{ gpm} \end{aligned}$$

Example 5-2 and Example 5-3 are the same problem, with Example 5-2 using velocity to calculate gpm and Example 5-3 using pressure. You can easily verify that 86 fps is equivalent to 50 psi by using the formula $Ve = 12.16\sqrt{P}$ to calculate velocity when given pressure. In fact, this was already done in Example 4-2 in Chapter 4.

FREEMAN'S FORMULA

$$gpm = .7854 \times D^2 \times \frac{1}{144} \times 12.16 \times \sqrt{P} \times 60 \times 7.48$$

$$gpm = 29.77 \times D^2 \times \sqrt{P}$$

The formula used to calculate gpm from pressure in Example 5-3 is a short version of $gpm = .7854 \times D^2 \times 1/144 \times 12.16 \times \sqrt{P} \times 60 \times 7.48$. By starting with the entire formula we can create a short version. We only need to multiply all the factors in this formula to get the constant, leaving D^2 and $\sqrt{P}$ as variables. By multiplying $.7854 \times 1/144 \times 12.16 \times 60 \times 7.48$ we get a constant of 29.7656, which when rounded to two decimal points is 29.77. The new, simpler formula for calculating gpm becomes $gpm = 29.77 \times D^2 \times \sqrt{P}$.

Example 5-4 Calculate the gpm from a 1¼-inch tip if it has a 50 psi nozzle pressure.

ANSWER This problem is the same as in Example 5-3.

$$\begin{aligned} gpm &= 29.77 \times D^2 \times \sqrt{P} \\ &= 29.77 \times 1\tfrac{1}{4}^2 \times \sqrt{50} \\ &= 29.77 \times 1.56 \times 7.07 \\ &= 328.339 \text{ rounded to } 328 \text{ gpm} \end{aligned}$$

Once again we have arrived at a flow of 328 gpm using a nozzle pressure. Because discharge in gpm is the same as we got in the long version of the gpm formula, we have just verified the validity and origin of the formula gpm $= 29.77 \times D^2 \times \sqrt{P}$.

The simplified formula is called Freeman's formula, after John Freeman, who first developed the formula in 1888. Technically Freeman's formula is slightly different from the one we derived in that he arrived at a constant of 29.71. Freeman's original formula is then gpm $= 29.71 \times D^2 \times \sqrt{P}$.

The difference between Freeman's $gpm = 29.71 \times D^2 \times \sqrt{P}$ and our gpm $= 29.77 \times D^2 \times \sqrt{P}$ is extremely small. The primary reason for the difference in the constant for calculating velocity from pressure is that Freeman probably rounded off the weight of a gallon of water to 62.5 pounds, instead of using 62.4 pounds.

If Freeman used 62.5 as the weight of a cubic foot of water, his formula for calculating velocity from pressure would have been $Ve = 12.14\sqrt{P}$ instead of our $Ve = 12.16\sqrt{P}$; this difference would account for a constant of 29.71.

Example 5-5 Using Freeman's original formula, find the gpm flow from a 1¼-inch tip at 50 psi nozzle pressure (see Figure 5-1).

ANSWER Freeman's original formula is $gpm = 29.71 \times D^2 \times \sqrt{P}$.

$$gpm = 29.71 \times D^2 \times \sqrt{P}$$
$$= 29.71 \times 1\tfrac{1}{4}^2 \times \sqrt{50}$$
$$= 29.71 \times 1.56 \times 7.07$$
$$= 327.67$$

The difference is due to rounding.

Variations of Freeman's Formula

We have already shown that the formula we derived for calculating gpm is slightly different from the original Freeman's formula, because we used a different weight of water than Freeman did. But we can be more technically correct because we have technological advantages that Freeman did not have, such as handheld calculators and desktop computers.

Another variation of Freeman's formula includes rounding the 29.71 to 30 for simplified multiplication, making the formula $gpm = 30 \times D^2 \times \sqrt{P}$. Rounding off in this fashion is not recommended because technology makes it so easy to carry out mathematical calculations with higher precision.

AIA formula formula for determining flow from hydrants

$gpm = 29.83 \times D^2 \times C \times \sqrt{P}$

A final formula for calculating flow from a discharge was developed by the American Insurance Association (AIA) for determining flows from hydrants. It is used to determine flow from a fire hydrant when testing hydrant capacity. The **AIA formula** is $gpm = 29.83 \times D^2 \times C \times \sqrt{P}$. It is used in place of Freeman's formula when determining the flow from a fire hydrant.

Discharge Coefficient

Regardless of which formula you use for calculating gpm, they will all be inaccurate unless you factor in the discharge coefficient or C factor. But first, what is the C factor?

All openings designed for water to flow through have some degree of inefficiency, that is, they all impede the flow of water to some extent. Some openings, such as hydrant butts, have a great deal of resistance to water flow and can have a discharge coefficient of as low as .7. This means that only 70% of the water calculated will actually flow through the opening. The amount of the discharge coefficient can vary from one manufacturer to another, and must always be taken into consideration, otherwise the flow calculated will not be accurate.

Table 5-1 *Discharge coefficients.*

Sprinkler, ½-inch	.75
Sprinkler, 17/32-inch	.95
Smooth-bore nozzle	.96 to .98
Underwriters playpipe	.97
Deluge or Monitor nozzle	.997

C factors for nozzles are usually very large, because they are designed for efficiency. For instance, the *C* factor for a monitor gun can be .997, which means that 99.7% of the calculated flow from a monitor nozzle will actually be delivered. Table 5-1 shows a sample list of discharge coefficients for various types of solid stream devices. For devices not on this list, consult the *Fire Protection Handbook* or check with the manufacturer.

$gpm = 29.77 \times D^2 \times C \times \sqrt{P}$

When calculating gpm using Freeman's formula, the *C* factor should always be included. With the *C* factor included Freeman's formula becomes, $gpm = 29.77 \times D^2 \times C \times \sqrt{P}$, where *C* is the discharge coefficient.

Example 5-6 Calculate the discharge from a 1¼-inch nozzle, at 50 psi nozzle pressure, with a *C* factor of .98.

ANSWER The *C* factor was obtained from Table 5-1.

$$\begin{aligned} \text{gpm} &= 29.77 \times D^2 \times C \times \sqrt{P} \\ &= 29.77 \times 1.25^2 \times .98 \times \sqrt{50} \\ &= 29.77 \times 1.56 \times .98 \times 7.07 \\ &= 321.77 \end{aligned}$$

In Example 5-4 we calculated the flow from a 1¼-inch tip at 50 psi nozzle pressure and got an answer of 328.339 gpm, which is only 6.6 gpm different. This difference is so small for smooth-bore nozzles that the *C* factor is often omitted. A word of caution is needed if you choose to omit the *C* factor from your calculations when working with smooth-bore nozzles: When calculating discharge from other types of devices, the *C* factor is critical and cannot be omitted. It makes more sense to include the *C* factor as a matter of habit than to forget it when it is needed.

For the remainder of this book, anytime we need to calculate the flow from a nozzle, the formula gpm = $29.77 \times D^2 \times \mathrm{C} \times \sqrt{P}$ is used. When calculating flow from a hydrant in Chapter 9, the formula gpm = $29.83 \times D^2 \times C \times \sqrt{P}$ is used.

Example 5-7 What is the flow from a 1½-inch monitor nozzle at 80 psi nozzle pressure?

ANSWER From Table 5-1 the *C* factor is .997

$$\text{gpm} = 29.77 \times D^2 \times C \times \sqrt{P}$$
$$= 29.77 \times 1.5^2 \times .997 \times \sqrt{80}$$
$$= 29.77 \times 2.25 \times .997 \times 8.94$$
$$= 597$$

Calculating Nozzle Diameter When gpm and Pressure Are Known

$$D = \sqrt{\frac{gpm}{29.77 \times C \times \sqrt{P}}}$$

By using Freeman's formula it is possible to calculate the nozzle diameter when we know gpm and pressure. This formula is sometimes used to calculate an "equivalent" nozzle diameter when working with fog nozzles. In order to find the nozzle diameter from Freeman's formula, the formula must be rearranged to read $D = \sqrt{gpm/(29.77 \times C \times \sqrt{P})}$.

Example 5-8 What size nozzle delivers 812 gpm at 80 psi nozzle pressure if the C factor is .997 (see Figure 5-2)?

ANSWER Hint: With a C factor of .997 we are looking for a monitor nozzle tip.

$$D = \sqrt{\text{gpm}/(29.77 \times C \times \sqrt{P})}$$
$$= \sqrt{812/(29.77 \times .997 \times \sqrt{80})}$$
$$= \sqrt{812/(29.77 \times .997 \times 8.94)}$$
$$= \sqrt{812/265}$$
$$= \sqrt{3.06}$$
$$= 1.75 \text{ or } 1\tfrac{3}{4}\text{-inch}$$

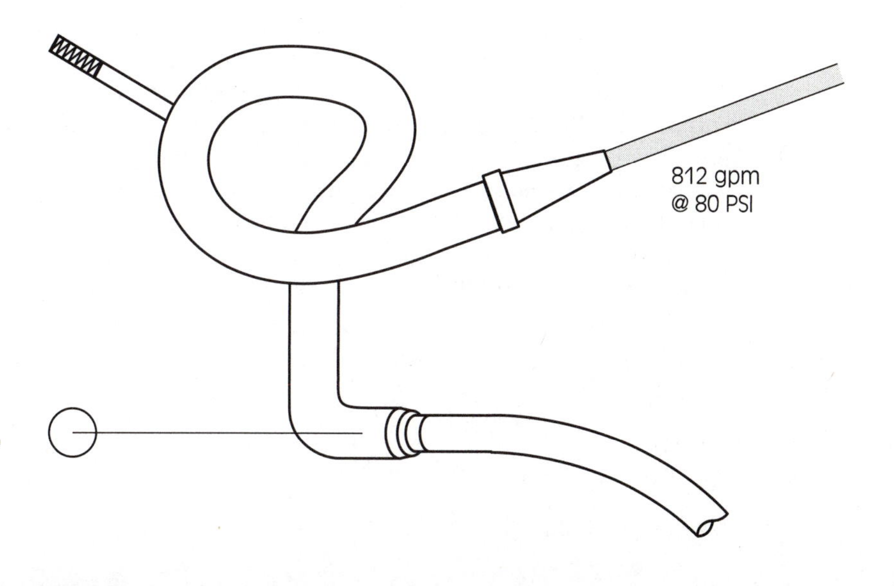

Figure 5-2 *Calculate the diameter of the tip.*

Table 5-2 *Decimal equivalents.*

Tip Size (in inches)	Decimal Equivalent
½	.5
17/32	.53
15/16	.938
1-1/8	1.125
1¼	1.25
1-3/8	1.375
1½	1.5
1-5/8	1.625
1¾	1.75
1-7/8	1.875

This particular example resulted in an exact tip size for *D*. This is not usually the case, not because the formula is not good or precise, but because of how we round off numbers. As much as we always want to be as precise as possible, some judgment must be used with this formula to determine the closest actual tip size to our answer. Additionally, the answer is always given in decimals and must be converted to a fraction. Table 5-2 gives the decimal equivalent of various tip sizes.

As mentioned previously, this formula can be used to calculate an equivalent smooth-bore tip size to correspond with the flow from a fog nozzle. With this knowledge, we can change a fog tip to an equivalent smooth-bore tip without changing the flow. Some people advocate the use of this formula to determine an equivalent tip size, as a means to determine what the flow from the fog nozzle will be if the nozzle pressure is varied. This, however, is not a good practice. The reach and effective pattern of the fog nozzle will be changed if the nozzle pressure and subsequent flow is altered. In short, the nozzle pressure on a fog nozzle should never be altered from what is recommended by the manufacturer.

Example 5-9 If a 2½-inch fog nozzle delivers 225 gpm at 100 psi nozzle pressure, what size smooth-bore tip will deliver a similar flow?

ANSWER Assume a nozzle pressure of 50 psi and *C* factor of .98.

$$D = \sqrt{GPM/(29.77 \times C \times \sqrt{P})}$$

$$= \sqrt{225/(29.77 \times .98 \times \sqrt{50})}$$

$$= \sqrt{225/(29.77 \times .98 \times 7.07)}$$

$= \sqrt{225 / 206.26}$

$= \sqrt{1.09}$

$= 1.04\text{-inch}$

Because this formula is used to calculate an equivalent tip size in inches, an exact fractional equivalent is not necessary. This formula has a very practical use in Chapter 9 when we calculate nozzle reaction for fog nozzles.

Calculating Nozzle Pressure When Nozzle Diameter and gpm Are Known

$$P = \left(\frac{gpm}{29.77 \times D^2 \times C}\right)^2$$

In Example 5-9 we were able to determine that it would take a 1.04-inch tip at 50 psi nozzle pressure to flow 225 gpm. The only problem is there is no 1.04-inch tip. Smooth bore hand line tips are usually sized in 1/8-inch increments. We still need to be able to find out how much pressure is needed on a 1-inch tip to flow 225 gpm. By rearranging Freeman's formula to find pressure, we get the formula $P = (gpm/[29.77 \times D^2 \times C])^2$. Now we can find what pressure is needed to deliver a flow of 225 gpm from a 1-inch tip.

Example 5-10 How much nozzle pressure is needed to flow 225 gpm from a 1-inch tip (see Figure 5-3)?

ANSWER Use a C factor of .98

$P = (gpm/[29.77 \times D^2 \times C])^2$

$= (225/[29.77 \times 1^2 \times .98])^2$

$= (225/29.17)^2$

$= (7.71)^2$

$= 59.4 \text{ psi}$

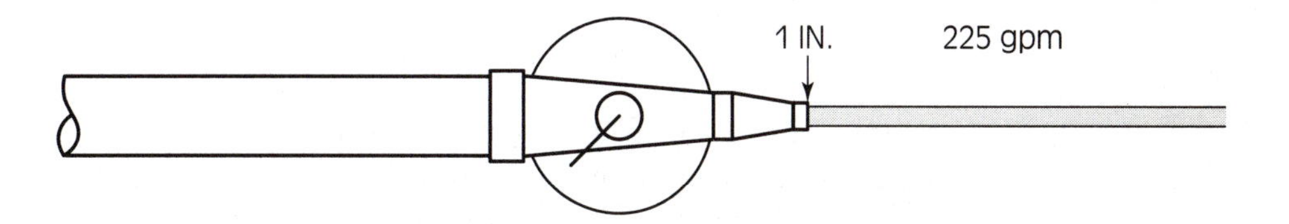

Figure 5-3 *Find the nozzle pressure.*

Example 5-11 Use Freeman's formula to verify the results of Example 5-9 and Example 5-10 (see Figure 5-4).

ANSWER $\text{gpm} = 29.77 \times D^2 \times C \times \sqrt{P}$

$= 29.77 \times 1^2 \times .98 \times \sqrt{59.4}$

$= 29.77 \times 1 \times .98 \times 7.71$

$= 224.94$

Once again the answer is off only a very small amount due to rounding.

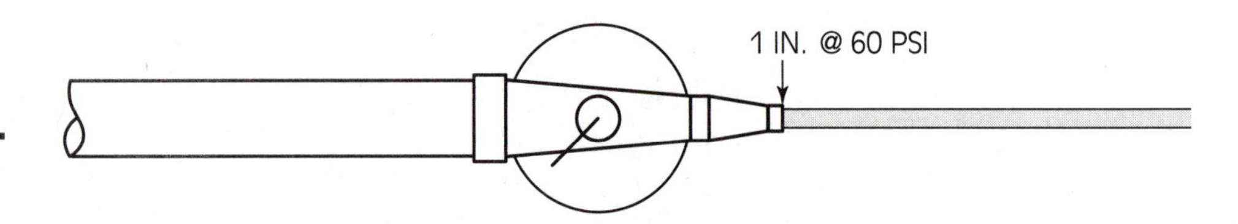

Figure 5-4 *Calculate the gpm.*

CALCULATING VELOCITY WHEN GPM AND NOZZLE DIAMETER ARE KNOWN

You have learned to use a variation of Freeman's formula to calculate the nozzle diameter and nozzle pressure when gpm are known. It is also possible to calculate the velocity of water as it exits a nozzle when the gpm are known, but to do so we need to go back one step to our original gpm formula $gpm = .7854 \times D^2 \times 1/144 \times Ve \times 60 \times 7.48$. This formula has an expression for velocity built in, *Ve*, and enables us to get an answer directly as velocity. Just as with Freeman's formula, we need to manipulate the formula a little to make velocity the unknown. But in order to make the process a little simpler, we use the simplified version of the formula for calculating gpm from velocity, $\text{gpm} = 2.448 \times D^2 \times Ve$.

$$Ve = \frac{gpm}{2.448 \times D^2}$$

If we begin with $\text{gpm} = 2.448 \times D^2 \times Ve$ we need to rearrange the formula to solve for *Ve*. The resulting formula is $Ve = gpm/(2.448 \times D^2)$. With this formula, we can calculate the velocity of water as it leaves the nozzle, as long as we know the gpm and diameter of the nozzle.

Example 5-12 If a flow meter indicates a flow of 250 gpm and the tip size is 1¼-inch, what is the velocity of the water as it leaves the nozzle (see Figure 5-5)?

ANSWER $Ve = gpm/(2.448 \times D^2)$

$= 250/(2.448 \times 1.25^2)$

$= 250/(2.448 \times 1.56)$

$= 250/3.82$

$= 65.45$ fps

Example 5-13 Verify Example 5-12.

ANSWER We verify our result by converting the velocity to pressure and then plugging it into Freeman's formula.

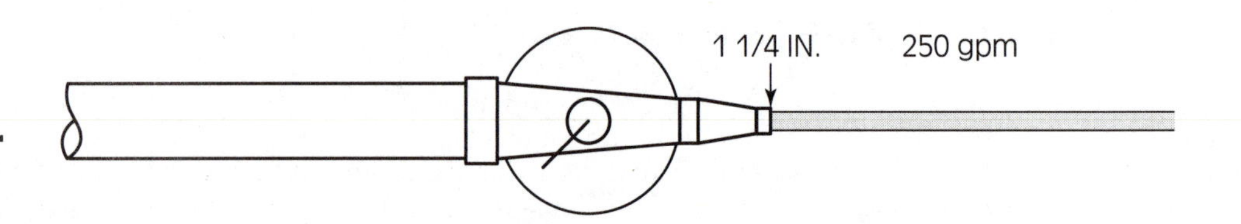

Figure 5-5 *Calculate the velocity.*

$$P = (Ve/12.16)^2$$
$$= (65.96/12.16)^2$$
$$= (5.42)^2$$
$$= 29.38 \text{ psi}$$

Now use Freeman's formula to find gpm

$$gpm = 29.77 \times D^2 \times \sqrt{P}$$
$$= 29.77 \times 1.25^2 \times \sqrt{29.38}$$
$$= 29.77 \times 1.56 \times 5.42$$
$$= 251.7$$

Once again the answer is off only a very slight amount due to rounding. The *C* factor was not used here because it was not used in the formula in Example 5-12.

CALCULATING GPM FROM SPRINKLER HEADS

The invention of the sprinkler system is probably the most important discovery, in terms of saving property, in the history of the fire service. For sprinkler systems to be effective, it is necessary that firefighters understand them, so they can be utilized to their fullest. The first step in understanding sprinkler systems is to be aware that they are fairly simple in design and operate according to the laws of hydraulics contained in this book.

Sprinkler heads are designed to operate at a minimum of 7 psi pressure. At that pressure, a sprinkler with a ½-inch opening should flow 15 gpm. As the pressure increases, so will the flow, just like any other nozzle. For now we can use Freeman's formula to calculate the flow from a sprinkler head at 7 psi pressure.

Example 5-14 Apply Freeman's formula to verify a 15 gpm flow from a ½-inch sprinkler head at 7 psi (see Figure 5-6).

ANSWER The *C* factor for a ½-inch sprinkler, from Table 5-1, is .75.

$$gpm = 29.77 \times D^2 \times C \times \sqrt{P}$$
$$= 29.77 \times .5^2 \times .75 \times \sqrt{7}$$
$$= 29.77 \times .25 \times .75 \times 2.65$$
$$= 14.79 \text{ rounded to } 15$$

Remember, different size sprinkler heads have different *C* factors. It is important to use the correct *C* factor in order to get the correct flow.

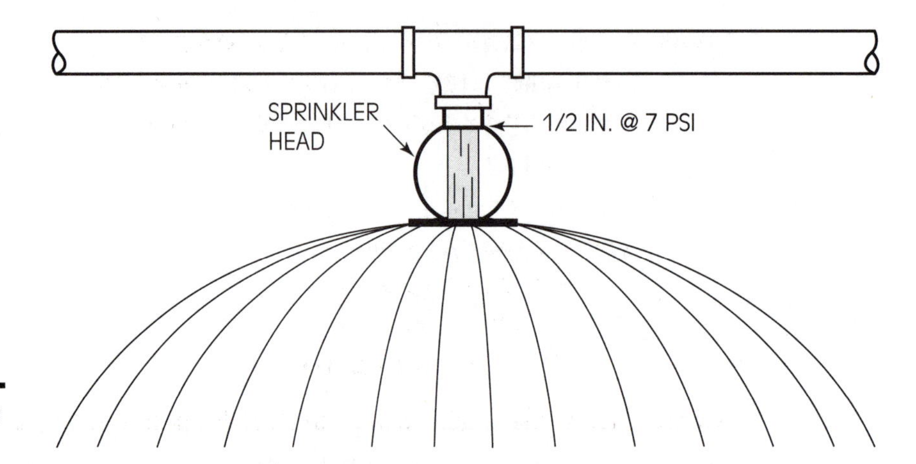

Figure 5-6 *Calculate the gpm.*

APPLICATION ACTIVITIES

Use your knowledge of Freeman's formula and its variations to solve the following problems.

Problem 1 How much water will flow, in gpm, from a 1¾-inch monitor nozzle at a velocity of 109 fps (see Figure 5-7)?

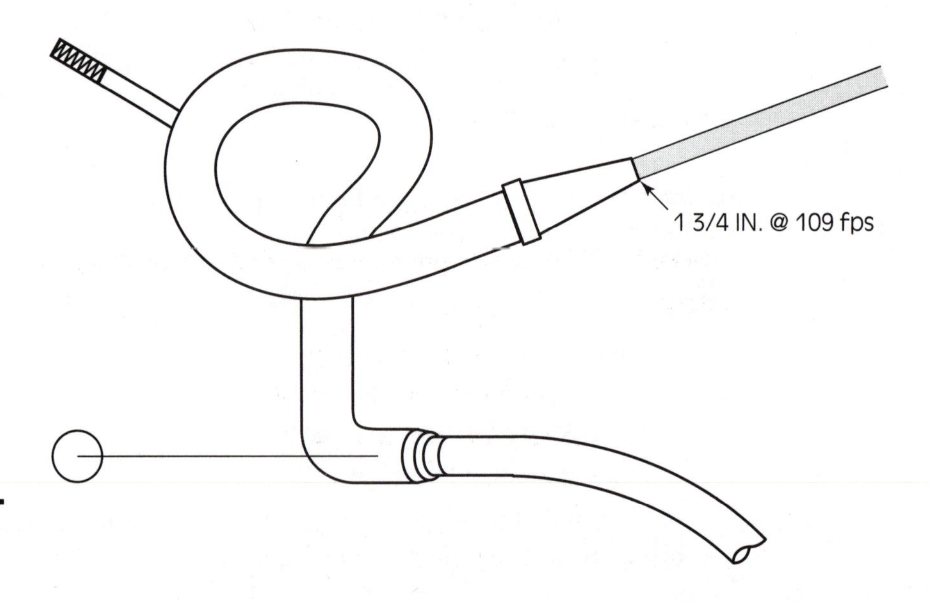

Figure 5-7 *Calculate the gpm.*

ANSWER $gpm = .7854 \times D^2 \times 1/144 \times Ve \times 60 \times 7.48$

$= .7854 \times 1.75^2 \times 1/144 \times 109 \times 60 \times 7.48$

$= .7854 \times 3.06 \times 1/144 \times 109 \times 60 \times 7.48$

$= 811.2$

Or

$gpm = 2.432 \times D^2 \times Ve$

$= 2.432 \times 1.75^2 \times 109$

$= 2.432 \times 3.06 \times 109$

$= 811.2$ rounded to 811

Problem 2 How many gpm will a 15/16-inch tip flow at 50 psi nozzle pressure?

ANSWER From Table 5-1 use a *C* factor of .98.

$gpm = 29.77 \times D^2 \times C \times \sqrt{P}$

$= 29.77 \times 15/16^2 \times .98 \times \sqrt{50}$

$= 29.77 \times .897 \times .98 \times 7.07$

$= 181.31$ rounded to 181

Problem 3 What size tip is needed to flow 650 gpm from a monitor nozzle at 80 psi nozzle pressure?

ANSWER The *C* factor from Table 5-1 is .997.

$D = \sqrt{gpm/29.77 \times C \times \sqrt{P}}$

$= \sqrt{650/29.77 \times .97 \times \sqrt{80}}$

$= \sqrt{650/29.77 \times .997 \times 8.94}$

$= \sqrt{650/265.35}$

$= \sqrt{2.45}$

$= 1.565$ or $1\frac{9}{16}$-inch tip

Problem 4 What nozzle pressure is needed to flow 300 gpm from a $1\frac{1}{4}$-inch tip?

ANSWER Use a *C* factor of .98.

$P = (gpm/[29.77 \times D^2 \times C])^2$

$= (300/[29.77 \times 1.25^2 \times .98])^2$

$= (300/[29.77 \times 1.56 \times .98])^2$

$= (300/45.51)^2$

$= (6.59)^2$

$= 43.43$ psi

Problem 5 How much velocity is needed to flow 600 gpm through a 1½-inch tip?

ANSWER $Ve = gpm/(2.432 \times D^2)$

$= 600/(2.432 \times 1.5^2)$

$= 600/(2.432 \times 2.25)$

$= 600/5.47$

$= 109.68$ fps

Problem 6 How much water will a 17/32 sprinkler head flow at 15 psi pressure?

ANSWER The C factor from Table 5-1 is .95.

$gpm = 29.77 \times D^2 \times C \times \sqrt{P}$

$= 29.77 \times 17/32^2 \times .95 \times \sqrt{15}$

$= 29.77 \times .282 \times .95 \times 3.87$

$= 30.87$

Summary

This chapter completes the process of developing a formula that gives the flow of water in gallons of water in 1 minute. The final formula includes all the elements needed to get gpm from the formula $Q = A \times Ve$. It factors in the area of the opening in square inches ($.7854 \times D^2 \times 1/144$), converts pressure to velocity to give the flow in cubic feet per second ($12.16 \times \sqrt{P}$), calculates the volume in cubic feet per minute ($\times$ 60), then converts the cubic feet per minute into gallons per minute ($\times$ 7.48), and finally a *C* factor was added. All this is then factored down to a formula with just one constant and three variables, $GPM = 29.77 \times D^2 \times C \times \sqrt{P}$.

In addition to being used to make straight gpm calculations, this formula can be re-arranged to calculate either nozzle diameter or nozzle pressure. In short, assuming the *C* factor stays constant, as long as we have any two of the following, we can find the third: gpm, nozzle diameter, or nozzle pressure. By using a different version of the gpm formula ($gpm = 2.448 \times D^2 \times Ve$) the velocity of discharge can be found when gpm and nozzle diameter are known.

Freeman's formula can also be used to find flow from a sprinkler head if the pressure at the head is known. Here it is particularly important to know the *C* factor for the particular size sprinkler head in order to obtain useful results.

Review Questions

1. Give the specific reason each of the following is included in the formula for gpm: $.7854 \times D^2$, 1/144, *Ve*, 60, 7.48.
2. In the formula $gpm = 2.448 \times D^2 \times Ve$, from where was the constant 2.448 derived.?
3. In the formula $gpm = 2.448 \times D^2 \times Ve$, how is it possible to introduce pressure in place of velocity?
4. If a 1 1/8-inch smooth-bore tip is used on a hand line and has a velocity at the tip of 81.5 fps, what is the gpm flow?
5. How many gpm will flow from a 1 7/8-inch tip on a monitor nozzle if the nozzle pressure is 70 psi?
6. Why did John Freeman's original formula have a constant of 29.71 instead of 29.77?
7. What size tip will flow 247.69 gpm at 45 psi? (Do not forget to include the *C* factor. Assume a *C* factor of .98.)
8. At what pressure will a 15/16-inch tip flow 181.5 gpm? (Assume a *C* factor of .98.)
9. If a 1-inch tip is flowing 210 gpm, what is the velocity of the water as it leaves the tip?
10. If a ½-inch sprinkler head is flowing 20 gpm, what is the pressure at the sprinkler head?
11. How many gpm will flow from a 1½-inch tip if it has a velocity at the tip of 85.98 fps?
12. How many gpm will flow from a 1½-inch tip if it has a nozzle pressure of 50 psi and a *C* of .99?
13. What tip velocity is necessary to achieve a flow of 325 gpm from a 1¼-inch tip?

List of Formulas

Calculating gpm when diameter and velocity are known:

$$gpm = 2.448 \times D^2 \times Ve$$

Calculating gpm when diameter and pressure are known:

$$gpm = 2.448 \times D^2 \times 12.16 \times \sqrt{P}$$

Basic gpm formula:

$$gpm = .7854 \times D^2 \times \frac{1}{144} \times 12.16 \times \sqrt{P} \times 60 \times 7.48$$

Freeman's formula:

$$gpm = 29.77 \times D^2 \times \sqrt{P}$$

AIA formula:

$$gpm = 29.83 \times D^2 \times C \times \sqrt{P}$$

Freeman's formula with coefficient of discharge:

$$gpm = 29.77 \times D^2 \times C \times \sqrt{P}$$

Finding diameter when gpm and pressure are known:

$$D = \sqrt{\frac{gpm}{29.77 \times C \times \sqrt{P}}}$$

Finding pressure when gpm and diameter are known:

$$P = \left(\frac{gpm}{29.77 \times D^2 \times C}\right)^2$$

Finding velocity when gpm and diameter are known:

$$Ve = \frac{gpm}{2.448 \times D^2}$$

Reference

Cote, Arthur E., P.E., Editor-in-Chief, *Fire Protection Handbook*, 18th ed. Quincy, MA: National Fire Protection Association, 1997.

Friction Loss

Learning Objectives

Upon completion of this chapter, you should be able to:

- Understand the causes of friction loss.
- Understand the four laws of hydraulics governing friction loss.
- Given gpm, calculate friction loss for various size hose.
- Given friction loss and gpm, calculate the friction loss conversion factor.
- Given friction loss conversion factor and length of any size hose, calculate an equivalent length of hose for the same gpm in any other size hose.
- Given friction loss in any size hose, calculate the friction loss in any other size hose for the same gpm.
- Given friction loss and conversion factor, calculate the gpm for a specified hose size.

INTRODUCTION

In Chapter 3 we talked at length about energy as it pertains to hydraulics. Although energy cannot be lost or destroyed, it does often change into less useful forms. One of the most prevalent and unforgiving causes of energy conversion is friction.

Friction loss plays a major role in hydraulics. We must always compensate for it in calculating engine pressure, when calculating hose lays, and when determining how much water we can flow. Fortunately, it is easy to calculate and charts can be made with friction loss at various flows and in different size hose.

In this chapter we learn the principles of friction loss, how to calculate friction loss, and how to adjust hose size and length based on a thorough understanding of friction loss. In addition, we learn how to calculate the gpm when the hose size and friction loss is known. Finally, we learn to find equivalent length of different size hose, or equivalent friction losses for a given gpm by using conversion factors.

The formula used in this chapter to find friction loss can find a friction loss for virtually any amount of gpm through any size hose. Just because you can calculate a friction loss does not mean you should actually attempt to pump that much water through the hose. Practically speaking, when the flow in a given size hose reaches a friction loss of 50 psi per 100 feet, the capacity of the hose has been reached. At pressures higher than this, even small diameter hose can become difficult to maneuver. Excessively high friction loss also leads to excessive engine pressure, which can reduce the capacity of the pump. (More on pump capacity in Chapter 7.)

■ Note
When the flow in a given size hose reaches a friction loss of 50 psi per 100 feet, the capacity of the hose has been reached.

While reading this chapter, keep in mind that the formulas identified for calculating friction loss are approximate. As the velocity of a fluid increases, it reacts to friction differently, so no single formula can precisely calculate friction loss. For practical purposes, the results of these formulas should be considered accurate, because they are the best we have.

FACTORS AFFECTING FRICTION LOSS

friction
the resistance to movement of two surfaces in contact

Friction can be defined as the resistance to movement of two surfaces in contact. For example, if you rub your hand across the surface of this book, you will feel a slight resistance. That resistance is friction trying to keep your hand from sliding across the surface of the book. The conversion of useful energy into nonuseful energy due to friction is called *friction loss*.

In hydraulics, friction is created when water (or other fluids) rubs against the inside of hose or pipe and against itself. Friction loss in hose is usually expressed in terms of loss of pressure per 100 feet. Several factors affect friction loss, including the roughness of the hose, roughness of hose appliances, and restrictions to the flow of the water. Restrictions can include protruding gaskets, sharp bends in hose, and excess or incorrect use of hose appliances.

Viscosity

viscosity
resistance to flow

Another factor affecting friction loss is the viscosity of the liquid. **Viscosity** is the resistance to flow of a liquid. The viscosity of a liquid determines how much friction loss is created inside the liquid itself as it tries to flow. In general, the thicker the liquid, the greater the viscosity or resistance to flow. As viscosity increases, so does friction loss within the liquid.

To better understand viscosity, think of the comparison of maple syrup to corn syrup. Maple syrup is very watery and easily flows off your pancakes; it has a very low viscosity. Corn syrup, however, is thick and when put on pancakes stays put; the corn syrup has a very high viscosity.

From the foregoing discussion of friction loss, it is evident that friction in hose has multiple causes. The condition of the hose, the condition of the appliances, the hose lay itself, and other factors affect friction loss and cause one of two friction loss scenarios: laminar flow or turbulent flow.

Laminar Flow

laminar flow
the smooth and orderly flow of water, with layers, or cores, of water gliding effortlessly over the next layer of water

Laminar flow represents a best-case scenario with the least amount of friction loss. In laminar flow, the flow of the water is smooth and orderly, with layers, or cores, of water effortlessly gliding over the next layer of water, much as depicted in Figure 6-1. Figure 6-1 shows the inner core of water, A, smoothly passing through core B, which in turn is smoothly passing through core C. Each core moves progressively faster as it gets further away from the rough walls of the hose. In Figure 6-1, core A is moving the fastest.

Laminar flow is found where hose has a relatively smooth lining, the hose is laid straight, and hose appliances are in good condition. Most of the friction loss is created by the friction of the water against the walls of the hose. In laminar flow, because the flow of water is orderly and in layers or cores, there is little friction of water against itself.

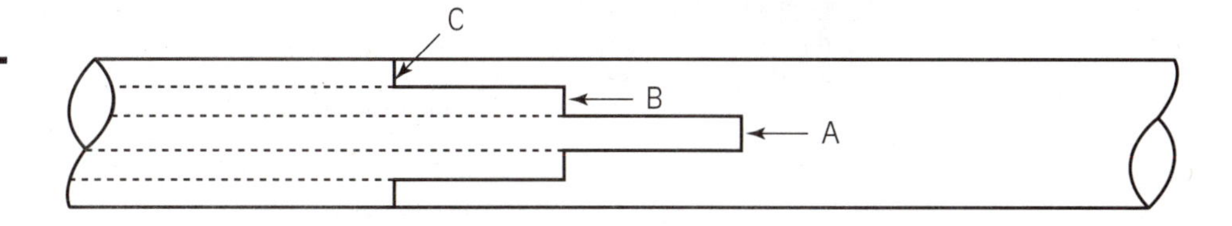

Figure 6-1 *Laminar flow: Water flows in smooth orderly layers.*

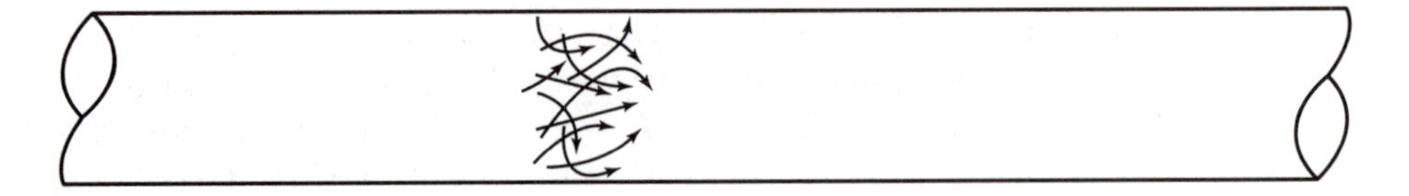

Figure 6-2 *Turbulent flow: Water flow is disorganized.*

Turbulent Flow

turbulent flow the flow of water is disorganized and random

Turbulent flow results when water flows in a disorganized, random manner. Figure 6-2 represents the flow of water under turbulent conditions, where a great deal of the friction loss is due to water rubbing against itself.

Rough hose, protruding gaskets, excessive bending, and kinking of hose all contribute to turbulence in the hose. In fact, it can logically be assumed that in most situations familiar to the fire service, turbulent flow is present to some degree. To reduce the effects of turbulent flow, deliberate efforts need to be taken when laying hose to prevent kinking and excessive bends. Reasonable efforts must also be made to keep pump pressures as low as possible.

FOUR LAWS GOVERNING FRICTION LOSS

Within the discipline of hydraulics we have already encountered principles that govern the behavior of pressure. Friction loss has four laws governing how friction behaves. By knowing these laws, we are able to calculate and understand friction loss. There is one difference from the principles of pressure we learned and these four laws of friction loss. The principles of pressure applied to all applications of hydraulics, while some of these laws of friction loss apply only to hose.

Law 1 *Friction loss varies directly as the length of the hose, provided all other conditions are equal.* Law 1 tells us that as the length of the hose line increases, so does the total friction loss (see Figure 6-3). The friction loss increases proportionally to the change in length of hose. If the length of the line is double, the friction loss also doubles. If the length of the line is increased by a factor of three, the friction loss increases by a factor of three.

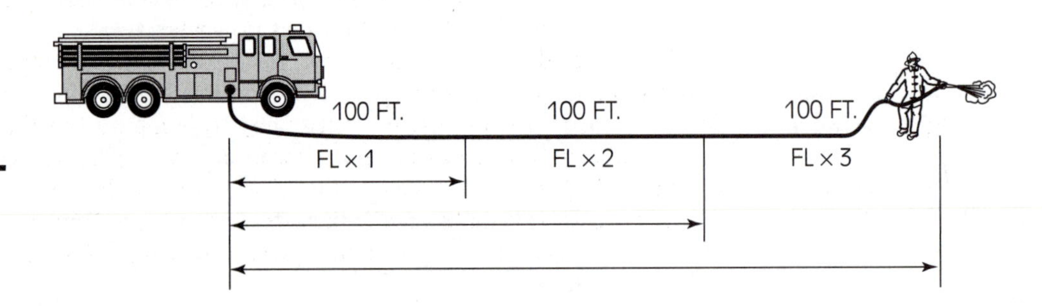

Figure 6-3 *Friction loss varies directly as the length of the hose.*

Law 2 *In the same size hose, friction loss varies approximately as the square of the velocity.* Two important principles make up Law 2. The first principle is that friction loss is a function of velocity, not gpm. This technicality is important to understand because gpm and velocity are directly related. If gpm is increased by a factor of two, the velocity is also increased by a factor of two. But, since friction is the result of movement of the water against the hose, and itself, velocity is technically the cause of friction loss.

The second principle tells us that the friction loss varies by the square of the change in velocity. The fact that friction loss changes exponentially is a direct result of the flow of water becoming more turbulent as pressure is increased. This principle is a perfect illustration of the previous discussion of laminar versus turbulent flow. As velocity is increased, the turbulence of water increases, causing a disproportionate increase in friction loss.

$$Fm = \left(\frac{Ve_2}{Ve_1}\right)^2$$

This change in friction loss, due to the change in velocity, can be expressed as a mathematical formula, $Fm = (Ve_2/Ve_1)^2$, where Fm is the friction loss multiplier, Ve_1 is the reference or original velocity, and Ve_2 is the new velocity.

Example 6-1 If the friction loss in 100 feet of hose is 5 psi, what will the new friction loss be if the velocity is doubled?

ANSWER In this example, Ve_1 is 1 because this is the reference velocity, and Ve_2 is 2 because the velocity is doubled.

$Fm = (Ve_2/Ve_1)^2$
$= (2/1)^2$
$= (2)^2$
$= 4$, which means the friction loss will be four times as great after the velocity is doubled.

The new friction loss is 5 psi × 4, or 20 psi.

Law 3 *For the same discharge, friction loss varies inversely as the fifth power of the diameter of the hose.* Just as in Law 2, this law also has multiple points to make. The first is that, for a given gpm, friction loss varies inversely to the size of the hose. That is to say, as the hose gets larger, there is less friction loss. Conversely, if the hose gets smaller, the friction loss is greater.

The second point is that friction loss varies to the fifth power of the diameter of the hose. More simply put, if we want to compare friction loss for a given gpm in various hose sizes, we only need to compare the diameter to the fifth power. This can be expressed in the formula, $Cf = D_1^5/D_2^5$, where Cf is the conversion factor, D_1 is the diameter of the hose with which you are making the comparison, and D_2 is the hose for which you are trying to find the conversion factor. This law, however, has one serious flaw; it only calculates a conversion factor for rubber-lined hose.

$$Cf = \frac{D_1^5}{D_2^5}$$

Example 6-2 If a given gpm in 2½-inch hose has a friction loss of 10 psi, what will be the friction loss in 3-inch hose for the same gpm?

ANSWER $Cf = D_1^5/D_2^5$

$= 2.5^5/3^5$

$= 97.66/243$

$= .4$

The friction loss for the same gpm in 3-inch hose is 10 psi × .4 or 4 psi.

Law 4 *For a given velocity, the friction loss in hose is approximately the same no matter what the pressure may be.* The single most important issue in Law 4 is that friction loss in hose is only approximately independent of pressure. Remember, it was already stated that velocity is responsible for friction loss. In reality, as pressure changes, it is possible for friction loss to change, even if only to a small degree, for two reasons.

The first reason is that as pressure in hose increases, it actually causes the hose diameter to enlarge slightly. It is impossible to factor this diameter change into friction loss calculations because, technically, the actual size of the hose due to pressure can change every foot or so because friction loss reduces the pressure. The effect of this diameter increase would be to reduce the actual friction loss. Additionally, as pressure increases, the rubber lining of fabric-covered hose can assume the texture of the fabric, causing a rougher surface and increased friction loss.

The second reason it is said that friction loss is *approximately* independent of pressure is that as hose is pressurized, it elongates. After the hose is charged, there is technically more hose to generate more friction loss. Again, this condition is impossible to factor into friction loss calculations because all hose does not behave alike, but the serious student of hydraulics should be aware of this behavior.

Even considering the theoretical change in friction loss due to these factors, they are not sufficient to cause a change in friction loss that is critical to us. Again recall that friction loss formulas are technically only approximate.

CALCULATING FRICTION LOSS

The National Board of Fire Underwriters, now known as the Insurance Services Office, developed the formula used today to calculate friction loss in hose. Originally the formula was $FL = 2Q^2 + Q$, where FL is friction loss, and Q is gpm flow in hundreds. In the late 1960s the formula was altered slightly to account for improved hose manufacturing processes that reduced the friction loss. Today we use the revised formula $FL = 2Q^2$, known as the Underwriters formula. The Underwriters formula is used to find friction loss in 2½-inch hose *only* and is calculated for 100 feet of hose. The answer obtained by use of the Underwriters formula is in units of psi, because we are compensating for loss of energy due to friction, and the energy put into the hose is in units of psi.

To use the Underwriters formula, we only need to know the gpm in hundreds. The gpm in hundreds is easily calculated from gpm by moving the decimal point two places to the left. For example, 250 gpm in hundreds is 2.5. The 2.5 would then replace Q in the equation.

$FL\ 100 = 2Q^2$

Before we begin doing friction loss calculations, we are going to make a slight change to the formula for friction loss. Instead of using the formula $FL = 2Q^2$, we are going to modify it slightly to read $FL\ 100 = 2Q^2$. FL 100 reminds us that this friction loss is only for 100 feet of hose. Also, in Chapter 10 the engine pressure formula uses FL for the total friction loss in a line. By using FL 100 here we can keep the two friction loss figures separate later.

Example 6-3 What is the friction loss in 2½-inch hose for 375 gpm?

ANSWER 375 gpm in hundreds is 3.75.

$$\begin{aligned} FL\ 100 &= 2Q^2 \\ &= 2 \times 3.75^2 \\ &= 2 \times 14.06 \\ &= 28.12 \text{ psi} \end{aligned}$$

$FL\ 100 = Cf \times 2Q^2$

Today's fire service uses a variety of hose sizes and we need a means of calculating friction loss for hose in sizes other than 2½-inch. To calculate the friction loss for hose other than 2½-inch we begin by calculating the friction loss for 2½-inch and then converting it to the friction loss for the hose in question. At this point Law 3 becomes important. By comparing other hose to 2½-inch hose, conversion factors can be generated for any size rubber-lined hose. The formula for calculating friction loss for hose other than 2½-inch then becomes, $FL\ 100 = Cf \times 2Q^2$, where Cf is the conversion factor. In short, this formula calculates friction loss for 2½-inch hose ($2Q^2$) and then multiplies it by a conversion factor to find friction loss for the same gpm in another size hose.

Example 6-4 What is the friction loss in 3-inch hose for 375 gpm?

ANSWER In Example 6-2, the conversion factor for 3-inch hose was calculated to be .4.

$$\begin{aligned} FL\ 100 &= Cf \times 2Q^2 \\ &= .4 \times 2 \times 3.75^2 \\ &= .4 \times 2 \times 14.06 \\ &= 11.25 \text{ psi for 375 gpm in 3-inch hose} \end{aligned}$$

By use of Law 3 we can calculate the conversion factor to compare friction loss for any two hose sizes, as long as they are rubber-lined hose. Because the fire service today uses more than just rubber-lined hose, we also need conversion factors for these other hoses.

Empirical Method for Calculating Conversion Factors

$$Cf = \frac{FL\ 100}{2Q^2}$$

Because Law 3 only directly pertains to rubber-lined hose, we need a method to calculate the conversion factor for all hose. For that we look to the formula we just developed for calculating friction loss, $FL\ 100 = Cf \times 2Q^2$. Just as we have done several times already in this book, we simply rearrange the formula to find for Cf. The empirical formula for calculating the conversion factor then becomes $Cf = FL\ 100/2Q^2$.

The conversion factor for a given size hose is generally constant from manufacturer to manufacturer, so there is not generally a need to do empirical testing on all hose purchased. However, from time to time bad hose does get sold or bought and verifying the conversion factor by doing random sampling is sensible. The conversion factor formula gives us conversion factors for hose that does not fit Law 3, as well as hose that does. The good thing about the conversion factor formula is that we only need to develop the conversion factor for a single flow in the test hose. The derived conversion factor will then be good for all flows in that size hose.

To use the formula $Cf = FL\ 100/2Q^2$ we need to know two things: (1) the gpm and (2) the friction loss for the gpm. This process may seem a little backwards, but if the following procedure is followed the formula will work with a high degree of precision.

This procedure is fairly simple. Start by laying a single line of the hose you are going to test. Make the line several hundred feet (at least 300 feet) long so you will end up with an averaged friction loss. Flow enough water to truly test the hose. For hand line hose, test at the midpoint of the advertised flow range. For supply line, test at 150 gpm per 1 inch of hose diameter at a minimum. The friction loss will be the engine pressure minus the nozzle pressure. (When testing supply line, subtract the intake pressure of pumper 2 from the discharge pressure of pumper 1 to obtain the friction loss.) Divide the calculated total friction loss by the amount of hose, in hundreds, to get the friction loss per 100 feet. Next calculate the gpm flow based on the nozzle pressure and tip size. Finally, plug these figures into the formula $Cf = FL\ 100/2Q^2$ and calculate the conversion factor.

When conducting a test to determine conversion factor, use a smooth-bore tip so you can take a pitot reading. The pitot reading serves two purposes: (1) it gives a more accurate nozzle pressure than using a gauge at the base of the nozzle, and (2) the pitot reading can be used to calculate the exact gpm flow during the test.

Example 6-5 What is the conversion factor for 180 gpm, where the engine pressure is 198.2 psi, the line is 300 feet long, and the nozzle pressure is 47.3 psi (see Figure 6-4)?

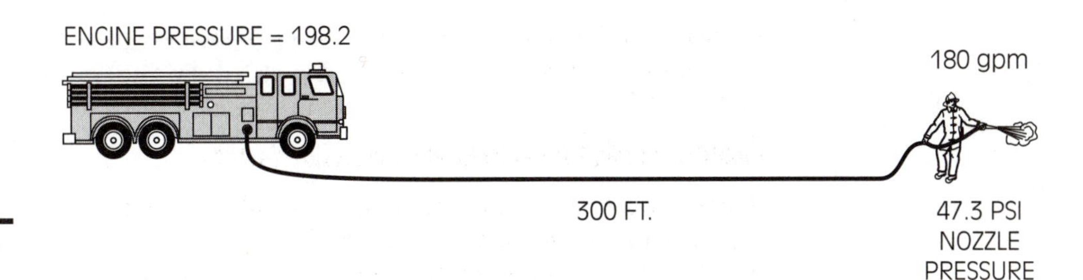

Figure 6-4 *Find the conversion factor.*

ANSWER First, subtract the nozzle pressure from the engine pressure and divide by three to obtain the friction loss per 100 feet. 198.2 – 47.3 = 150.9 ÷ 3 = 50.3. Then, solve the equation

$$Cf = FL\ 100/2Q^2$$
$$= 50.3/2 \times 1.8^2$$
$$= 50.3/2 \times 3.24$$
$$= 50.3/6.48$$
$$= 7.76$$

The conversion factor was correctly found to be 7.76, which corresponds to the conversion factor for 1 3/4-inch hose in Table 6-1.

Table 6-1 *Conversion factors for common hose sizes.*

Hose Diameter	Conversion Factor
¾ inch	500
1 inch	75
1½ inch	12
1½ inch (linen)	25.6
1¾ inch with 1½-inch couplings	7.76
2 inch	4
2 inch (linen)	6.25
2½ inch (linen)	2.13
3 inch with 2½-inch couplings	.4
3½ inch	.17
4 inch	.1
5 inch	.04
6 inch	.025

Table 6-1 contains the conversion factors for the most common size hose. All hose, unless otherwise noted, is rubber lined. For hose of type or size not contained in Table 6-1, you should run your own test to determine the correct conversion factor.

FINDING EQUIVALENT HOSE LENGTH

$L_2 = L_1 \times \left(\frac{Cf_1}{Cf_2}\right)$

At times it is helpful to determine how much of one size hose is equivalent to another size hose. For example, how much 3-inch hose is equivalent to 300 feet of 2½-inch hose. To make these comparisons, it is helpful to have a formula that will directly compare any one size hose to any other size hose. That formula is $L_2 = L_1 \times (Cf_1/Cf_2)$. L_1 is the known length of hose, and L_2 is the length of the hose we are looking for. The conversion factors used in this formula are the same as those in Table 6-1, with Cf_1 the conversion factor of the length of the hose we already know, and Cf_2 the conversion factor for the hose we are looking for. If any comparisons are made to 2½-inch hose, the conversion factor for 2½-inch hose is 1.

This formula tells us how much of another size hose will give us the same total friction loss, if the flow remains constant, compared to our reference hose. For example, if there is *X* amount of total friction loss in 300 feet of 2½-inch hose flowing an unspecified gpm, how much 3-inch hose will have *X* amount of total friction loss flowing the same gpm?

Example 6-6 How much 3-inch hose is equivalent to 300 feet of 2½-inch hose?

ANSWER Remember, the *Cf* for 2½-inch hose is 1, and the *Cf* for 3-inch hose is .4.

$$L_2 = L_1 \times (Cf_1/Cf_2)$$
$$= 300 \times (1/.4)$$
$$= 300 \times 2.5$$
$$= 750 \text{ feet of 3-inch hose}$$

750 feet of 3 inch hose is equivalent to 300 feet of 2½-inch hose.

Example 6-7 How much 1¾-inch hose is equivalent to 150 feet of 1½-inch hose?

ANSWER From Table 6-1, Cf_1 is 12 and Cf_2 is 7.76.

$$L_2 = L_1 \times (Cf_1/Cf_2)$$
$$= 150 \times (12/7.76)$$
$$= 150 \times 1.546$$
$$= 231.9 \text{ feet of 1¾-inch hose.}$$

Obviously hose this size must be in 50-foot lengths, so this answer would have to be rounded down to 200 feet.

FINDING EQUIVALENT FRICTION LOSS

$FL_2\ 100 = FL_1\ 100 \times \left(\frac{Cf_2}{Cf_1}\right)$

Being able to compare friction loss between two different size hose is another method of comparing hose efficiency. To do this we need to know the friction loss for the reference hose. Then we need to find the friction loss for the comparison hose. The formula to make the comparison is $FL_2\ 100 = FL_1\ 100 \times (Cf_2/Cf_1)$. In this equation, FL_1 is the friction loss we already know, FL_2 is the friction loss we are looking for, Cf_1 is for the hose we know the friction loss of, and Cf_2 is for the size hose for which we are calculating friction loss.

If we were comparing friction loss for a given gpm, we could easily make the comparison using the Underwriters formula. The formula, $FL_2\ 100 = FL_1\ 100 \times Cf_2/Cf_1$ allows us to make friction loss comparisons even when we do not know the gpm. This formula can be helpful when doing maximum flow problems comparing various hose size (which are done in Chapter 11.)

Example 6-8 If a 4-inch supply line has a friction loss of 15 psi, how much friction loss will there be in 3-inch hose for the same gpm?

ANSWER From Table 6-1, Cf_1 is .1 and Cf_2 is .4.

$$\begin{aligned} FL_2\ 100 &= FL_1\ 100 \times (Cf_2/Cf_1) \\ &= 15 \times (.4/.1) \\ &= 15 \times 4 \\ &= 60 \text{ psi} \end{aligned}$$

Another useful application of this formula is to find a factor to compare friction loss between any two sizes of hose. This formula is useful in making quick comparisons without having to go through the entire process for each friction loss adjustment. To calculate the comparison factor, simply use a friction loss of 1 for FL_1 100. Rather than create a whole new formula just to calculate the comparison factor, we will use $FL_2\ 100 = FL_1\ 100 \times (Cf_2/Cf_1)$. Just remember the answer is actually a comparison factor and not a friction loss.

Example 6-9 Calculate the friction loss conversion factor to compare friction loss in 1½-inch hose to that of 1¾-inch hose.

ANSWER From Table 6-1, Cf_1 is 12 and Cf_2 is 7.76

$$\begin{aligned} FL_2\ 100 &= FL_1\ 100 \times Cf_2/Cf_1 \\ &= 1 \times 7.76/12 \\ &= 1 \times .647 \\ &= .647 \text{ or a comparison factor of } .647 \end{aligned}$$

Just multiply the known friction loss for the 1½-inch hose by .647 and you have the friction loss for the same gpm in 1¾-inch hose. If you happen to have the friction loss in 1¾-inch hose, divide it by .647 to get friction loss for the same flow in 1½-inch hose. In other words, friction loss for a given gpm in 1¾-inch hose is .647 times the friction loss in 1½-inch hose.

Example 6-10 If 1½-inch hose has an *FL* 100 of 37.5 for 125 gpm, what would the friction loss in 1¾-inch hose be for the same flow?

ANSWER Simply multiply the *FL* 100 for the 1½-inch hose by .647.

$37.5 \times .647 = 24.26$

FL 100 for 1¾-inch hose flowing 125 gpm is 24.26 psi.

CALCULATING GPM FROM FRICTION LOSS

$$Q = \sqrt{\frac{FL\ 100}{(Cf \times 2)}}$$

There are times when it is necessary to find gpm when the only information we have is the friction loss. Under these conditions, it is possible to find the gpm by using the Underwriters formula, $FL\ 100 = Cf \times 2Q^2$. By rearranging the formula to solve for Q, the formula becomes $Q = \sqrt{FL\ 100/(Cf \times 2)}$. The conversion factor in this formula corresponds to the size hose being used.

Example 6-11 If 5-inch hose has a friction loss of 15 psi, how much water is it flowing?

ANSWER The conversion factor for 5-inch hose from Table 6-1 is .04.

$$\begin{aligned} Q &= \sqrt{FL\ 100/(Cf \times 2)} \\ &= \sqrt{15/(.04 \times 2)} \\ &= \sqrt{15/.08} \\ &= \sqrt{187.5} \\ &= 13.69 \text{ or } 1{,}369 \text{ gpm (remember, } Q \text{ is in hundreds)} \end{aligned}$$

CONVERSION FACTORS FOR PARALLEL HOSE LAYOUTS

There may be times when pumpers, master stream devices, or sprinkler and standpipe systems are supplied by more than one hose from the same pumper called parallel lines. If the hose used is all the same size, this situation does not present a problem, because each line will be flowing an equal share of the water being supplied. The friction loss for the appropriate share of water will be the same for all the lines. For example, if two 2½-inch lines are supplying a monitor nozzle from the same pumper, each line will be delivering exactly one half of the water. If the monitor is flowing 600 gpm, then each line is flowing 300 gpm. When calculating friction loss for this set up, it will be calculated for a flow of 300 gpm.

There may be times when hoses of different diameters are used to supply the same device (see Figure 6-5). When pumping from one pumper through parallel

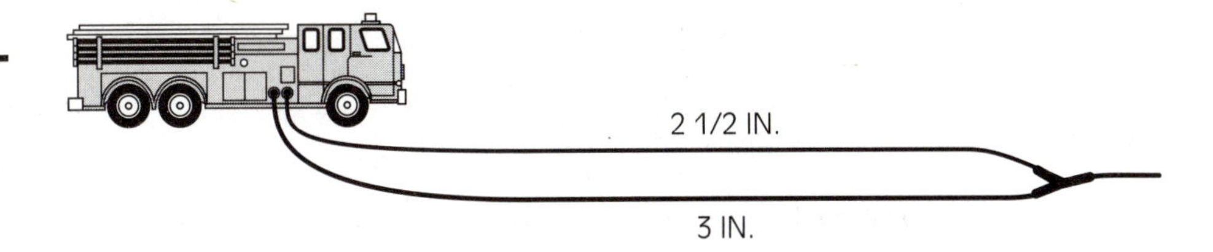

Figure 6-5 *Parallel lines of unequal diameter. FL 100 is the same for each hose line.*

lines to a device, another pumper, or a system of some kind, the pressures in the parallel lines will equalize at the point where both lines have the same friction loss. At any given friction loss, the larger hose will have the greater flow. Calculating the friction loss is impossible under such situations unless you know exactly how much water is flowing through each size hose, and you will not know how much water is flowing unless you know the friction loss.

The solution to this problem is to develop conversion factors for possible multiple line evolutions. This can be done the same way we figured the conversion factor under Empirical Method for Calculating Conversion Factors. Instead of using a single line, use two equal length lines of two different diameters. Pump from one pumper to a second pumper that is delivering a precise amount of water. You can then calculate the friction loss per 100 feet for the gpm being supplied. Finally, use the formula $Cf = FL\ 100/2Q^2$ to determine the conversion factor for the multiple line layout as done previously for the single line.

There is also another method that does not require an actual flow test to determine the conversion factor. Start by selecting a friction loss at random and then find the flow for each size hose at that friction loss. You now have the gpm and friction loss to plug into the conversion factor formula.

Example 6-12 What is the conversion factor for parallel lines, one 2½-inch and one 3-inch?

ANSWER In Example 4 we determined that 3-inch hose has a friction loss of 11.25 psi when flowing 375 gpm. This gives us a convenient place to start. Now let us find how much water is flowing in 2½-inch hose for the same friction loss. Remember the conversion factor for 2½-inch hose is 1.

$$\begin{aligned} Q &= \sqrt{FL\ 100/Cf \times 2} \\ &= \sqrt{11.25/1 \times 2} \\ &= \sqrt{11.25/2} \\ &= \sqrt{5.625} \\ &= 2.37 \text{ or } 237 \text{ gpm} \end{aligned}$$

The flow will be 375 gpm for the 3-inch hose and 237 gpm for the 2½-inch hose for a total flow of 612 gpm. Now calculate the conversion factor.

$Cf = FL\ 100/2Q^2$

$= 11.25/2 \times 6.12^2$

$= 11.25/2 \times 37.45$

$= 11.25/74.9$

$= .15$

By inserting a conversion factor of .15 into the formula $FL\ 100 = Cf \times 2Q^2$, the friction loss for any flow through parallel lines of one 2½-inch hose and one 3-inch hose can be found. This procedure can be applied to any parallel line situation to determine the conversion factor. Just remember to total up all the flows at the same friction for each size hose, even if some of the lines are the same diameter.

APPLICATION ACTIVITIES

Problem 1 What is the friction loss for 200 gpm in 2-inch, rubber-lined hose?

ANSWER The conversion factor for 2-inch hose in Table 6-1 is 4.

$FL\ 100 = Cf \times 2Q^2$

$= 4 \times 2 \times 2^2$

$= 4 \times 2 \times 4$

$= 32$ psi

Problem 2 While testing a sample of hose to verify the conversion factor, you determine the hose has a friction loss of 18 psi at 300 gpm. What is the conversion factor? What size is the hose?

ANSWER $Cf = FL\ 100/2Q^2$

$= 18/2 \times 3^2$

$= 18/2 \times 9$

$= 18/18$

$= 1$

A conversion factor of 1 indicates 2½-inch hose.

Problem 3 Find the length of 1¾-inch hose that is equivalent to 500 feet of 2½-inch hose (see Figure 6-6).

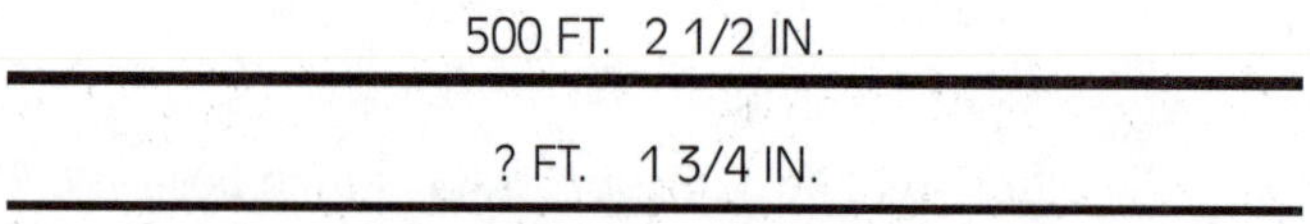

500 FT. 2 1/2 IN.

? FT. 1 3/4 IN.

Figure 6-6 *Find the equivalent length of 1¾-inch hose.*

ANSWER From Table 6-1, the conversion factor for 1¾-inch hose is 7.76.

$$L_2 = L_1 \times (Cf_1/Cf_2)$$
$$= 500 \times (1/7.76)$$
$$= 500 \times .129$$
$= 64.5$ rounded down to a single 50-foot length of hose.

Problem 4 If 3-inch hose has a friction loss of 20 psi for a given flow, how much friction loss will 3½-inch hose have for the same flow?

ANSWER From Table 6-1, the conversion factor for 3-inch hose is .4 and the conversion factor for 3½-inch hose is .17.

$$FL_2\ 100 = FL_1\ 100 \times Cf_2/Cf_1$$
$$= 20 \times .17/.4$$
$$= 20 \times .425$$
$= 8.5$ psi per 100 feet of hose

Problem 5 Calculate a friction loss conversion factor to compare the friction loss in 4-inch hose to friction loss in 3-inch hose.

ANSWER From Table 6-1, the conversion factor for 3-inch hose is .4, and for 4-inch hose is .1. Remember to use 1 as FL_1.

$$FL_2\ 100 = FL_1\ 100 \times (Cf_2/Cf_1)$$
$$= 1 \times (.4/.1)$$
$$= 1 \times 4$$
$$= 4$$

The answer tells us that 3-inch hose has 4 times the friction loss of 4-inch hose.

Problem 6 Calculate the amount of water flowing in 3½-inch hose if it has a friction loss of 45 psi.

ANSWER The conversion factor for 3½-inch hose is .17.

$$Q = \sqrt{FL\ 100/Cf \times 2}$$
$$= \sqrt{45/.17 \times 2}$$
$$= \sqrt{45/.34}$$
$$= \sqrt{132.35}$$
$= 11.5$ or 1,150 gpm (remember, Q is in hundreds)

Problem 7 Go back to Example 6-7 and prove that 231.9 feet of 1¾-inch hose is equivalent to 150 feet of 1½-inch hose.

ANSWER Calculate the friction loss for a given flow in both 1½-inch hose and 1¾-inch hose. Then multiply the *FL* for the 1½-inch hose by 1.5 and the *FL* for the same flow in the 1¾-inch hose by 2.319.

First, calculate friction loss for 125 gpm in 1½-inch hose.

$FL\ 100 = Cf \times 2Q^2$

$= 12 \times 2 \times 1.25^2$

$= 12 \times 2 \times 1.56$

$= 37.44$ psi

Total friction loss is 37.44 times 1.5, or 56.16 psi

Now calculate the friction loss for 125 gpm in 1¾-inch hose.

$FL\ 100 = Cf \times 2Q^2$

$= 7.76 \times 2 \times 1.25^2$

$= 7.76 \times 2 \times 1.56$

$= 24.21$ psi

Total friction loss is 24.21 times 2.319, or 56.14 psi.

The two answers are only off by .02 due to rounding. We now have proof that Example 6-7 is correct.

Problem 8 Calculate a conversion factor for a parallel lay of one 3-inch hose and one 3½-inch hose.

ANSWER We already know that at 11.25 psi friction loss 3-inch hose will flow 375 gpm. First calculate how much water will flow in 3½-inch hose with 11.25 psi friction loss.

$Q = \sqrt{FL\ 100/Cf \times 2}$

$= \sqrt{11.25/.17 \times 2}$

$= \sqrt{11.25/.34}$

$= \sqrt{33.09}$

$= 5.75$ or 575 gpm

375 gpm for 3-inch hose and 575 gpm for 3½-inch hose totals 950 gpm.

$Cf = FL\ 100/2Q^2$

$= 11.25/2 \times 9.5^2$

$= 11.25/2 \times 90.25$

$= 11.25/180.5$

$= .062$

Summary

Because friction represents the loss of energy in a system, we need to compensate for it. The measurement of needed compensation is called friction loss. Friction is the by-product of two surfaces rubbing together. It can be orderly, that is, laminar, or disorganized, that is, turbulent. Even the viscosity of the liquid contributes internal friction to the liquid.

There are four laws that govern the behavior of friction within hose. By understanding these laws we can best influence friction loss to obtain the best water flow. After all, the ultimate goal of hydraulics is getting water from the source to the fire most efficiently.

Because the basic friction loss formula is designed around 2½-inch hose, it is necessary to develop conversion factors to adjust both friction loss and hose length to hose other than 2½-inch. Where the conversion factor cannot be calculated by use of Law 3, an empirical method to calculate the correction factor was developed.

Finally, given the friction loss, it is possible to calculate the gpm flow. Later, in Chapter 11, the formula $Q = \sqrt{FL\ 100/Cf} \times 2$ becomes useful in doing maximum flow calculations.

Review Questions

1. Define friction loss.
2. Define viscosity.
3. What is laminar flow?
4. What is turbulent flow?
5. If you were testing 2½-inch hose with a flow of 150 gpm and a friction loss of 4.5 psi, and then increase the flow to 300 gpm, how could you determine the new friction loss without using the friction loss formula? What is the new friction loss?
6. Using Law 3, how much less friction loss will you have if the hose size is doubled but the flow stays the same?
7. Assume hose size does not vary due to pressure; why is friction loss not affected by pressure?
8. What is the formula for calculating friction loss for hose other than 2½-inch?
9. What is the friction loss for 150 gpm in 1¾-inch hose?
10. Go back to question 5 and use the friction loss formula to verify the answer.
11. You have been given the task of testing new hose for your department. During the test you have 500 feet of supply line laid from the hydrant. The discharge pressure at the pumper on the hydrant is 125 psi and the intake at the second pumper is 89 psi. During the test, the second pumper is flowing a 1½-inch wagon pipe ($C = .99$) at 80 psi tip pressure. What is the *Cf*, and what size hose is it?

12. During your testing of the hose, you are asked to consider the possibility of changing to 2-inch hose. If you replaced your 300 foot 1¾-inch attack line and ran a test with 2-inch hose at the same flow, how much more 2-inch hose could you carry in place of the 1¾-inch hose?

13. As you test the hose, the question of using 1¾-inch hose in place of 2½-inch is investigated. You know that some gpm flows in 2½-inch have friction losses as high as 20 and 25 psi. How much friction loss would there be in the 1¾-inch hose to compare with 20 psi in the 2½-inch hose?

14. After calculating the friction loss in question 13, you have decided that the friction in the 1¾-inch hose is prohibitive. What was the gpm?

List of Formulas

Finding the friction loss multiplier:

$$Fm = \left(\frac{Ve_2}{Ve_1}\right)^2$$

Finding the conversion factor:

$$Cf = \frac{D_1^5}{D_2^5}$$

Friction loss per 100 feet of 2½-inch hose:

$$FL\ 100 = 2Q^2$$

Friction loss per 100 feet of hose other than 2½-inch:

$$FL\ 100 = Cf \times 2Q^2$$

Finding conversion factor when friction loss and quantity of flow is known:

$$Cf = \frac{FL\ 100}{2Q^2}$$

Equivalent length formula:

$$L_2 = L_1 \times \left(\frac{Cf_1}{Cf_2}\right)$$

Equivalent friction loss formula:

$$FL_2\ 100 = FL_1\ 100 \times \left(\frac{Cf_2}{Cf_1}\right)$$

Finding quantity flow when friction loss and conversion factor are known:

$$Q = \sqrt{\frac{FL\ 100}{(Cf \times 2)}}$$

Pump Theory

Learning Objectives

Upon completion of this chapter, you should be able to:

- Identify the types of pumps familiar to the fire service.
- Understand the operation of positive displacement pumps.
- Understand the operation of nonpositive displacement pumps.
- Identify the parts and function of nonpositive displacement pumps.
- Identify the operational difference between multi- and single stage pumps.

INTRODUCTION

No course on hydraulics is complete without a chapter on pump theory. The pump generates the pressure (energy) we have been talking about to this point. In this chapter, the operation of pumps is examined in detail. Differences between various types of pumps are examined and advantages and disadvantages between types of pumps are explored.

Pumps can be generally divided into two different types: positive displacement and nonpositive displacement. Each has its role in today's fire service, but they are generally not interchangeable. In addition to these two larger types, positive and nonpositive displacement pumps, each can be further divided into subtypes of pumps.

The method of operation of all positive displacement pumps is similar. However, subtypes of nonpositive displacement pumps have different means of creating pressure.

Because the nonpositive displacement pump is the primary and most familiar pump used in the fire service today, the inner workings of both single stage and multistage pumps are explained. Specific terminology concerning the naming of pump components and how these components are essential to the creation of pressure are detailed.

A BRIEF HISTORY OF FIRE PUMPS

The first fire pumps were hand-operated units that had to be drawn to the fire by the firefighters. Once at the fire, the firefighters who had just pulled the unit to the fire then became the motive power to operate the pump. With as few as two or as many as fifteen on a side, firefighters would line up on each side of the "pumper" and alternately pump, discharging water to a hose or monitor device (see Figure 7-1). While some of the firefighters pumped, others would form an old-fashioned bucket brigade to fill the water tub on some models. Eventually the idea of using horses to pull the fire engine caught on and the apparatus became horse drawn. In the early 1800s, the motive power began to switch from manpower to steam power. By the mid-1800s, after the invention of the gasoline engine, it was first used to supply power to operate pumps that were still drawn to the fire by horses. Eventually as gasoline engines improved, they replaced the horses and the rest, as they say, is history.

The early fire engines used piston pumps, the leading edge of technology for their time. But as time passed and apparatus eventually became mechanized (run exclusively by gasoline engines), rotary gear pumps began to show up on fire engines. And still later, rotary gear and piston pumps gave way to the centrifugal pump. Just when the transition from piston and rotary pumps to centrifugal pumps took place is impossible to identify. As early as 1912 Seagrave was using centrifugal pumps almost exclusively, and Ahrens-Fox did not stop producing piston pumps until 1952. Today, except for special applications, the centrifugal pump is the pump of choice for the fire service.

Figure 7-1 *An 1853 James Smith hand pumper, owned by the Fire Museum of Maryland. Photo courtesy of Curt Elie.*

It is interesting to note that on some of the early mechanized fire apparatus, pumps of only 300- or 400-gpm capacity were not unusual. Today it is common to find pumps that have a capacity of as much as 2,000 gpm.

TYPES OF PUMPS

Pumps are divided into two primary categories: positive and nonpositive displacement pumps. Each has specific advantages and disadvantages, which is why they both fill a specific niche in the fire service today. In order to understand how each category of pump serves the needs of the fire service, it is important that the operating principles of each be thoroughly understood.

Positive Displacement Pumps

positive displacement pump
a pump that operates on the principle of discharging a fixed quantity of fluid with each cycle of the pump

The **positive displacement pump** operates on the principle of discharging a fixed quantity of fluid with each cycle of the pump. The quantity discharged per cycle is the same, regardless of how fast or slow the pump is operating. Obviously, the faster the pump operates, the larger the total volume of fluid discharged.

Most positive displacement pumps are also incapable of taking advantage of any incoming pressure. The only pressure that is discharged from one of these pumps is what the pump itself generates. It is also impossible for water to flow through rotary type positive displacement pumps if the pump is not engaged.

There are also some advantages to positive displacement pumps. They have the ability to develop higher pressure than their nonpositive displacement counterparts. In fact, where high pressures are needed for special applications, positive displacement pumps are used. Additionally, because there is no continuous pathway through the pump, they also have the ability to pump air. This is critically important when drafting, as we shall learn in Chapter 8.

The positive displacement pump is not just a single kind of pump, but a class of pumps. Positive displacement pumps can be piston pumps, rotary gear pumps, rotary vein pumps, and more. They all operate by the same general principle of discharging a fixed amount of fluid with each cycle.

Piston Pumps The piston pump is the oldest type pump. Its principles of operation are very simple and readily illustrate the general operation principles of positive displacement pumps. Basically, the piston pump operates much like a two-cycle piston engine. On the intake stroke, the piston draws fluid into the cylinder. Then, on the discharge stroke that same volume of fluid is discharged.

If you examine Figure 7-2 closely, it is easy to understand in more detail how the pump works. As the piston retreats from the top of the cylinder, it draws in a fixed volume of fluid determined by the volume of the cylinder, causing the intake valve to open and the discharge valve to close. After the piston has retreated enough to allow the cylinder to completely fill with fluid, it changes direction and begins to push the fluid out. This movement causes the intake valve to close and the discharge valve to open, allowing the fluid to discharge.

Previously it was stated that most positive displacement pumps cannot take advantage of incoming pressure or allow water to run through the pump unless the pump is operating. The piston pump is the exception to both of these conditions. The valves on the piston pump are activated by a differential in pressure. At draft, the pressure on the piston side of the intake valve is reduced as the piston retreats, allowing the atmospheric pressure to push the valve open and admit water and at

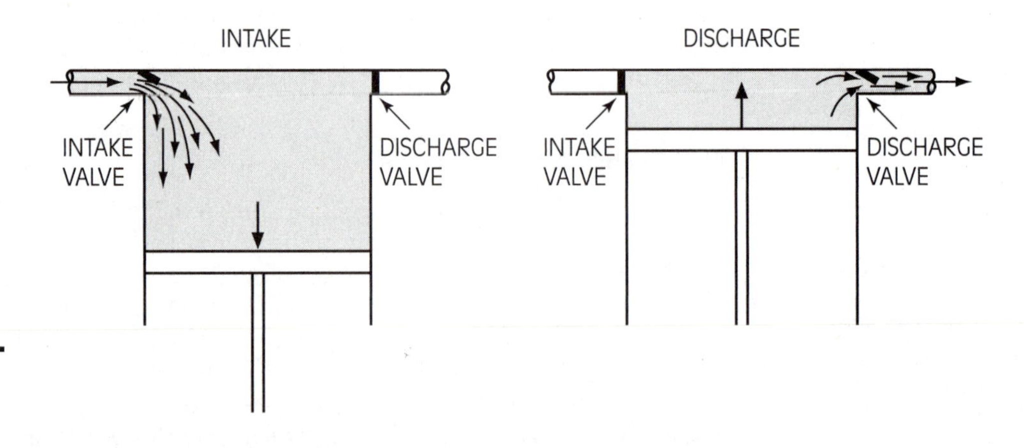

Figure 7-2 *The piston pump.*

the same time close the discharge valve. As the piston advances, the pressure on the piston side of the intake valve is greater and it closes the intake valve and opens the discharge valve. If the water on the intake side of the pump has any pressure on it, such as from a fire hydrant, the pressure of the water coming in is sufficient to open the intake valve and admit water into the piston, even if the pump is not operating. Once the piston fills with water, the pressure opens the discharge valve and discharges water to the discharge manifold. When the pump is operating with a positive intake pressure, this condition is met each time the piston retreats. As the piston begins to advance, the intake pressure is already present so any pressure the piston generates is in addition to the intake pressure.

It is important to understand that because there is a fixed volume of fluid to be displaced, as the piston discharges the fluid there must always be an open discharge while the pump is in operation. Because water is noncompressible, a positive displacement pump operating without any means of discharging water will cause something to break. The part that breaks can be the cylinder, piping to the discharge, or the connecting rod to the piston. At the very least, the power source will stall out.

Piston pumps can be built as single-acting pumps or double-acting pumps. The pump illustrated in Figure 7-2 is a single-acting pump. That is, it only pumps fluid on one stroke of the piston. But if a second set of valves were placed at the other end of the cylinder, a double-acting pump would be created (see Figure 7-3). A single cylinder could then discharge almost twice as much fluid in a single cycle of the pump.

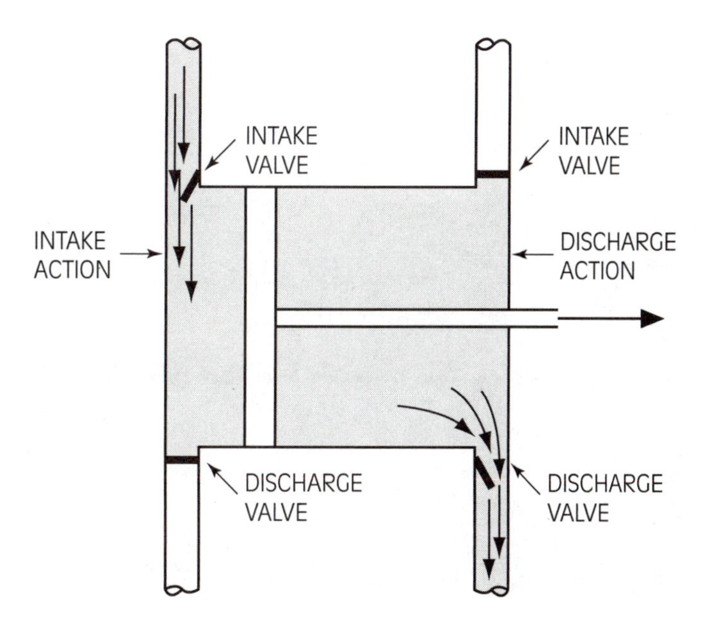

Figure 7-3 *Dual action piston pump.*

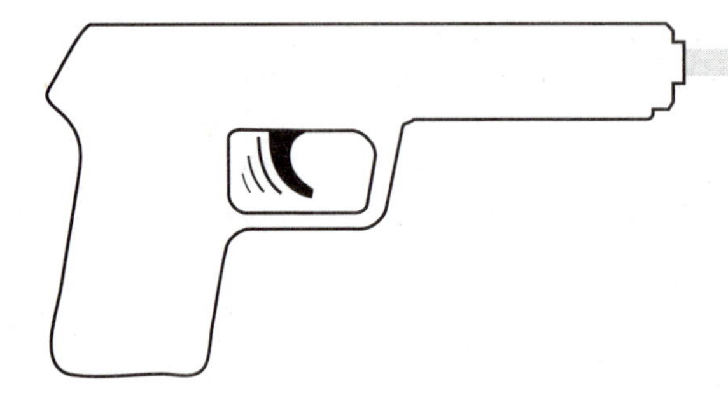

Figure 7-4 *Water pulse effect.*

Because the piston pump discharges fluid on only half of the cycle of the pump, the fluid comes out in surges, similar to how water discharges from a water pistol (see Figure 7-4). For firefighting this action is unacceptable. To remedy this, piston pumps used a combination of solutions. First, there were multiple cylinders. To reduce the surging, the cylinders all discharged at different times, equally spaced apart. Obviously the more cylinders, the less the effect of the surges. However, the surges were not completely eliminated. Another strategy to reduce the effects of the surge is to discharge all the fluid into a common discharge manifold. There the surges are further dampened before the fluid is finally discharged. This surge-dampening manifold is the shiny brass or chrome pressure dome evident on so many old pumpers (see Figure 7-5).

slippage
the tendency of water on the discharge side of the pump to slip back to the intake side

One of the advantages of the piston pump over other positive displacement pumps is that piston pumps do not have slippage. **Slippage** is the tendency of water

Figure 7-5 *The pressure manifold reduces pulsations (1897 American Fire Engine Co.)*

on the discharge side of the pump to slip back to the intake side. Because there are valves that open and close as the piston operates, it is not possible to have slippage.

Rotary Pumps Piston pumps are no longer used in the fire service, but a close cousin, rotary pumps, are very much in use. The rotary pump, like the piston pump, is a positive displacement pump. It discharges a fixed volume of fluid with each cycle, and there is no clear water path through the pump. The most common rotary pumps are rotary gear (see Figure 7-6), rotary vane (see Figure 7-7), and rotary lobe (see Figure 7-8). Some manufacturers used the rotary gear and rotary lobe pumps in place of piston pumps until nonpositive displacement pumps became common. The single disadvantage of rotary pumps is that as pressures become higher, it is possible for fluid to slip past the sides of the gears back to the intake side of the pump, that is, slippage can occur.

The following description of the operation of the rotary gear pump is common of rotary pumps. Fluid enters the pump through an intake manifold (see Figure 7-6) and is then captured by the gears of the pump. The gear teeth, which are turning away from the center of the pump at the intake, pick up a fixed volume of fluid between the teeth. As the teeth approach the discharge, they are turning toward the center of the pump and the teeth engage, forcing the fluid out the discharge.

Rotary vane pumps operate much like the rotary gear pumps, only instead of the fluid being forced out by gears meshing, the fluid is forced out as the vanes retract. The vanes retract because the hub with the vanes is placed eccentric to the pump casing. As the hub rotates, the vanes extend, trapping fluid and carrying it

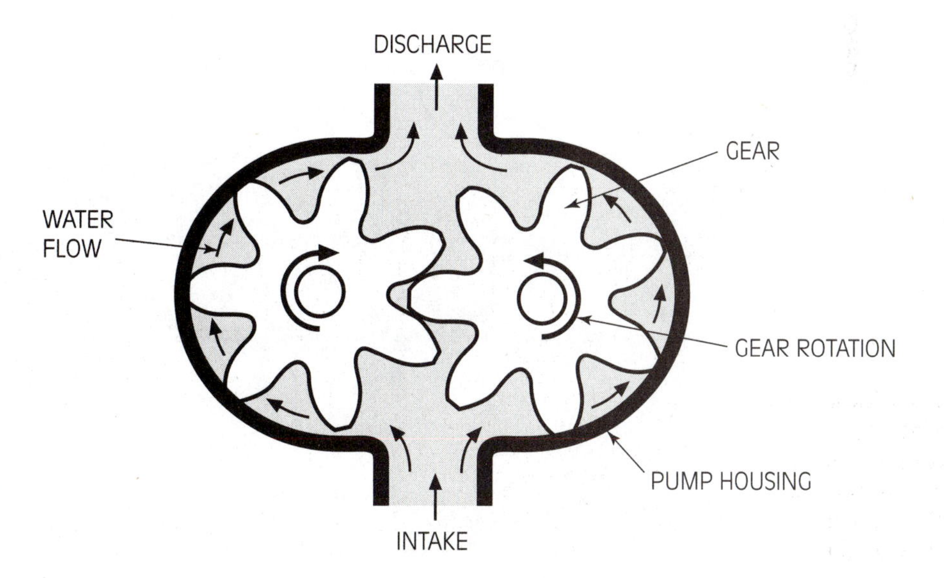

Figure 7-6 *Rotary gear pump.*

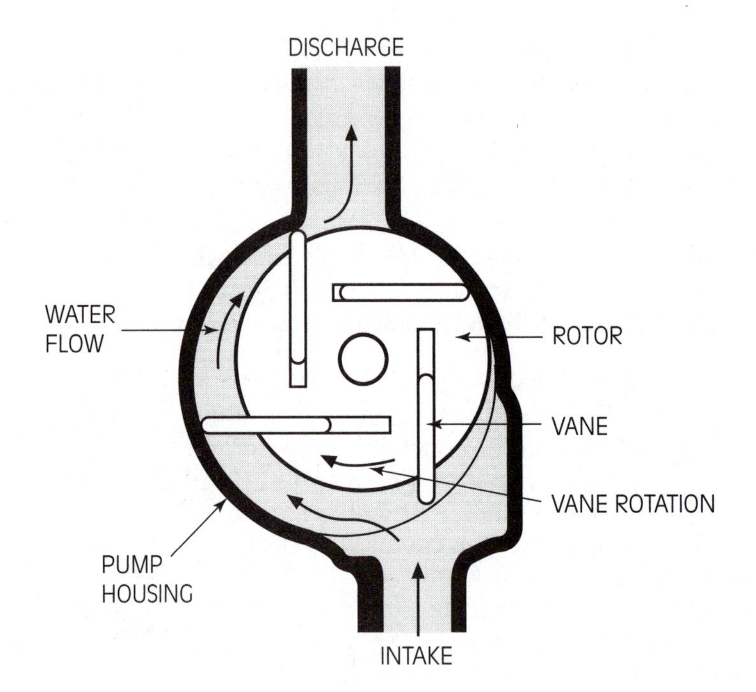

Figure 7-7 *Rotary vane pump.*

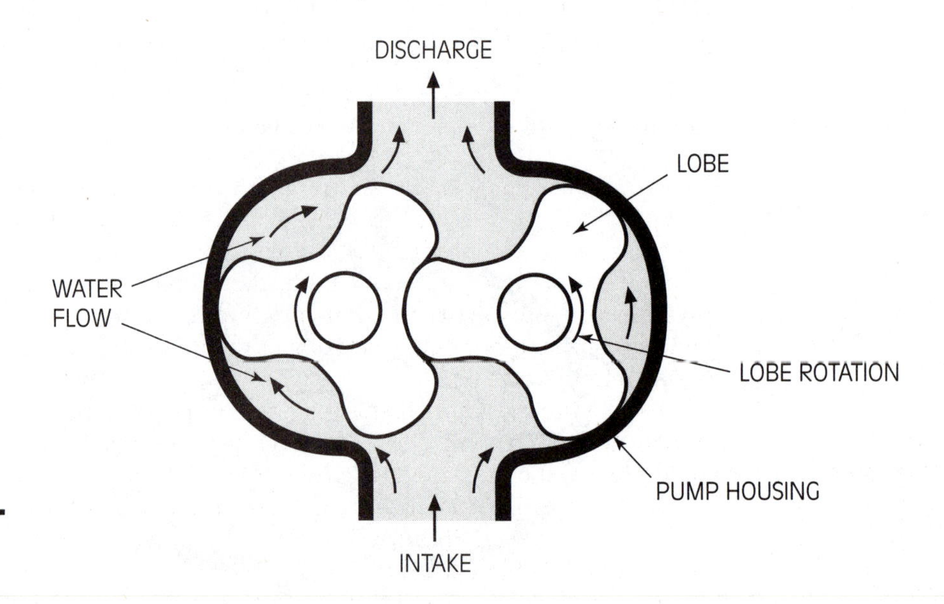

Figure 7-8 *Rotary lobe pump.*

to the discharge side of the pump. On the discharge side, the vanes retract and the fluid is forced out the discharge by the next volume of fluid being brought around by the vanes (see Figure 7-7).

The rotary pumps still in use in the fire service are used primarily as priming pumps. Because they are positive displacement, they can pump air, a feat impossible for nonpositive displacement pumps. There are some special application, high-pressure pumps of the positive displacement type.

Nonpositive Displacement Pumps

The second type of pump, the nonpositive displacement pump, is the type primarily used in the fire service today. It is used not only on fire apparatus, but also where pumps are needed to provide pressure for water-based fire protection systems. In short, wherever pumps are mentioned in the fire service, 99.9 percent of the time it is nonpositive displacement pumps being mentioned.

Nonpositive displacement pumps are often referred to as centrifugal pumps, a name given to them a long time ago when it was thought that there was a force called *centrifugal force*. We now know that centrifugal force does not exist, but the term is still used. Later in the chapter in the section "Radial flow pump," there is an explanation of how the nonpositive displacement pump imparts energy to water.

Just like there was not only one type of positive displacement pump, several types of nonpositive displacement pumps are broadly classified as kinetic energy pumps. In the fire service, we use only the subcategory of kinetic energy pumps classified as centrifugal pumps. These centrifugal pumps come in three types: radial flow, mixed flow, and axial flow. Of the three types of pumps, the radial flow pump is the workhorse of the fire service, however the other pumps deserve mention. All nonpositive displacement, centrifugal pumps have one thing in common: The part of the pump that actually imparts the energy to the liquid is called the **impeller**. The impellers for each type of centrifugal pump are different, due to their different means of imparting energy to the liquid. If the three types of centrifugal pumps were placed on a continuum, the radial flow pump would come first, operating at the lowest revolutions per minute (RPM), and the axial flow pump would be on the other end, operating at the highest rpm. The mixed flow pump would be in the middle.

impeller
the part of the nonpositive displacement pump that actually imparts the energy to the liquid

nonpositive displacement pump
a pump that does not have any form of valves or other restriction that would prevent the free flow of liquid

Unlike their positive displacement counterparts, **nonpositive displacement pumps** do not have any form of valves or other restriction in the pumps that would prevent the free flow of liquid. It is possible to open an intake and discharge on a nonpositive displacement pump and let liquid flow through the pumps without the pumps running, allowing the nonpositive displacement pump to take advantage of intake pressure.

Also unlike their positive displacement counterparts, nonpositive displacement pumps do not pump a fixed volume of liquid with each cycle of operation. Instead, the volume of liquid discharged is dependent on the resistance offered to the movement of the liquid. The pump exerts a force on the liquid that is constant for any given speed of the pump. If the force of resistance equals the force being created by the pump, the liquid will reach a state of equilibrium and the liquid will not flow. The pump will then churn the liquid and heat will be generated. However, if the force being generated is greater than the force of resistance, the water flows.

The primary disadvantage of the nonpositive displacement pump is that it cannot pump air. This disadvantage is serious where liquid has to be drafted (drawn) from a static supply, such as a pond or river. Once all the air is exhausted from the pump, the nonpositive displacement pump can pump liquid, even from a draft source, as long as air does not reenter the system. To overcome this inability to pump air, nonpositive displacement pumps that are required to draft are fitted with a small positive displacement pump. The sole purpose of this priming pump is to exhaust air from the main pump when operating from a draft. The National Fire Protection Association (NFPA) has a standard that addresses this need. NFPA 1901, *Automotive Fire Apparatus*, requires that all apparatus-mounted pumps be able to develop a vacuum equivalent to 22 inches of mercury. All apparatus-mounted, nonpositive displacement pumps must have a priming pump built in.

NFPA 1901

Standard for Automotive Fire Apparatus

Axial Flow Pump The axial flow pump is so named because the flow of liquid is in line with the axis of the impeller. In Figure 7-9 you can see how the impeller is on a shaft and the liquid flows in line with the impeller shaft. In the axial flow pump the impeller pushes the liquid through the pump, much the same way a propeller on a boat pushes the boat through the water, only in the axial flow pump, the pump is stationary, so the liquid has to move. The faster the impeller rotates, the more liquid and/or pressure there is.

radial flow pump
a pump in which the liquid is discharged out of the center (radially) through the impeller to its circumference

Radial Flow Pump The **radial flow pump** gets its name from the fact that liquid is discharged radially (out from the center) through the impeller to its circumference,

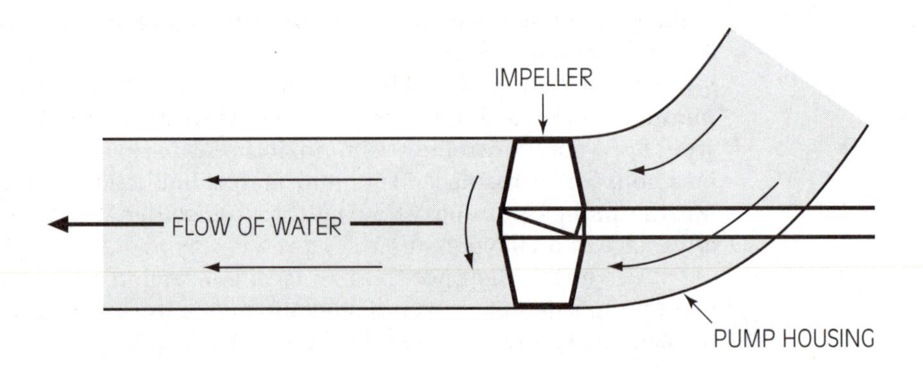

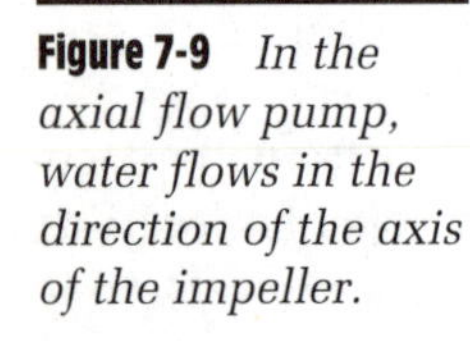
Figure 7-9 *In the axial flow pump, water flows in the direction of the axis of the impeller.*

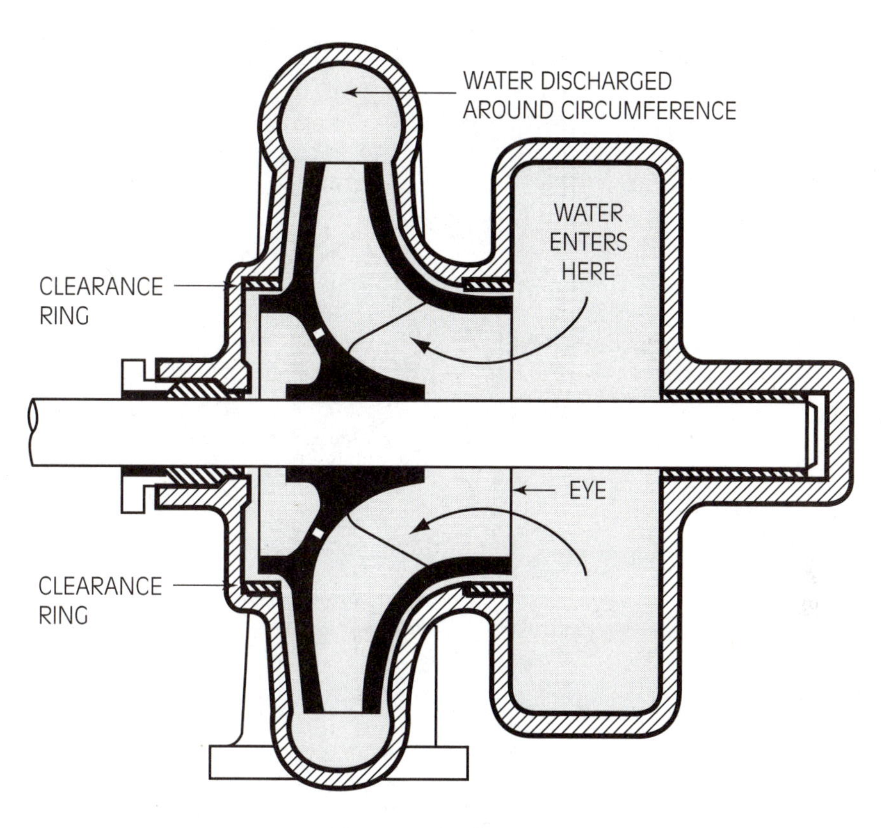

Figure 7-10 *In the radial flow pump, water flows radially through the impeller.*

as depicted in Figure 7-10. The radial flow pump is the apparatus-mounted pump so often referred to as a centrifugal pump.

Before discussing the process by which the radial flow pump pressurizes liquid, it is necessary to understand the critical parts of the pump. The moving part of the pump is called the impeller (see Figure 7-11). Liquid enters the impeller at the eye and flies to the outer circumference of the impeller as it rotates. This impeller is enclosed in a pump housing that has a volute shape with the impeller placed eccentric (off center) to the pump housing. The volute serves two purposes. First, it accommodates an increasing volume of liquid as the impeller rotates from point A in Figure 7-11 to point B. Second, after the impeller has imparted velocity to the liquid, the volute confines the accelerated liquid, converting velocity to pressure. Remember, the purpose of any pump is to add energy to the liquid it is pumping and that energy is in the form of pressure, velocity, or some combination of both. Impellers come in one of two types: single suction (see Figure 7-12) or double suction (see Figure 7-13)

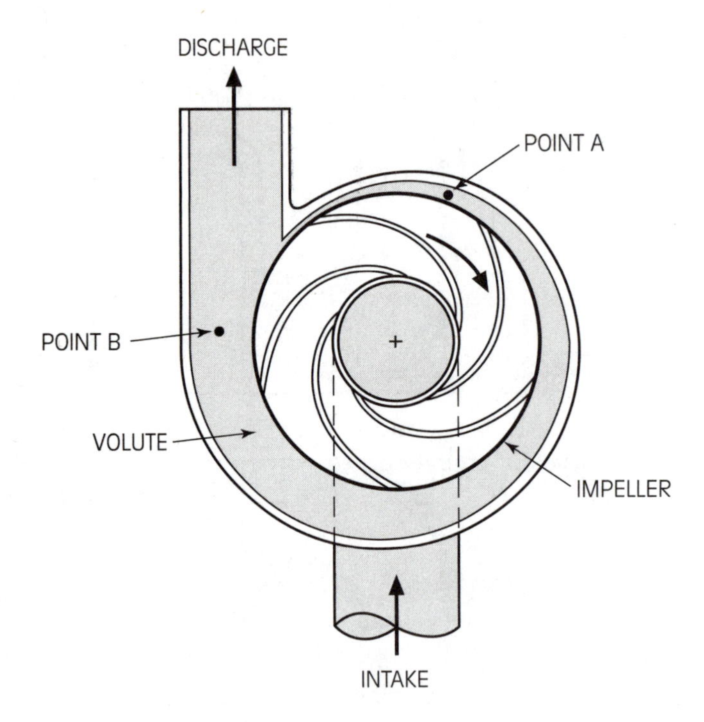

Figure 7-11 *The impeller is* eccentric *in the pump.*

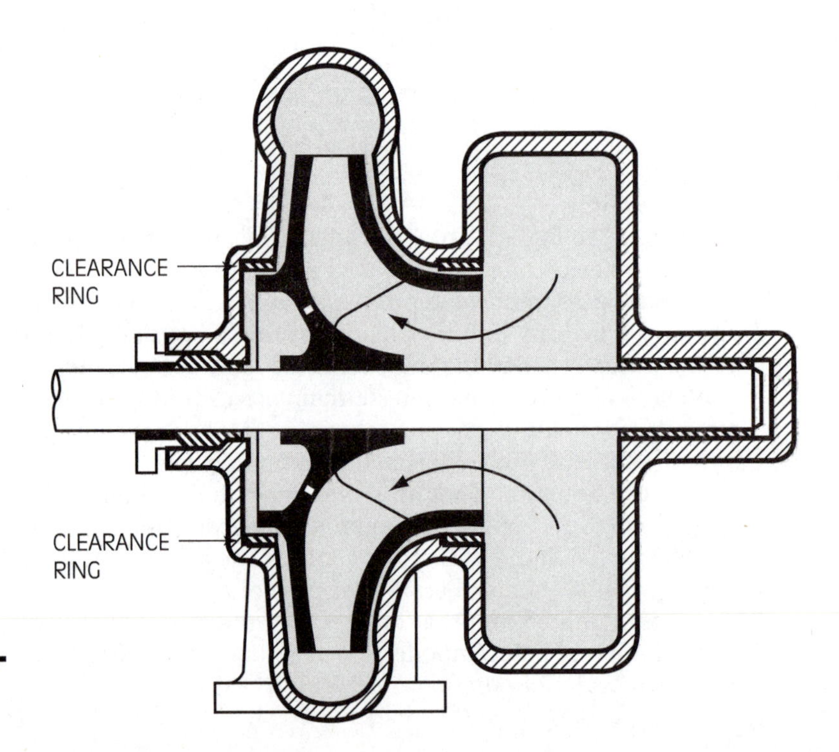

Figure 7-12 *Single suction impeller.*

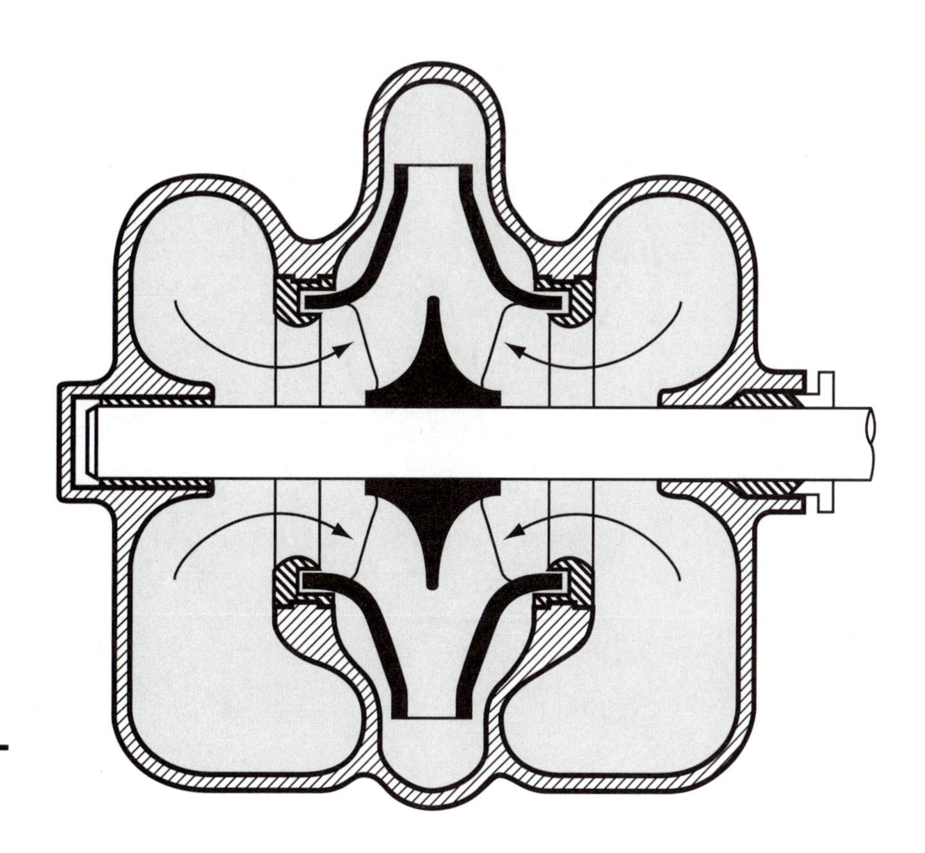

Figure 7-13 *Double suction impeller.*

eddying
water running contrary to the direction of the main flow

Between the walls of the impeller there are structures in the impeller called *vanes*. The purpose of the vanes is to maintain an orderly flow of liquid through the impeller by preventing eddying of the water. **Eddying** is water running contrary to the direction of the main flow, creating friction loss. The vanes also allow partial conversion of velocity to pressure. It is important to note that the vanes are curved away from the direction of rotation of the impeller so as to allow a natural movement of liquid through the impeller (see Figure 7-14).

The nonpositive displacement pump operates by allowing liquid to enter the pump by the force of an external pressure. That pressure either comes from the positive pressure of a municipal water system or the force of atmospheric pressure if drafting. (More on drafting in Chapter 8.) As the impeller rotates, liquid enters the impeller at the eye where an area of low pressure exists.

Next, the impeller adds velocity to the liquid by "throwing" the water to the edge of the impeller in the direction of the force being created by the pump. Figure 7-15 shows the direction of that force. The force is tangential to the rotation of the impeller. For instance, if you were to draw a circle on the impeller in Figure 7-15 so that it intersected droplet A, the direction of the travel of the liquid droplet would be on a tangent (a line that touches but does not intersect a

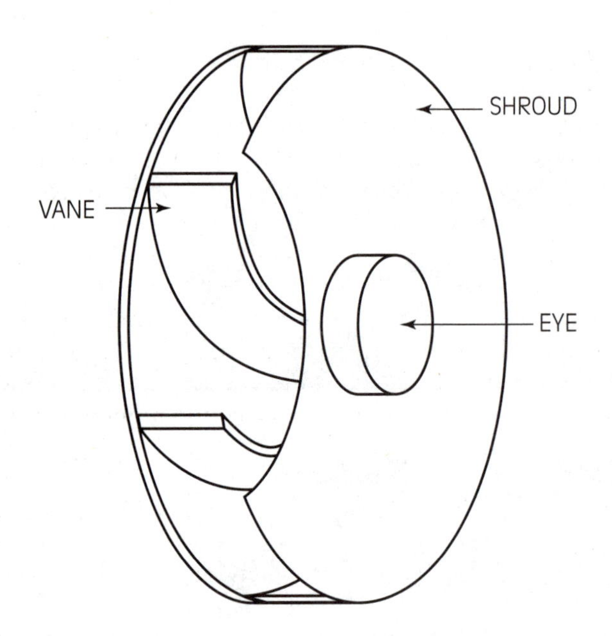

Figure 7-14 *The impeller.*

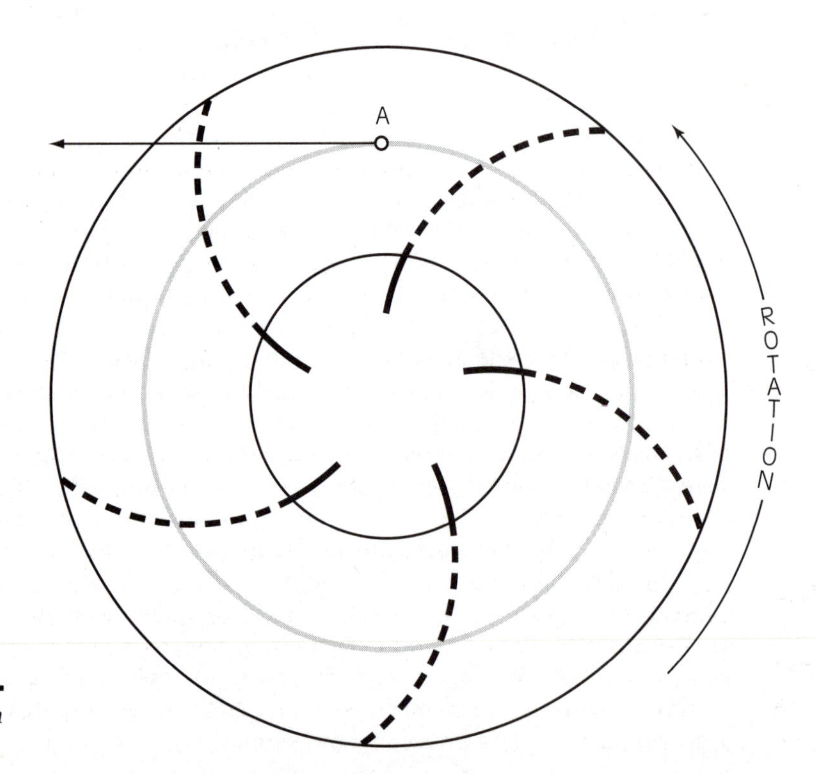

Figure 7-15 *Direction of force.*

circle) to the circle. The energy imparted to the liquid by the impeller is in the form of velocity.

We would expect the path of the water in the impeller to take a straight line as stated by Newton's first law of motion: "Every body continues in its state of rest or uniform speed in a straight line unless acted upon by a nonzero force." The path of the liquid through the impeller, however, is not a straight line. Due to the rotation of the impeller and the total volume of liquid, the liquid takes a spiral path. Because the outer edge of the impeller is travelling at a much faster rate than the edge of the inlet, additional velocity is imparted to the liquid as it approaches the outer edge. This results in a tangential velocity increase because the radius of the spiral increases as the liquid passes through the impeller. In short, liquid passing through an impeller changes direction numerous times, each time in the direction of the force as shown, resulting in the liquid spiraling through the impeller and gaining velocity.

Finally, after the liquid has been accelerated (energy added), the volute of the pump confines the liquid. By confining the liquid that has just been accelerated, the velocity of the water is reduced with a corresponding increase in pressure. This example shows Bernoulli's theory in operation.

In most instances dealing with pumps, energy simultaneously exists as pressure and velocity. However, because pressure is the more easily measured energy form, and the one that needs to be compensated for due to friction loss created by velocity, we universally refer only to pressure. Be aware, however, that there is velocity energy, even if only a minimal amount, any time water is flowing.

Thus far we have referred to the pressure added by the pumps as energy. In Chapter 3 we defined energy as the ability to do work. Before any action can take place, that is, before liquid can be moved through a distance, energy must be transformed into work. Since work is defined as force moving an object through some distance, in the pump's work is a combination of net pump pressure and how much liquid is being discharged. An expression for the work done by the pumps would look like this: 400 gpm @ 150 psi. In this expression the 400 gpm is the total amount of water flowing and 150 psi is the net pump pressure.

The Multistage Radial Flow Pump The multistage pump gets its name from the fact that it has more than one impeller where each impeller is a separate stage. Figure 7-16 shows the two different configurations of a two-stage radial flow pump. In Figure 7-16A, liquid enters the first stage from the intake, pressure is added by the first stage impeller and is then discharged to the second stage. The liquid is further pressurized by the second stage impeller before being discharged.

When a multistage pump is configured to route liquid first through one impeller and then the other (Figure 7-16A), it is capable of obtaining its highest operational pressure. This arrangement is called the *series* or *pressure* configuration. To better remember how the two-stage pump is configured when in series, remember that in series, each impeller pumps all the liquid and generates half the added pressure. The disadvantage to the series position is that it limits the pump to a maximum of about 70% of its rated capacity.

■ Note

In series, each impeller pumps all the liquid and generates half the added pressure.

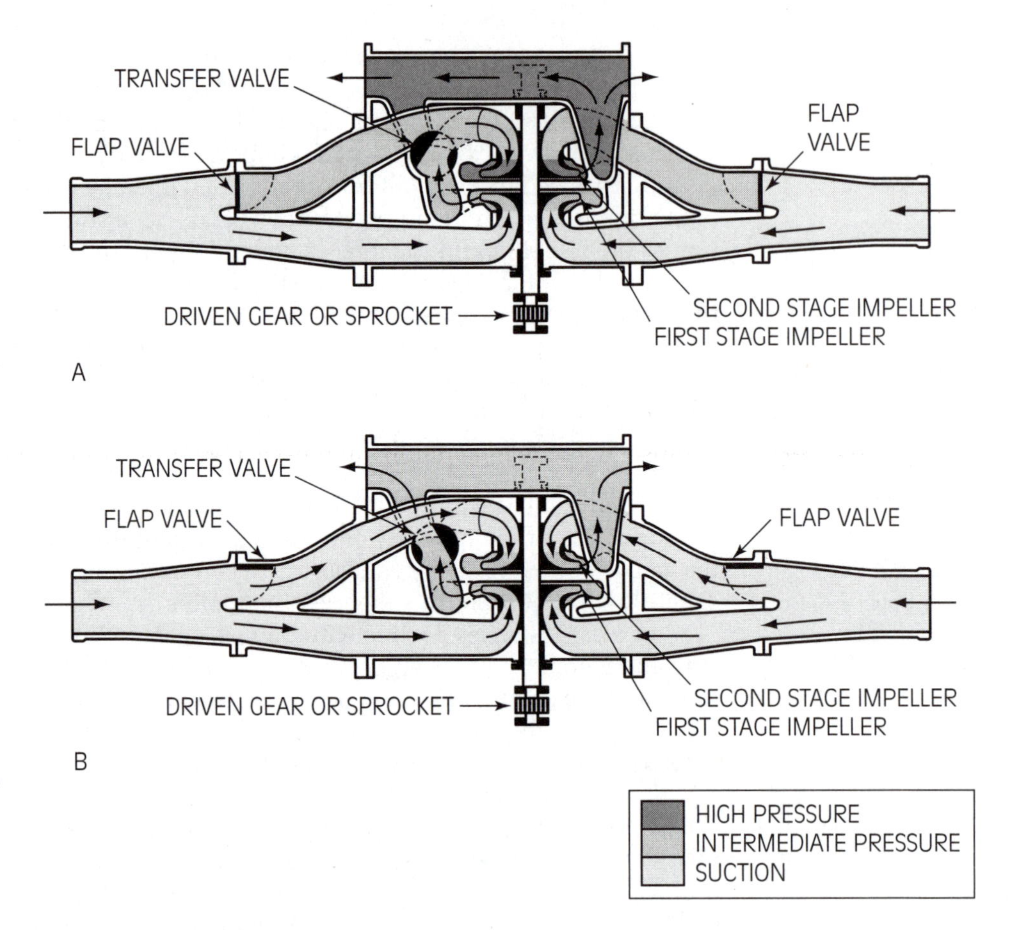

Figure 7-16 *The two-stage pump. Courtesy of Waterous Company.*

net pump pressure the pressure generated by the pump

gross pump pressure the total pressure discharged by the pump

Since the nonpositive displacement pump is capable of taking advantage of incoming pressure, where there is a positive intake pressure, the pumps do not work as hard. For example, if the pump is hooked up to a hydrant that has a residual pressure of 50 psi and the desired discharge pressure is 150 psi, the pump only needs to develop 100 psi of the final pressure. This pressure is defined as the net pump pressure. **Net pump pressure** is the pressure generated by the pump. The **gross pump pressure** is the total pressure discharged by the pump.

Example 7-1 How much pressure will each impeller of a two-stage pump develop if the residual intake pressure is 60 psi and the discharge pressure is 180 psi?

ANSWER To determine how much pressure each impeller will develop, start by subtracting the intake pressure from the discharge pressure. This figure is the total work done by the pump.

Discharge pressure – intake pressure = pressure generated by the pump

180 psi – 60 psi = 120 psi

Now divide the pressure generated by the pump by the number of stages, in this case two.

120 psi ÷ 2 = 60 psi

Each impeller (stage) will develop 60 psi of pressure.

Note
In parallel, each impeller pumps half of the total liquid but must generate all the added pressure.

The second configuration possible (see Figure 7-16 B) with the two-stage pump is called the *parallel* or *volume* configuration. In the parallel configuration, water enters each impeller at the same time directly from the intake manifold and is then discharged directly to the discharge manifold. In parallel, each impeller pumps half of the total liquid but must generate all the added pressure. In the parallel configuration the pump is capable of pumping its rated capacity, but cannot do it at high pressure.

Note: When two pressures come together, such as in the discharge manifold of a two-stage pump in parallel configuration, the pressures will combine equally without adding onto one or the other. If one pressure is higher than the other, the higher pressure will be the one to register if measured after the two pressures are combined.

With the two-stage pump it is necessary to have a mechanism that can configure the pump from the series configuration to the parallel configuration. The mechanism that transfers the pump from one configuration to the other is called the *transfer valve.* The transfer valve has only two positions in a two-stage pump, series or parallel. In the series position liquid is directed from the discharge of the first stage to the intake manifold of the second stage instead of being allowed to enter the discharge. Because this pressure is higher than intake pressure, due to pressure added by the first stage impeller, flap valves in the intake manifold close, preventing liquid from entering the second stage from the manifold. The liquid is then routed through the second stage to the discharge manifold.

When the transfer valve is in the volume position, liquid from the first stage is directed directly to the discharge manifold, which allows liquid from the intake manifold to open the flap valves to the second stage and enter the second stage directly. Liquid from the second stage is then discharged simultaneously with liquid from the first stage.

The Single-Stage Radial Flow Pump A second common design of the radial flow pump is the single-stage pump (see Figure 7-17). The single stage pump is easier to operate because the operator does not need to be concerned with selecting the proper position for the transfer valve. Since there is only one impeller on the impeller shaft, depending on capacity and design, a double suction impeller is sometimes employed. The impeller itself is big enough to allow for the full rated volume of the pump.

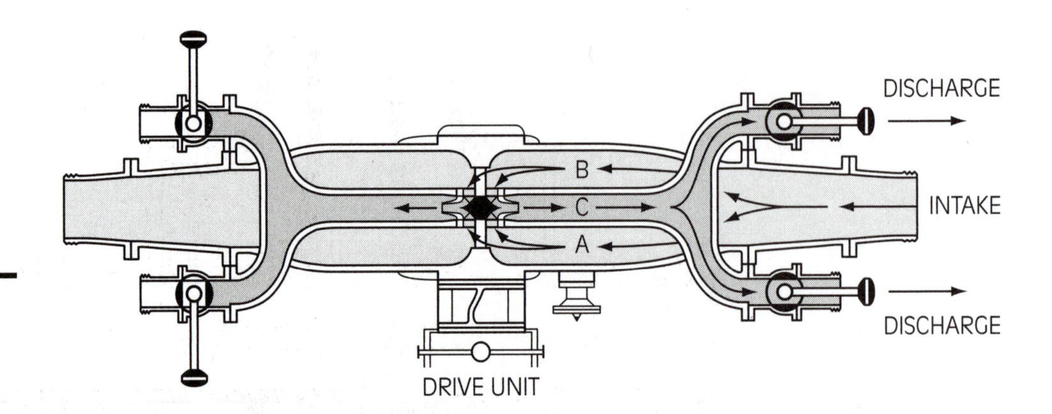

Figure 7-17 *Single-stage pump with double suction impeller.*

The disadvantage of the single-stage pump is that it is most efficient at high rpm. Diesel engines have the needed torque, and with proper gearing, can achieve the higher rpm. For this reason single-stage pumps did not achieve wide-spread use until the fire service had adopted the diesel engine.

end thrust
produced when the direction of flow of liquid is abruptly changed

End Thrust and Radial Hydraulic Balance Two conditions associated with radial flow pumps should be mentioned. The first is a condition known as end thrust. **End thrust** is produced when the direction of flow of liquid is abruptly changed. As water enters the impeller, it forces the impeller against the back wall of the pump housing before the water changes direction 90 degrees. This end thrust can harm the pump because it causes excessive wear if no compensation is provided. To minimize end thrust, clearance rings are provided at the rear of the impellers. Holes are also provided in the rear of the impeller, between the hub and the vanes, to admit water to the backside of the impeller to partially cushion the force of the water (see Figure 7-12). These two details provide partial hydraulic balance to reduce the forces associated with end trust. Any remaining force is then easily absorbed by the impeller support bearings.

Another method to compensate for end thrust is seen in Figure 7-18. By placing the two impellers of a two-stage pump on the same shaft, back-to-back, the end thrust of the first stage is cancelled out by the end thrust of the second stage. A double suction impeller is in hydraulic balance due to its design.

radial hydraulic balance
the equal discharge of liquid around the circumference of the impeller

The second condition associated with the radial flow pump is an issue of radial hydraulic balance. **Radial hydraulic balance** can be defined as the equal discharge of liquid around the circumference of the impeller. Because the volute of the pump accumulates more liquid as the discharge is approached, the load on the impeller is not uniform throughout its circumference. The result is a lack of hydraulic balance around the circumference of the impeller and can cause excess wear on the clearance rings called *bell mouthing*. To maintain radial hydraulic balance, one of two pump designs is used. First, a double volute design (see Figure

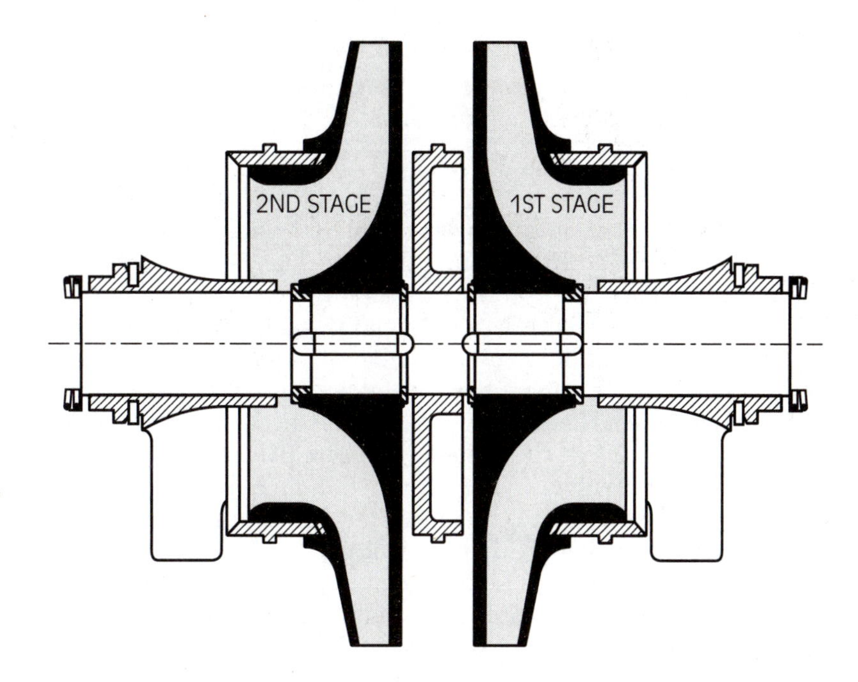

Figure 7-18 *Single suction impellers, back-to-back. Courtesy of Waterous Company.*

7-19) is sometimes employed. With the double volute design, liquid distribution is mirrored at 180 degrees allowing the design to be in radial hydraulic balance. The double volute design is generally limited to single-stage pumps, but even then the capacity and design of the pump determines if it is used.

A design to counter radial hydraulic balance where there are multiple stages involves alternating the direction of the volute for each stage on the same impeller

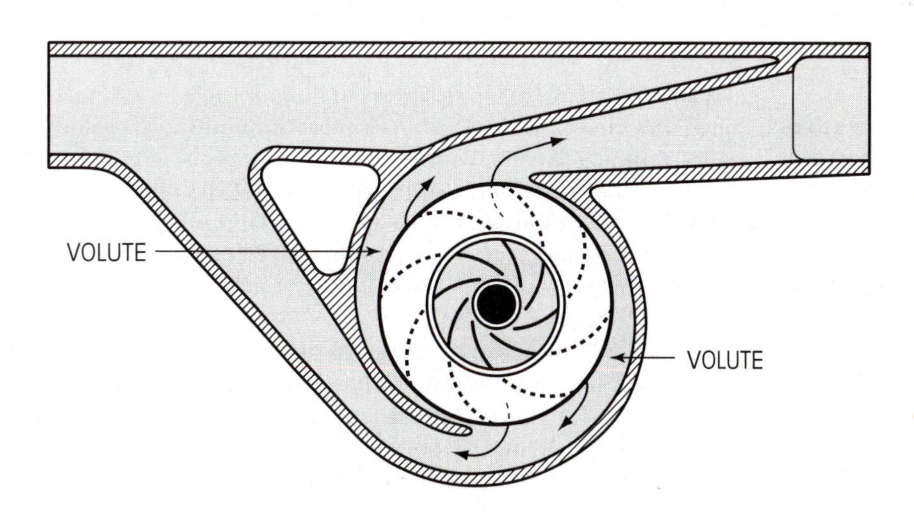

Figure 7-19 *Double volute.*

shaft. In the case of a two-stage pump, the volute of the first stage is in one direction and the volute of the second stage is rotated 180 degrees. Because both impellers are on the same shaft, they work together to maintain radial hydraulic balance.

Cavitation The operation of a radial flow pump is fundamentally simple. Only a few things can go wrong, other than mechanical breakdowns. The most serious operational mistake is allowing the pump to try to pump more water than it has available. Doing so causes a potentially serious condition known as cavitation.

cavitation
the formation of vapor (steam) bubbles in the impeller of the pump

Cavitation is the formation of vapor (steam) bubbles in the impeller of the pump. As the pressure on water is reduced, the temperature at which water boils is lowered. Under the right circumstances water can actually boil at ambient temperature inside the impeller of a pump. The circumstances under which this can happen involves trying to make the pump supply more water than is available.

Under normal operation, there is an area of low pressure (vacuum) at the eye of the impeller. With sufficient pressure coming into the impeller, the pressure at the eye remains high enough to prevent the formation of vapor bubbles. However, if there is insufficient pressure coming in, the pressure can fall below the vapor pressure of the water for its temperature and vapor bubbles will form. These vapor bubbles then travel to the interior of the impeller where they encounter higher pressures and instantly implode.

Cavitation can occur any time the pumps are operating and the impeller(s) are spinning too fast for the intake pressure. Mild cavitation may hardly be noticeable, but severe cavitation can cause the pumps to shake and make a sound similar to marbles being sucked through the pumps. Any cavitation can cause pitting of the interior impeller surfaces. Severe cavitation, if not stopped when first heard, can cause the impeller to break up and damage the pumps. Anytime cavitation is heard or even suspected, corrective steps should be taken immediately. The appropriate corrective step is to reduce the rpm of the pump.

rated capacity of a pump
the maximum amount of water a pump can deliver at 150 psi discharge and 10 feet of lift

Pump Capacity Pumps are designed to deliver their rated capacity at 150 psi discharge pressure from a 10 foot lift at draft, which allows us to define the **rated capacity of a pump** as the maximum amount of water a pump can deliver at 150 psi discharge and 10 feet of lift. If pressures higher than 150 psi are needed, the amount of water being pumped is reduced. This example is a direct application of Bernoulli's theorem of conservation of energy. We can either have a lot of water at a little pressure or a little water at a lot of pressure. The total energy in either evolution must be the same.

Table 7-1 illustrates the breakdown of the pump capacity versus maximum pressure. These figures apply to pumps at draft with a 10-foot lift. While these figures apply directly to draft situations, the principle itself applies to hydrant operations too. For hydrant operations, no absolute table of figures can be developed because the actual capacity of the pump varies depending on the capacity of the

Table 7-1 *Pump capacity and maximum pressure.*

Pressure	Capacity
0 to 150 psi	Up to 100 %
151 to 200 psi	Up to 70%
201 to 250	Up to 50%

water supply and the residual pressure of the hydrant. For calculating engine pressures and determining maximum lay and maximum flow problems Table 7-1 is used as a generalized guide.

Pumps are generally designed to deliver a maximum pressure of 300 psi. At this pressure, the maximum capacity is less than 50%.

Mixed Flow Pump Between the axial flow pump and the radial flow pump on our continuum is a third type of nonpositive displacement pump, called a *mixed flow pump* (see Figure 7-20). The mixed flow pump is a hybrid between the axial flow and the radial flow pump. It employs elements of both pumps to impart energy to water. It pushes the water through the pump to some extent, like its axial flow cousin. It also throws the water out of the impeller to some extent, like its radial flow cousin.

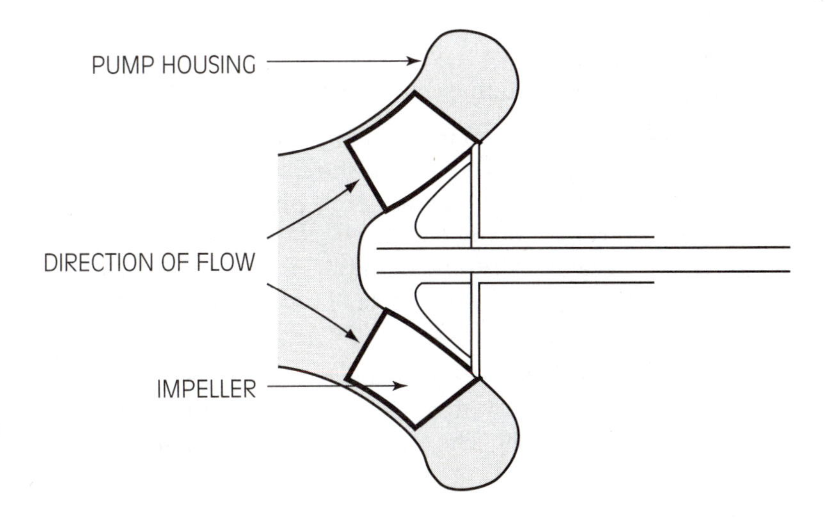

Figure 7-20 *Mixed flow pump.*

APPLICATION ACTIVITIES

Put your knowledge of pumps to work to solve the following problems.

Problem 1 A pumper connected to a hydrant has a residual intake pressure of 65 psi. If the pumper is discharging water at 235 psi, how much pressure is being added by each impeller of a two-stage pump in the series configuration?

ANSWER Because the pump is in the series position, each impeller adds half the pressure difference between the intake pressure and the discharge pressure.

Discharge pressure – intake pressure = pressure generated by the pump

235 – 65 = 170 psi

Each impeller adds half of the total pressure generated by the pump.

170 ÷ 2 = 85 psi

Each impeller generates 85 psi

Problem 2 At the same fire as the pumper in Problem 1 is another pumper that is supplying water to a large master stream device. If the pumper has a residual intake pressure of 45 psi, and is discharging at 150 psi, how much pressure is each impeller generating if the pump is in the parallel configuration?

ANSWER The pump is in the parallel configuration, so each impeller will develop the total pressure added by the pump.

Discharge pressure – intake pressure = pressure generated by the pump

150 – 45 = 105 psi

Each impeller develops 105 psi

Problem 3 Prove that the force that accelerates water in a radial flow pump is tangential to the rotation of the impeller.

ANSWER This problem is not all that hard to prove. The following experiment will prove the direction of the force that accelerates water in the impeller of a radial flow pump. Take a piece of string about 2 feet long and tie a small weight to one end. Twirl the string around just fast enough to keep the weight at the end of the string, scribing a circle. When the weight is at the top of the circle, let go of the string. What direction did the weight go in (see Figure 7-21)?

In Figure 7-21 you can see the circle scribed by the weight at the end of the string. When released, the weight will fly in the direction of force on the weight. The weight will fly out in the direction indicated by A.

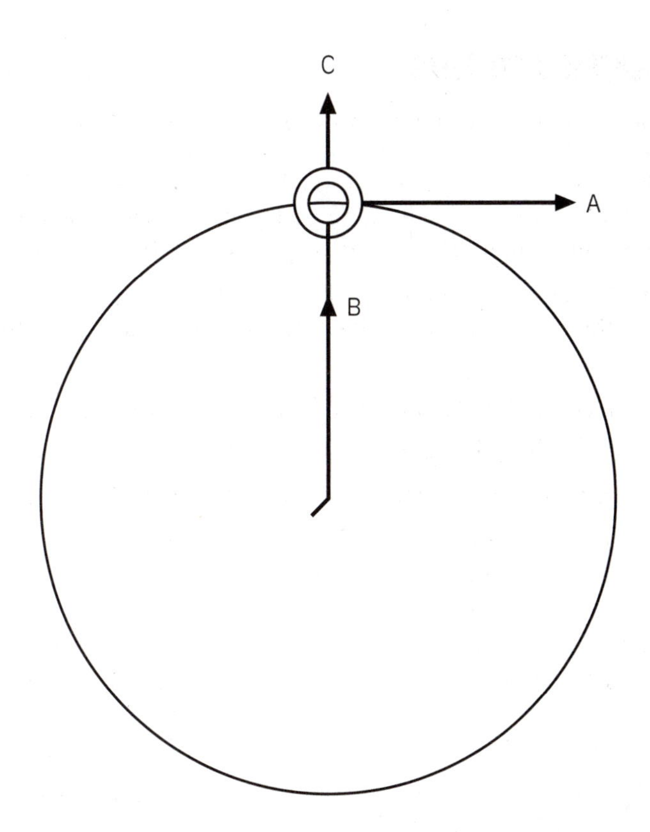

Figure 7-21 *Direction of force proof.*

It was mentioned earlier that the nonpositive displacement pumps are often referred to as centrifugal pumps. At one time it was mistakenly believed that a force (a center fleeing force or centrifugal force) pushed objects out from the center of a circle as indicated by arrow B in Figure 7-21. However, if there were centrifugal force acting in the direction of B, our weight would go straight out in the direction of C when the string was released. Instead the weight follows the path indicated by A. This force is *tangential* to the direction of rotation of the weight at the end of the string.

Summary

One of the more interesting aspects of the fire service is the evolution of the fire engine. At first it was drawn to the fires by the firefighters, but today internal combustion engines provide the motive force as well as the power to operate the pumps at the fire.

Pumps can be placed in one of two major categories: positive displacement or nonpositive displacement. Within each of these major categories, pumps can be further divided into subcategories of pumps.

Positive displacement pumps work by discharging a fixed quantity of fluid with each cycle of the pump. They are incapable of taking advantage of any incoming pressure. Additionally, they must always have an open discharge when operating, or damage can occur to the pump.

Nonpositive displacement pumps do not discharge a fixed volume of liquid for each revolution of the impeller. Instead, the volume of a nonpositive displacement pump depends on resistance offered to the movement of the liquid. This type pump is also capable of taking advantage of incoming pressure, which in turn reduces the workload on the engine providing the motive power to the pump.

Because the radial flow pump is the workhorse of the fire service, it is helpful to know, in depth, how it works. Being able to understand concepts such as end thrust and radial hydraulic balance is critical to a true working understanding of pump theory.

Review Questions

1. When did the nonpositive displacement pump first find a place in the U.S. fire service?
2. What general category of pump does not take advantage of incoming pressure?
3. When discussing the characteristics of pumps, reference is made to positive displacement pumps pumping fluid and nonpositive displacement pumps pumping liquid. Why is this distinction made?
4. What is the name of the moving part of a nonpositive displacement pump?
5. What is the name of the inlet of the item referred to in Question 4?
6. What is the purpose of the volute?
7. If a two-stage pump is discharging liquid at 200 psi, how much pressure is each impeller providing if the intake pressure is 60 psi and the pump is in the parallel configuration?
8. If during the operation of a pump it begins to cavitate, what action should be taken?
9. What is cavitation?
10. What is the name given to the general category of pump installed on all radial flow pumps as "priming pumps"?

11. Give an example of Bernoulli's principle at work in the radial flow pump.
12. Define eddying.
13. What is the name of the part of the discharge pressure generated by the pump?
14. What is the path of water through the two-stage, radial flow pump when it is in the series configuration?
15. What condition results when the direction of flow of liquid is abruptly changed in a radial flow pump?

List of Formulas

Finding net pump pressure:

Discharge pressure – intake pressure = pressure generated by the pump

Reference

NFPA 1901, *Standard for Automotive Fire Apparatus.* Quincy, MA: National Fire Protection Association, 1999.

Theory of Drafting

Learning Objectives

Upon completion of this chapter, you should be able to:

- Understand the laws of physics that permit pumps to draft water.
- Be able to calculate the amount of work a pump has to perform to get water into the pumps from a draft source.
- Convert inches of mercury, feet of lift, or psi.
- Interpret the static and dynamic reading on the intake gauge in terms of feet of lift or friction loss.
- Understand basic pump test procedures.
- Understand pump capacity limitations as discharge pressure increases.

INTRODUCTION

This chapter goes hand-in-glove with Chapter 7. You cannot fully understand pump theory without also understanding how water is drafted into a pump. Pump theory and the theory of drafting are divided into two separate chapters to make understanding them a little simpler.

In Chapter 2 we discussed at length the concepts of force and pressure. A thorough understanding of pressure is a prerequisite to the study of this chapter. When talking about drafting, it is necessary to understand how the pressure of the atmosphere is responsible for getting water into the pumps. Unless you know force and pressure this is not possible. It is also necessary to be able to calculate the maximum theoretical lift at atmospheric pressure and the amount of energy loss realized on the intake side of a pump while drafting. From a positive water supply, a fire hydrant, this is not a factor because the pressure of the incoming water is independent of the pump. However, when drafting, the pump must not only perform enough work to discharge water from the pump, it must also perform work to get water into the pump.

Finally, Chapter 3 discussed conservation of energy. In this chapter you will see it applied to a real world situation. As the pump discharge pressure is increased, the capacity of the pump must go down. After all, the amount of energy the pump imparts to the liquid it is pumping is finite, and it can be in the form of a lot of water at low pressure, or a little water at a lot of pressure. This concept is explained in the section Testing Pumps.

DRAFTING

Under most circumstances when we think of pumping water we usually think of a pumper hooked up to a fire hydrant connected to a municipal water system. The municipal water supply supplies water under pressure to the pump making it possible for the pump to simply add any additional pressure, thus taking advantage of the incoming pressure. But this situation is not universal; some places have no municipal water supply and must rely on fire apparatus to draft water from static water sources.

drafting
the process of removing water from a static source, such as a river, by the influence of the atmosphere

Drafting is the process of removing water from a static source, such as a river, by the influence of the atmosphere. In order to remove the water from the static source, some outside factor must be utilized to get the water into the pump. Water itself has no tensile strength, so it is not possible to "suck" the water into the pump. In fact, the only way water will enter the pumps is if it is pushed in. It does not matter whether the source of supply is a municipal system or a pond; something has to push the water into the pump.

From a municipal water system, pressure to force water into the pump is created by pumps or elevation (more on this in Chapter 12), but from a draft we must rely on a more natural source of pressure, the atmosphere. Understand that the pressure read from the municipal water supply is actually relative pressure. When the relative pressure is reduced to zero, we still have atmospheric pressure left. In order to draft water into our pumps we need to rely on atmospheric pressure to generate the intake pressure.

Water will not automatically flow into our pumps just because we place hard-suction hose into a static water source, because while there is a source of positive pressure outside the pump in the form of atmospheric pressure, the same atmospheric pressure exists inside the pump at the same time. In short, pressure inside and outside the pump is in equilibrium and water will not flow (see Figure 8-1). Remember pump pressure gauges are calibrated on a relative scale and read zero when there is actually an atmospheric pressure present. In order to draft, all we have to do is eliminate the condition of equilibrium.

priming the pump reducing the pressure inside the pump, allowing atmospheric pressure to fill the pump with water

In order for atmospheric pressure to force water into our pumps, we must reduce the pressure inside the pump to a pressure below atmospheric pressure. **Priming the pump** is the term given to the process of reducing the pressure inside the pump, allowing atmospheric pressure to fill the pump with water. For non-positive displacement pumps to be able to draft water, a small positive displacement priming pump is integrated into the pump housing. For every pound of pressure we reduce the internal pressure of the pump, the atmosphere will push the water up the intake 2.31 feet (see Figure 8-2). Initially the pressure reduction

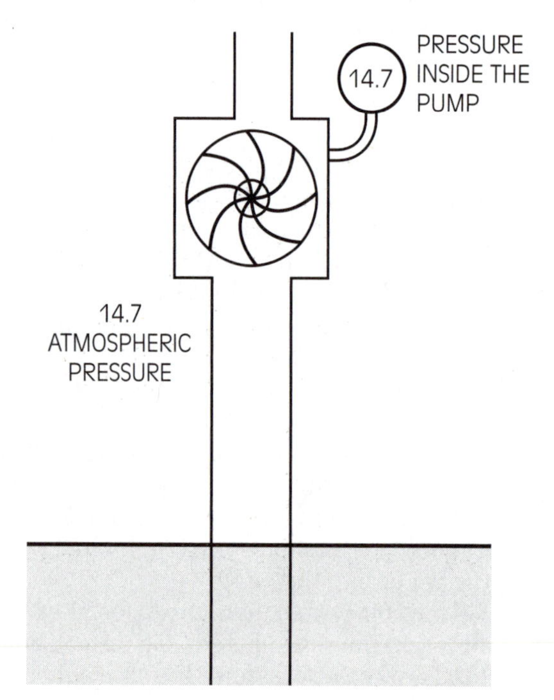

Figure 8-1 *Pressure in equilibrium.*

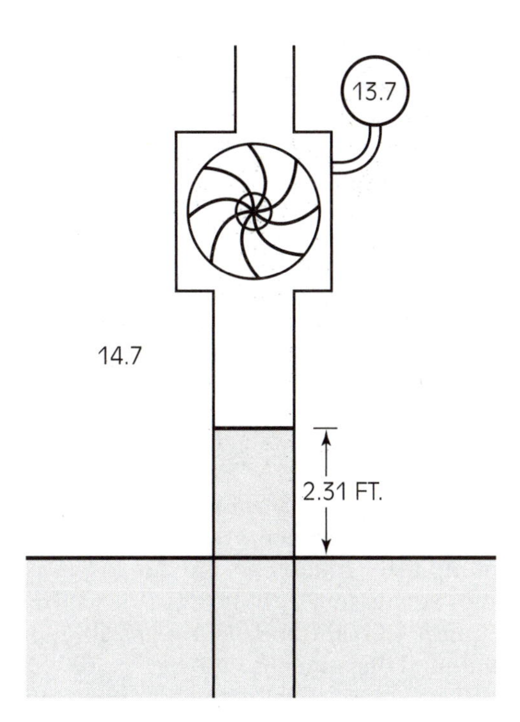

Figure 8-2 *Pressure in the pump is reduced by 1 psi.*

is accomplished by the activation of the priming pump, but once water is flowing, the pump must do enough work not only to pump the water, but also to maintain the intake pressure.

The pressure that is indicated on the intake gauge after the pump has been primed is referred to as a vacuum because it is below atmospheric pressure. Reading the vacuum created by the pump presents a small problem, because all the gauges on the pump panel are calibrated on the relative pressure scale. The intake gauge, however, is a compound gauge and is capable of reading pressures above and below atmospheric. Vacuum is always displayed as inches of mercury (Hg, the symbol for the chemical element mercury, is the customary abbreviation for mercury). This means that any time we are discussing drafting, the performance of the pump in getting water into the pump will always be in inches of Hg. Since we are familiar with referring to pounds of pressure and feet of water, we need to convert the inches of Hg to either pressure or feet of water. Table 8-1 is a pressure-mercury-water conversion chart and can be used to convert inches of Hg to pressure or feet of water, or to go from pressure or feet of water to inches of mercury. As we go through this chapter we will refer back to this chart many times and will teach its use as examples arise rather than as an abstract concept at this point.

Table 8-1 *Pressure-mercury-water conversion chart.*

Pressure	Mercury	Water
.433 psi	.882 inches	1 foot
.489 psi	1 inch	1.13 feet
1 psi	2.037 inches	2.31 feet
14.7 psi	29.94 inches	33.96 feet

Limitations

The process of drafting has several factors that affect a pump's ability to draft water, even under ideal circumstances. Some of these limitations are due to the mechanics of the pump itself, and others are due to the laws of physics.

In theory, pumps should ideally be able to evacuate every bit of pressure from the interior of the pump. In reality this is not possible. In order for the pump to operate, clearances around the impeller shaft have to be loose enough that the shaft can turn without restriction, yet tight enough to prevent air and water leaks. To meet this dual requirement, the pumps cannot be tight enough to prevent all air leakage when drafting. The seals are tight enough to allow the priming pump to evacuate the majority of air from the pumps, but an absolute vacuum is not possible. Also, leakage at gates and valves is a factor in allowing unwanted air into pumps. The older the pump, the more air leakage, so it is necessary to maintain pumps in top condition and be alert for deteriorating performance.

The atmosphere is also a limiting factor in determining how high we can lift (draft) water. Because the atmospheric pressure pushes the water into the pump, if that pressure varies, so will the ability to lift water. Two conditions determine the atmospheric pressure: (1) the elevation and (2) weather conditions. In general, atmospheric pressure will be reduced by .5 psi for every 1,000 feet of elevation we go up from sea level. As example, this means that at 2,000 feet above sea level the normal atmospheric pressure will only be 13.7 psi. This pressure limits the theoretical and actual height a pump can lift water. The second factor, weather conditions, affects drafting regardless of the normal atmospheric pressure. As weather patterns change, the atmospheric pressure rises and falls. As the pressure increases, the amount of lift possible increases; but as the pressure falls, the amount of lift possible decreases. You can find out how the weather is affecting drafting conditions by watching the daily news. Weathermen routinely report the barometric pressure as weather patterns change. The barometric pressure they are referring to is the current atmospheric pressure.

In Table 8-1, it should be noted that 29.94 are the maximum inches of mercury, assuming an atmospheric pressure of 14.7 at sea level. In reality this pressure can increase or decrease to some extent based on weather conditions.

There are two final factors to be considered with pumps and drafting. The first is the friction loss on the intake side of the pump. Here we are generally talking about friction loss in the hard suction hose and strainer. Just as hose has friction loss, hard sleeves and strainers also have a loss associated with them. This loss in the hard suction hose and strainer has to be overcome by the atmospheric pressure.

Finally, the water temperature is also a factor in drafting water. When water is hot, it has a tendency to vaporize. So when attempting to draft, as the pressure inside the sleeve and pump is reduced, hot water would rather vaporize than stay in liquid form. This does not mean that hot water can not be drafted, but it is a factor and may require that the pumper be placed closer, vertically, to the source of water.

CALCULATING LIFT

One of the most important calculations we need to perform in relation to drafting is determining lift, which includes two different calculations: (1) calculating maximum theoretical lift and (2) calculating actual lift.

Regardless of the calculated theoretical lift or calculated actual lift, there are practical limitations to drafting. As a rule of thumb, in order to obtain the maximum capacity of the pump, a lift of 10 feet should not be exceeded. A lift of 15 feet reduces the capacity of the pump to 70 percent, and a lift of 20 feet reduces the capacity of the pump to 60 percent.

Calculating Theoretical Lift

To calculate the maximum theoretical lift we need to know two things, what the atmospheric pressure is and how much lift each pound of pressure provides. The second factor has already been provided to us, both in Chapter 2 and in Table 8-1. Each pound of pressure is capable of lifting water 2.31 feet. Multiply this amount by the atmospheric pressure and you will find the maximum theoretical lift. In Chapter 2 the formula $H = 2.31 \times P$ was introduced. This formula can be used here. Insert the atmospheric pressure in place of P, and the maximum theoretical lift will be H.

Example 8-1 What is the maximum theoretical lift at sea level?

ANSWER At sea level the atmospheric pressure is 14.7 psi.

$$\begin{aligned} H &= 2.31 \times P \\ &= 2.31 \times 14.7 \\ &= 33.96 \text{ feet} \end{aligned}$$

Example 8-2 What is the maximum theoretical lift at Mile High Stadium in Denver, Colorado?

ANSWER Mile High Stadium is approximately 5,000 feet above sea level, therefore the atmospheric pressure is 2.5 psi less than at sea level.

$$H = 2.31 \times P$$
$$= 2.31 \times 12.2$$
$$= 28.2 \text{ feet}$$

As already mentioned, a good source for up-to-date information on the atmospheric pressure due to weather condition is the weather service. However, the weather people usually give pressure in terms of inches of mercury so it is necessary for us to convert the inches of mercury into pressure. To convert inches of mercury to psi, we only need to multiply the inches of mercury by .489 psi. (In Table 8-1, 1 inch of mercury is equal to .489 psi.) This creates a new formula, $P = .489 \times Hg$. In this formula .489 is a constant that represents the pressure equivalent of 1 inch of Hg, *Hg* represents the barometric pressure or reading in inches of mercury, and *P* is the pressure equivalent of the barometric reading.

$P = .489 \times Hg$

Example 8-3 What is the maximum theoretical lift if a storm is moving through and the weather service reports a barometric pressure of 28.5 inches of mercury?

ANSWER First find the pressure equivalent of 28.5 inches of mercury.

$$P = .489 \times Hg$$
$$= .489 \times 28.5$$
$$= 13.94 \text{ psi}$$

Maximum theoretical lift is:

$$H = 2.31 \times P$$
$$= 2.31 \times 13.94$$
$$= 32.2$$

It is also possible to make a direct calculation from inches of mercury to feet of lift by multiplying the barometric pressure by 1.13 feet, the equivalent lift of water to each inch of Hg. This requires another new formula, $H = 1.13 \times Hg$. In this formula 1.13 is a constant for the elevation of water that represents a reading of 1 inch of mercury from Table 8-1, *Hg* represents the barometric pressure or reading in inches of mercury, and *H* is the theoretical lift.

$H = 1.13 \times Hg$

$$H = 1.13 \times Hg$$
$$= 1.13 \times 28.5$$
$$= 32.2 \text{ feet}$$

Calculating Actual Lift

Because the atmospheric pressure is responsible for pushing water up the sleeve into the pump, we are able to determine actual lift from the Hg reading on the intake gauge. This simple calculation involves multiplying the inches of Hg indicated on the intake gauge by 1.13 feet using the formula $H = 1.13 \times Hg$. It should be noted that the reading on the intake gauge must be a static reading. At draft, the **static intake reading** is the reading on the intake gauge, in inches of Hg, after the pumps have been primed, but before water is flowing.

static intake reading the reading on the intake gauge, in inches of mercury, after the pumps have been primed, but before water is flowing

Example 8-4 How high is the actual lift if the static intake reading at draft is 6 inches of mercury (see Figure 8-3)?

ANSWER One inch of mercury is equal to 1.13 feet of lift.

$$H = 1.13 \times Hg$$
$$= 1.13 \times 6$$
$$= 6.78 \text{ feet}$$

Reverse Lift Calculations

Since we now know how to find the lift when given the Hg reading, it can be useful at times to be able to find the Hg reading when we know the lift. It is just a matter of multiplying the feet of lift by .882 inches of Hg. This uses the formula $Hg = .882 \times H$, where .882 is a constant that represents the Hg equivalent to 1 foot

$Hg = .882 \times H$

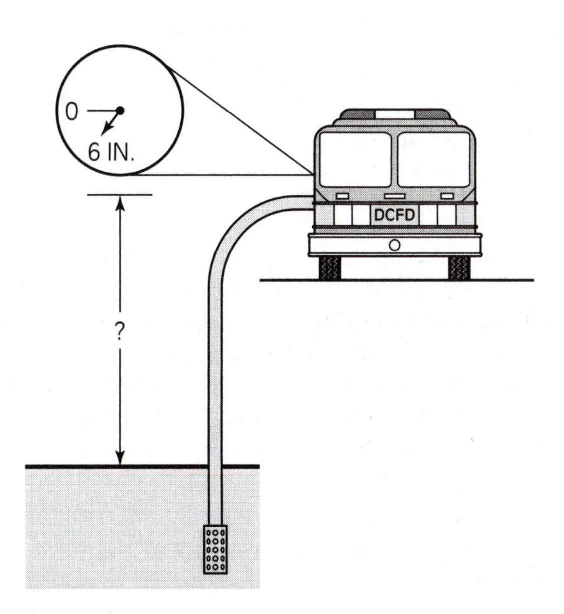

Figure 8-3 *Static reading = feet of lift.*

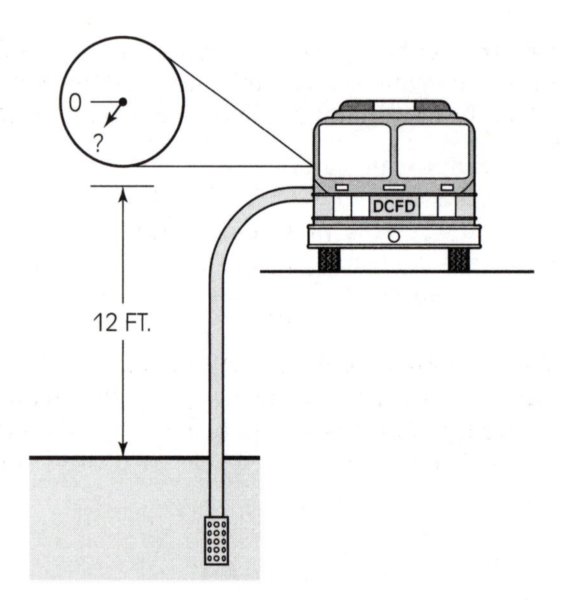

Figure 8-4 *What will the reading be on the intake gauge?*

of water from Table 8-1, H is the height of the lift of water, and Hg is the equivalent Hg reading.

Example 8-5 If a specific draft situation requires that the pump be located 12 feet above the surface of the water, how much Hg should be reading on the intake gauge after water has entered the pump (see Figure 8-4)?

ANSWER Essentially you are just calculating the equivalent in Hg of 12 feet of lift. From Table 8-1, a foot of water is equivalent to .882 inches of Hg.

$$Hg = .882 \times H$$
$$= .882 \times 12$$
$$= 10.58 \text{ inches}$$

Remaining pressure = Atmospheric pressure − (.433 × H)

It is also possible to calculate the atmospheric pressure remaining in the pump after enough pressure has been evacuated to allow water to enter the pump. To do this it is necessary that we use the formula Remaining pressure = Atmospheric pressure − (.433 × H), where atmospheric pressure is the ambient pressure, .433 is the pressure created by 1 foot of elevation, H is the height or lift, and remaining pressure is the pressure remaining in the pump.

Example 8-6 How much pressure will remain in the pump in Example 8-5 after water has entered the pump?

ANSWER You can start either by calculating the pressure equivalent of 12 feet and subtract it from atmospheric pressure, or if the mercury reading is given, calculate the equivalent pressure and subtract it from the atmospheric pressure.

$$\begin{aligned}\text{Remaining pressure} &= 14.7 - (.433 \times H)\\ &= 14.7 - (.433 \times 12)\\ &= 14.7 - 5.2\\ &= 9.5 \text{ psi}\end{aligned}$$

Remaining pressure = Atmospheric pressure – (.489 × *Hg*)

By altering the formula a little to read Remaining pressure = Atmospheric pressure – (.489 × *Hg*), we can calculate the remaining pressure when the Hg reading is known. The only thing that changes in the formula is the constant of .489 is used in place of .433, and we use the Hg reading instead of *H*.

Example 8-7 How much pressure will be remaining if there are 10.58 inches of Hg?

ANSWER

$$\begin{aligned}\text{Remaining pressure} &= 14.7 - (.489 \times Hg)\\ &= 14.7 - (.489 \text{ x } 10.58)\\ &= 14.7 - 5.2\\ &= 9.5 \text{ psi}\end{aligned}$$

It is also possible to calculate how far water will be drawn into the pumps if the intake reading is known. This calculation is done by using the formula $H = 1.13 \times Hg$. If we know the intake reading in inches of Hg we can easily convert it to feet of lift.

Example 8-8 In Example 8-5, how far up the hard suction hose will water be if the intake gauge reads 8 inches of Hg and the pumper is 12 feet above the surface of the water (see Figure 8-5)?

ANSWER Find the feet of lift equivalent to 8 inches of Hg.

$$\begin{aligned}H &= 1.13 \times Hg\\ &= 1.13 \times 8\\ &= 9.04 \text{ feet}\end{aligned}$$

In this problem it is not necessary to know the pumper was 12 feet above the water. It was put in here just to illustrate that the lift obtained by evacuating air to the point of 8 inches of Hg will not get water to the pump.

If the intake gauge in Example 8-8 reads only 8 inches of Hg, water will only be lifted to the 9.04 foot level. With water at this height, the pressure inside and outside the suction hose will be in equilibrium at the surface of the water source. that is, pressure at point A and point B in Figure 8-5 are equal.

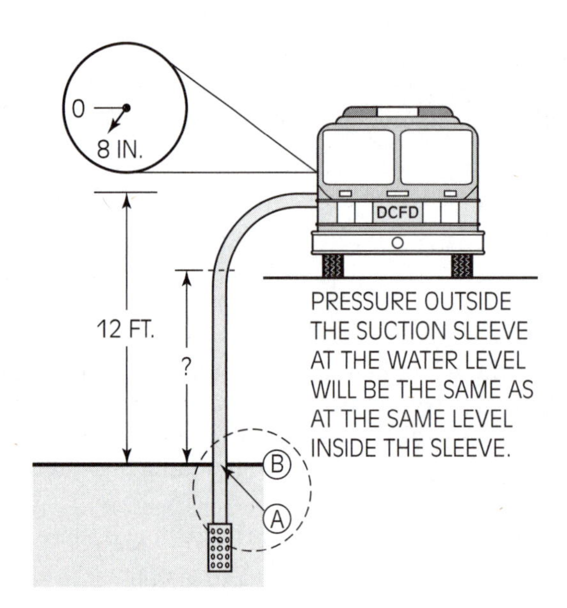

Figure 8-5 *Pressure at A is equal to pressure at B.*

CALCULATING INTAKE LOSS

dynamic intake reading includes lift, friction loss in the suction hose, and friction loss in the strainer

FL = (dynamic reading – static reading) × .489

When water is flowing, we get a dynamic reading on our intake gauge. At draft, the **dynamic intake reading** includes lift, friction loss in the suction hose, and friction loss in the strainer. To determine the actual loss on the intake side of the pump, we need two pieces of information: (1) the static reading and (2) the dynamic reading. We have already defined the static reading as a measurement of the lift, so to find the friction loss on the intake side of the pump we subtract the static reading from the dynamic reading, giving the formula *FL* = (dynamic reading – static reading) × .489. The terms of this formula are self-explanatory and the .489 converts inches of Hg to pressure. The answer is the amount of loss on the intake side of the pump due to friction.

Example 8-9 How much friction loss is there on the intake side of a pump that has a static reading of 8 inches of Hg and a dynamic reading of 20 inches of Hg?

ANSWER *FL* = (dynamic reading – static reading) × .489

= (20 – 8) × .489

= 12 × .489

= 5.87 psi

The good thing about calculating the intake loss in this fashion is that it is not theoretical or an approximation, it is an actual measured loss.

CALCULATING TOTAL WORK

In Chapter 7 we defined the work done by the pump as a combination of net pump pressure and how much liquid is being discharged. When calculating the work done by a pump at draft, we define work exactly the same with one addition. When drafting we need to account for the pressure on the intake side of the pump. Because the pump must create a vacuum on the intake side, the pressure the pump generates to get the water into the pump becomes part of the net pump pressure. In short, the work the pump does includes reducing the pressure on the intake side so the atmospheric pressure can overcome both lift and intake losses. Additionally, because there is no positive pressure coming in, the entire discharge pressure is part of the net pressure.

We have already spent time separating lift from intake losses by interpreting the static and dynamic intake reading. To calculate the intake pressure, for purposes of determining the net pump pressure, simply use the dynamic reading. It includes the pressure requirement for both the lift and intake losses.

Example 8-10 What is the total work being done by a pump at draft if it is flowing *x* gpm, the discharge pressure is 145 psi, and the intake is reading 18 inches of Hg? The amount of water the pump is flowing is represented by x.

ANSWER Add the pressure equivalent of 18 inches of Hg to the discharge pressure.

$$P = .489 \times Hg$$
$$= .489 \times 18$$
$$= 8.8 \text{ psi}$$

Now add the intake loss in psi to the discharge:

$$\text{Work done} = (\text{Intake loss} + \text{discharge pressure}) \text{ @ } x \text{ gpm}$$
$$= (8.8 + 145) \text{ @ } x \text{ gpm}$$
$$= 153.8 \text{ psi @ } x \text{ gpm}$$

In this problem the 153.8 psi represents both the net and the gross pump pressure.

TESTING PUMPS

Pump tests are not an everyday occurrence, but knowledge of the principles and procedures of pump tests can lend additional insight into how pumps operate. Because tests can vary to a certain degree depending on the pump capacity, only the general concepts are presented here. These tests should be conducted on pumps annually and after any major repair.

Pump tests should be done from draft because the true performance of the pump is easier to evaluate. The water source should be 4 to 8 feet below the level of the pump and no more than 20 feet of hard suction hose and a strainer is to be used per intake. (Larger capacity pumps may require as many as four hard suction intakes to flow the capacity of the pump.) The point where the sleeve is placed in the water should have at least 4 feet of water and the strainer needs to be covered by at least 2 feet of water.

It is important that the water supply not be aerated and not over 90°F. Recall from earlier in this chapter that the water temperature can affect a pump's ability to draft. Here is a practical application of that principle. The air temperature should also be between 0°F and 100°F, and the barometric pressure should be 29 inches of Hg minimum.

The exact configuration of any hose lay depends on the capacity of the pump and which test is being conducted. Care should be taken when calculating flows from nozzles to know and use the correct nozzle coefficient so the exact flow will be known. If the correct coefficients are not used, erroneous results will invalidate the test.

Three separate flow tests actually make up a pump test. Because a pump cannot pump full capacity at maximum pressure, the tests are designed to test the pump's ability at three different intervals. The first test is done at full capacity but only 150 psi net pressure. This area is where most pumps do their work most of the time, so this test is run for 20 minutes. The second test is the 70 percent capacity test at 200 psi. The final test is the 50 percent capacity test and is done at 250 psi. These last two tests are run for 10 minutes each. Table 8-2 summarizes the three tests by pressure, capacity, and duration. The first two columns of the table are critical to know and are put to practical application later when calculating maximum flow and maximum lay. The right-hand column states the position of the transfer valve if the pump is a two-stage pump.

Table 8-2 applies to all radial flow fire service pumps. As stated in earlier chapters, as the needed pressure increases, the capacity of the pump decreases. Conversely, as the need for water decreases, the pressure available is increased. Table 8-2 is a direct use of Table 7-1 to a real world application.

If the pump being tested is a two-stage pump, it is important that the transfer valve be in the correct position. For the full capacity test, the pump will be in the parallel position. For the 70 percent capacity test, the pump can be in either

Table 8-2 *Pump test criteria.*

Test	Pressure	Duration	Transfer valve
Full capacity	150 psi	20 minutes	Parallel
70%	200 psi	10 minutes	Series or parallel
50%	250 psi	10 minutes	Series

series or parallel, whichever position works best. But for the 50 percent capacity test, the pump must be in the series position.

In addition to the flow tests, a vacuum test is conducted separately. The vacuum test requires that all intake valves be opened and capped, and all discharge valves must be closed and uncapped. The pump is then required to draw 22 inches of Hg and cannot drop more than 10 inches in 5 minutes. Due to reductions in atmospheric pressure as elevation increases, the maximum vacuum may be reduced by 1 inch of Hg for every 1,000 feet of elevation. Another direct application of the laws of physics applied to fire pumps.

NFPA 1911

Standard for Service Tests of Fire Pump Systems on Fire Apparatus

More details and specifications of pump tests are contained in NFPA 1911, *Standard for Service Tests of Fire Pump Systems on Fire Apparatus.*

APPLICATION ACTIVITIES

It is now time to put your knowledge of drafting to work.

Problem 1 What is the maximum theoretical lift at an elevation of 3,000 feet?

ANSWER The atmospheric pressure will be 1.5 psi less than at sea level.

$$\text{Lift} = 2.31 \times \text{pressure}$$
$$= 2.31 \text{ x } 13.2$$
$$= 30.49 \text{ feet}$$

Problem 2 A barometric pressure of 30.36 inches of Hg is equivalent to how much atmospheric pressure?

ANSWER From Table 8-1, 1 inch of Hg is equal to .489 pounds of pressure.

$$\text{Pressure} = .489 \times \text{inches of Hg}$$
$$= .489 \times 30.36 \text{ inches of Hg}$$
$$= .489 \times 30.36$$
$$= 14.8 \text{ psi atmospheric pressure}$$

Problem 3 A pumper at draft shows a static intake reading of 10 inches of Hg, how much lift is the pump overcoming (see Figure 8-6)?

ANSWER From Table 8-1, an inch of Hg is equivalent to 1.13 feet of lift.

$$\text{Lift} = 1.13 \times \text{inches of Hg}$$
$$= 1.13 \times 10 \text{ inches of Hg}$$
$$= 1.13 \times 10$$
$$= 11.3 \text{ feet}$$

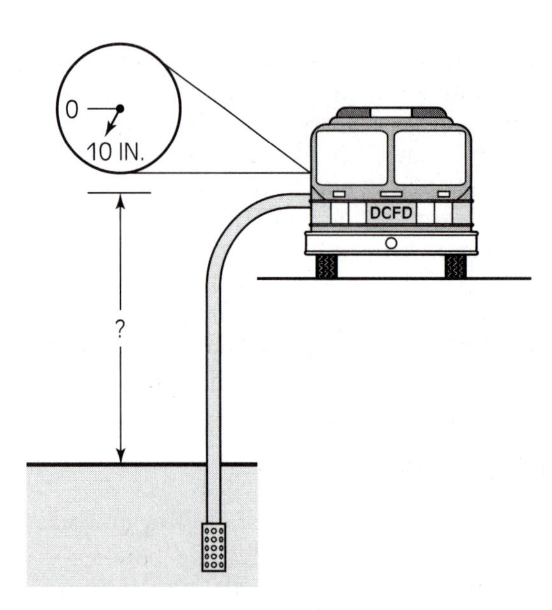

Figure 8-6 *What is the lift?*

Problem 4 If you had a sealed glass tube, 20 feet tall and 6 inches in diameter, and it had water 6 feet up the inside of the tube, how much pressure is there in the tube (see Figure 8-7)?

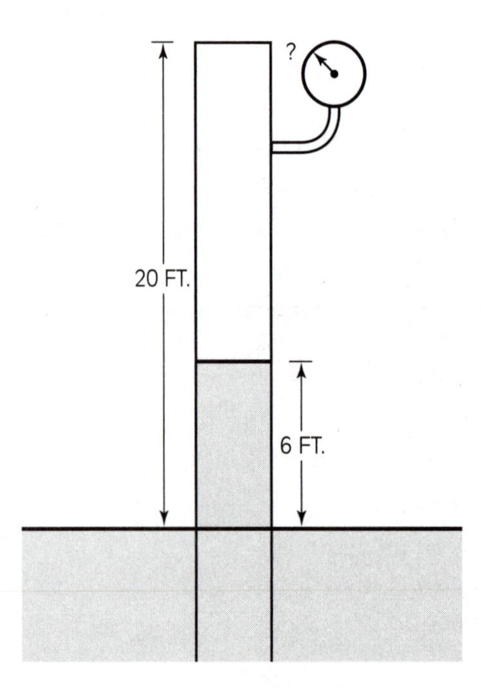

Figure 8-7 *Find the pressure in the tube.*

ANSWER Find how much pressure it would take to push the water 6 feet up the tube and subtract it from the atmospheric pressure. Recall the formula $P = .433 \times H$ from Chapter 2.

$= .433 \times H$

$= .433 \times 6$ feet

$= .433 \times 6$

$= 2.598$ psi

Internal pressure = Atmospheric pressure − 2.598 psi

$= 14.7$ psi − 2.589 psi

$= 14.7 - 2.598$

$= 12.1$ psi

Problem 5 A pumper operating at draft has a static intake reading of 11 inches of Hg and a dynamic reading of 18 inches of Hg, how much intake loss is there?

ANSWER FL = (dynamic reading − static reading) × .489

$= (18 - 11) \times .489$ psi

$= 7 \times .489$

$= 3.4$ psi

Problem 6 Prove that in Example 8-8, the pressure at point A and the pressure outside the pump are in equilibrium (see Figure 8-8).

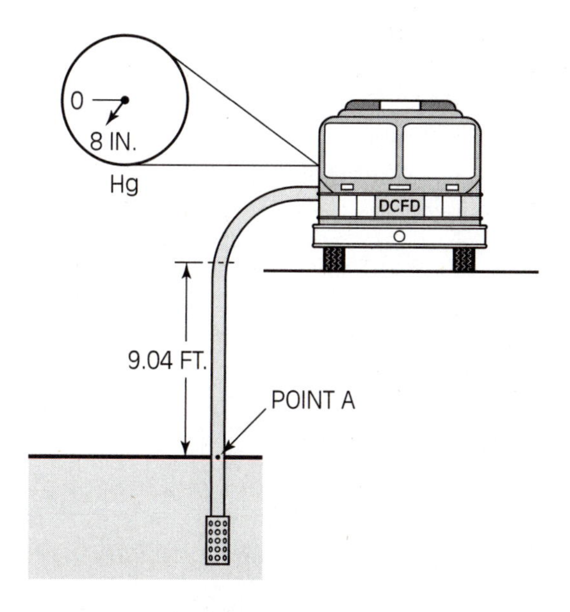

Figure 8-8 *Is this system really in equilibrium?*

ANSWER Finding this answer requires three steps. First, calculate the pressure remaining in the pump, and then calculate the pressure exerted by a column of water 9.04 feet tall. Finally, add them together.

$$\begin{aligned}\text{Remaining pressure} &= 14.7 - (.489 \times Hg)\\ &= 14.7 - (.489 \times 8)\\ &= 14.7 - (3.9)\\ &= 10.8 \text{ psi}\end{aligned}$$

Now find the pressure exerted by a column of water 9.04 feet tall:

$$\begin{aligned}P &= .433 \times H\\ &= .433 \times 9.04\\ &= 3.9 \text{ psi}\end{aligned}$$

10.8 + 3.9 = 14.7 psi. We now know the pressure at point A in Figure 8-8 is the same as atmospheric pressure; the system is in equilibrium.

Summary

The concepts and principles of drafting are firmly rooted in the laws of physics, as exemplified by the need for the atmospheric pressure to push the water into the pumps at draft.

By understanding the relationships between pressure, Hg as a measure of pressure, and lift of water, we can calculate how much pressure is needed to get water into the pumps. When we add this pressure to the discharge pressure, we have the net pump pressure for a draft situation. By calculating the flow at a given net pressure, we can establish the amount of work being done by the pumps.

In testing the pumps, we can see further evidence of how physics is so important to the operation of a pump. We see it in the need for the water to be less than 90°F, we see it in limiting lift during pump test to a maximum of 8 feet, and we see it in the requirement for a specified barometric pressure. The requirement for three separate tests, each at a different capacity and pressure, is a direct application of Bernoulli's principle in a real life situation.

At this point, between Chapter 7 and Chapter 8, the serious student of hydraulics should have a firm understanding of how pumps work. By demystifying the operation of the black box, the pump can be better utilized as a tool and its limits safely explored.

Review Questions

1. How does water enter the pumps when drafting if there is no "intake" pressure?
2. Define drafting.
3. What factors limit a pump's ability to draft?
4. Why is it impossible to obtain a perfect vacuum in a pump?
5. One foot of lift is equivalent to what reading in Hg?
6. How much Hg would be equivalent to exactly .5 psi?
7. Calculate the amount of lift indicated by a Hg reading of 7.5 inches.
8. What is the maximum theoretical lift at 4,000 feet elevation?
9. If the static intake reading is 5.5 inches of Hg and the dynamic reading is 23 inches of Hg, what is the friction loss on the intake side of the pump?
10. How much Hg must be reading on the intake gauge in order to get water into the pump if the lift is 12 feet? Atmospheric pressure is 1 atmosphere.
11. Go back to Question 10 and calculate the amount of pressure left in the pump.

12. What is the maximum amount of hard suction hose allowed, for each intake, during a pump test?
13. Why is the water temperature limited to a maximum 90°F during the pump test?
14. Why is the pump test divided into three separate tests?
15. When conducting the vacuum test part of the pump test, what is the minimum draw of mercury the pump must be able to achieve?

List of Formulas

Finding pressure from inches of mercury:

$$P = .489 \times Hg$$

Finding height of water from inches of mercury:

$$H = 1.13 \times Hg$$

Reverse lift calculation:

$$Hg = .882 \times H$$

Finding remaining pressure when height is known:

Remaining pressure = atmospheric pressure – (.433 × H)

Or

Finding remaining pressure when inches of mercury are known:

Remaining pressure = atmospheric pressure – (.489 × Hg)

Finding friction loss on intake side of pump at draft:

FL = (dynamic reading – static reading) × .489

Reference

NFPA 1911, *Service Testing Fire Pump Systems on Fire Apparatus*. Quincy, MA: National Fire Protection Association, 1997.

Fire Streams

Learning Objectives

Upon completion of this chapter, you should be able to:

- Determine what properties are essential for an effective fire stream.
- Calculate vertical and horizontal range of streams.
- Calculate the nozzle reaction for smooth-bore and fog nozzles.
- Calculate back pressure and distinguish it from friction loss.

INTRODUCTION

The development of an effective fire stream is arguably the most neglected aspect of firefighting today. In many instances, firefighters think that aiming water at the building is sufficient. This attitude is far from acceptable. Even under the best circumstances, friction of the water against the air will degrade an otherwise ideal fire stream.

Just what makes an average fire stream into an effective fire stream? For any fire stream to be considered effective, it must meet the following four goals:

1. *It must flow sufficient gpm to absorb the BTUs being generated.* Recall that in Chapter 1 we calculated the amount of water needed to absorb the BTUs generated by a fire. If the water cannot absorb the BTUs being given off, the fire simply will not go out until all the fuel is consumed.
2. *Water must be applied at the correct point(s).* Even if an adequate volume of water is being put on the fire, if it is put in the wrong place, the fire will not go out. It is often more effective to put multiple smaller streams in service than a few larger ones.
3. *Water must be applied in the correct form.* Regardless of the fire situation, the most effective conversion of water into steam comes when water is applied in the form of small droplets. From the information given in Chapter 1, we can calculate that water that is allowed to vaporize will absorb 7.7 times more heat than water that is only allowed to get to 212°F without vaporizing. Only where extreme reach or penetration is critical should fog nozzles not be used.

   Special note: When applying water in the form of fog during an interior attack, caution must be used to avoid steam burns.
4. *Water should be applied for the shortest possible time.* If the first three goals are met, fire streams that are operated for too long can cause excessive water damage. In larger firefighting operations, this goal is often difficult to achieve, but on smaller fires it is well within our means to eliminate damage from excess water.

This book is not about tactics, so we do not discuss proper placement or selection of streams. Instead, we discuss how to get the most out of the fire stream, so when it is properly placed it will have maximum effectiveness.

EFFECTIVE FIRE STREAMS

How important is an effective fire stream? Consider the following example. On 23 February 1991 a fire broke out on the twenty-first floor of One Meridian Plaza in Philadelphia, Pennsylvania. The fire burned for hours, consuming everything on floors 21 through 29, and, sadly, taking the lives of three courageous firefighters in the process. Philadelphia's bravest were unable to slow down this inferno despite their most valiant efforts. Eventually, for fear the building was going to collapse, all firefighters were withdrawn from the building and the fire was allowed to burn. However, the tenant on the thirtieth floor had required a sprinkler system be installed prior to occupying the tenant space. When the fire got to the thirtieth floor, it was stopped by the activation of just ten sprinkler heads. The exact flow of each head is not known, but even if they delivered a maximum flow per head of about 25 gpm, they still only delivered a total of 250 gpm. That is the flow from one 2½-inch hand line. The point is, they were able to (1) deliver sufficient water, (2) in the right form, (3) at the right spot to stop the spread of fire. This example alone validates the foregoing goals 1, 2, and 3.

We have already mentioned four goals that help to define an effective fire stream. However, there are also some very specific criteria that aid in defining the effectiveness of a fire stream. These criteria are not new—John Freeman first defined them in 1888—but they are as meaningful today as they were then. As you read and study the following characteristics of a good fire stream, keep in mind that the variables in each criterion can change with each fire. These criteria were written specifically for solid stream nozzles, but the broader principles apply to all fire streams.

An effective solid stream is one that displays the following criteria:

- At the limit named, has not lost continuity of stream by breaking into showers of spray. Reason: A broken stream will not reach the fire with the desired mass. If too much water is lost due to breakup, the effectiveness of the stream is compromised.
- Up to the limit named, appears to discharge nine-tenths of its volume of water inside a 15-inch circle and three-quarters of its volume inside a 10-inch circle. This further defines criterion 1. Because it is impossible for the stream not to experience some breakup due to friction with air, this criterion defines the amount of breakup allowed.
- Is stiff enough to attain, in a fair condition, the height or distance named even though a fresh breeze is blowing. Wind affects the quality of the stream. Directing a stream into a breeze can decrease its range, whereas directing it with the wind can increase the range. A wind from the side has an undetermined effect on the stream, but criteria 1 and 2 still apply.
- At a limit named, it will, with no wind, enter a room through a window opening and just barely strike the ceiling with force enough to spatter well. As previously stated, we want the stream to stay together as long as possible. When the stream gets to the fire we want it to break up so it will more efficiently absorb BTUs. By specifying that the stream break up at the fire, we are attempting to break the water down into smaller particles so it will more readily absorb heat.

Range of Streams

The range of a fire stream is dependent on the following five factors, four of which can be controlled by the firefighter:

1. *Nozzle pressure.* All else being equal, the nozzle pressure determines the reach of the stream. The higher the nozzle pressure, the greater the reach of the stream, to the point where the stream is overpumped and breaks up.
2. *Nozzle diameter.* All else being equal, the nozzle diameter determines the reach of the stream. The larger diameter nozzles have a greater flow and a greater reach.
3. *Angle of stream.* All else being equal, the angle of the stream determines the reach of the stream. An angle of 32 degrees above horizontal provides the maximum horizontal reach.
4. *Wind.* As already mentioned in criteria 3, the wind can increase or reduce the range of a stream.
5. *Pattern.* When using a fog nozzle, as the angle of the pattern increases, the reach decreases.

Of the five factors, the nozzle pressure is the most critical. It is easy to overpump or underpump a nozzle. If we provide too much pressure on a nozzle, we can cause the stream to break up, and if we underpump it, we can deliver too little water and reduce the range of the stream. Most nozzles can tolerate some degree of increased pressure before the stream begins to break up, but no nozzle will deliver the required flow if it is underpumped, regardless of the condition of the stream.

Another critical factor in determining the range of a stream is the angle. As already mentioned, an angle of 32 degrees will provide the maximum horizontal range. However, an angle of 32 degrees is not generally an effective angle for firefighting. To be an effective fire fighting stream the water must enter the building and then be deflected off the ceiling in order to break up the stream to achieve maximum heat absorption. The most effective angle for fire fighting is 45 degrees. At 45 degrees, the stream gets an acceptable horizontal range as well as a good angle for deflecting water off the ceiling of the room it enters. The best penetration and deflection off the ceiling is achieved when the stream is directed just above the lowest edge of the opening (see Figure 9-1). An advantage of the 45-degree angle is that it is easy to set up. In order to get the 45 degree angle, the nozzle is simply placed as far from the building as the stream will reach up the building. The third floor is considered to be the highest story into which a stream can be effectively directed from the ground.

$HR = \frac{1}{2} NP + 26^*$

*Add 5 for each 1/8 inch of nozzle diameter more than 3/4 inch.

The maximum horizontal range can be approximately calculated by the formula $HR = \frac{1}{2} NP + 26^*$, where *HR* stands for horizontal range, and *NP* is the nozzle pressure. The asterisk is there to indicate that for each 1/8 inch the nozzle diameter is more than 3/4 inch, another 5 is added to the 26. The answer is given in feet.

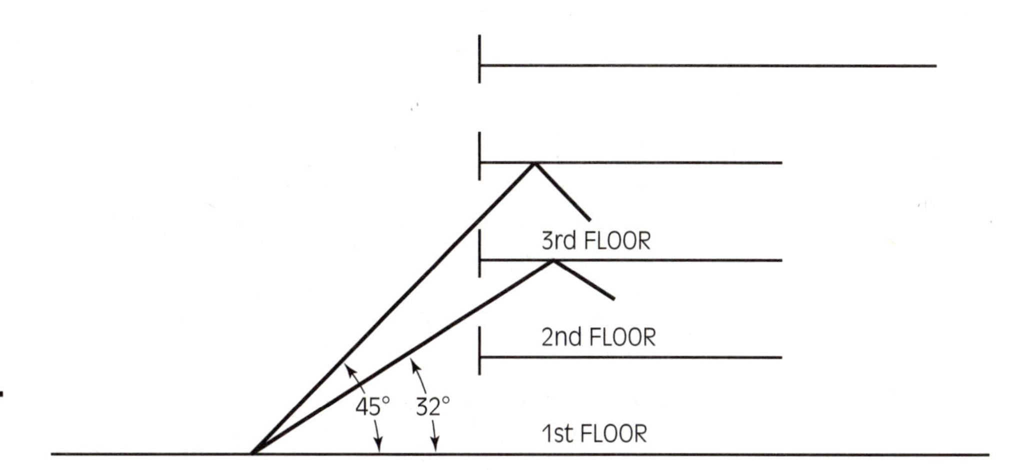

Figure 9-1 *Angle of streams.*

Example 9-1 What is the horizontal range of a nozzle with a 1¼-inch tip and operating at 50 psi nozzle pressure?

ANSWER Add 5 for each ⅛ inch over ¾ inch = 20

$HR = \frac{1}{2}\ NP + 26^*$

$= \frac{1}{2}\ 50 + 46$

$= 25 + 46$

$= 71$ feet

The vertical range is also affected by the factors in the list, however we can do a couple of things to increase our vertical range. First, a stream operating from on top a pumper has the advantage of being about 8 additional feet off the ground. Where vertical range is critical, it is a definite advantage to leave master stream devices on apparatus. Another way to gain vertical range is to use an elevated stream, such as a ladder pipe. Just how much additional range is possible depends on the type of elevated stream and its design length. When operating into the upper floors or area of a building, it is still necessary to place the aerial device so the stream enters the building at a 45 degree angle. An example of placement for a 100-foot aerial ladder using a ladder pipe assembly would be to place the inside edge of the turntable a distance from the building equal to one-half of the desired reach of the stream. The ladder should then be elevated to an angle of 70 degrees and extended until the nozzle is about 10 to 15 feet below the level of the opening. The stream is then directed into the building so it just clears the windowsill, giving an angle of about 45 degrees (see Figure 9-2).

$VR = \frac{5}{8}\ NP + 26^*$

The vertical range can be calculated by the formula $VR = \frac{5}{8}\ NP + 26^*$, where *VR* stands for vertical range, and *NP* is the nozzle pressure. Again the asterisk tells us that 5 must be added for each ⅛ inch the nozzle diameter is more than ¾ inch.

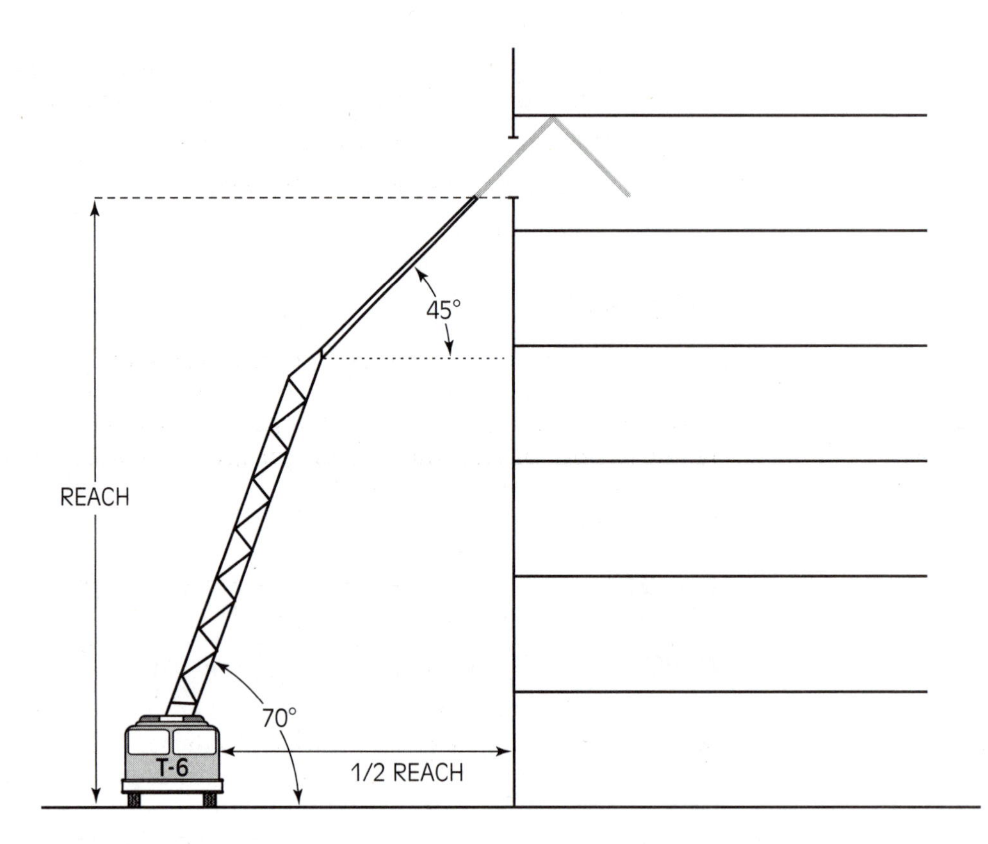

Figure 9-2 *Proper placement of elevated stream from aerial ladder.*

Example 9-2 What is the maximum vertical reach of a stream placed on the ground if the nozzle is $1\frac{1}{2}$ inch in diameter and has a nozzle pressure of 80 psi?

ANSWER Add 5 for each $\frac{1}{8}$ inch over $\frac{3}{4}$ inch = 30

$$VR = \tfrac{5}{8}\,NP + 26^{*}$$
$$= \tfrac{5}{8}\,80 + 56$$
$$= 50 + 56$$
$$= 106 \text{ feet}$$

If the stream is elevated in any way, simply add the height of the elevation to the answer obtained from the formula.

Example 9-3 What is the maximum vertical reach of a ladder pipe using a $1\frac{1}{2}$-inch tip at 80 pounds, if the ladder is extended 80 feet?

ANSWER The reach of the stream was determined in Example 9-2. Use 80 as the vertical reach advantage of the aerial device. (This amount is not exactly correct, but since the angle can vary somewhat for each situation, the extension of the aerial device is close enough for practical purposes.)

VR = 106 feet for the stream

Extension of the aerial device = 80

Actual reach = 186 feet

THE NOZZLE

The nozzle has several purposes. First, it serves as a means to direct the water onto the fire. Second, it often is associated with a way to shut down the stream. Only master stream devices, which are the same as nozzles, do not have a cut off. Finally, the nozzle converts the pressure energy in the water to velocity energy. This third point is sufficiently important and complicated enough to warrant a more thorough explanation.

Recall from Chapter 7 that water has both pressure and velocity energy as it flows through the hose. The nozzle must convert the pressure to velocity so there is no pressure in the water as it exits the nozzle (see Figure 9-3). This conversion is critically important because of Principle 2 in Chapter 2, "Pressure in a fluid acts equally in all directions." When the nozzle is able to convert all the pressure into velocity, we have a nice smooth stream as depicted in Figure 9-3. However, if the nozzle is overpumped and the nozzle is unable to convert all the pressure to velocity, the stream will look more like the one in Figure 9-4 as it exits the nozzle,

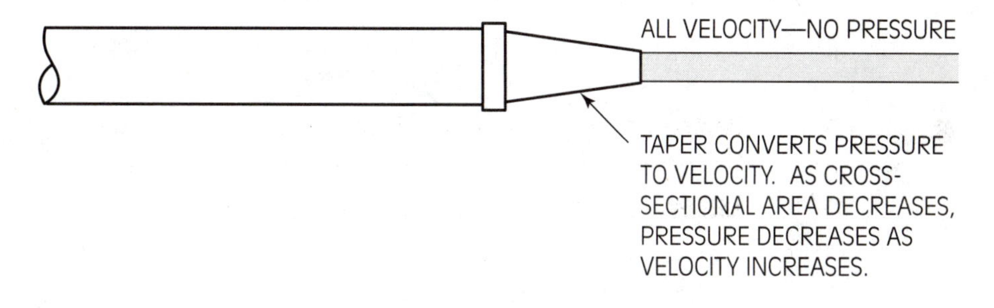

Figure 9-3 *The nozzle converts pressure to velocity.*

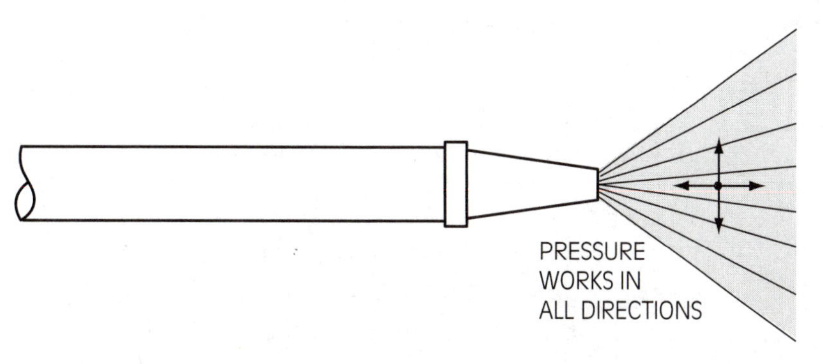

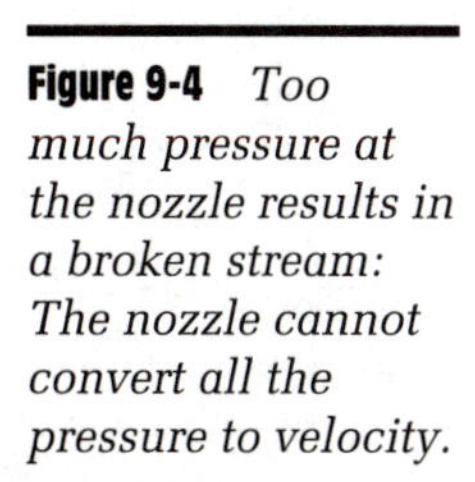

Figure 9-4 *Too much pressure at the nozzle results in a broken stream: The nozzle cannot convert all the pressure to velocity.*

because the pressure inside the stream that is not converted to velocity is trying to push out in all directions after the stream has left the nozzle. Where there is no pressure in the stream as the water exits the nozzle this cannot happen.

Types of Nozzles

smooth-bore nozzle a nozzle only capable of producing a solid stream of water

CVFSS nozzle one that is capable of producing any pattern from a straight stream to 100 degree fog

There are two basic types of nozzles: (1) smooth-bore, or straight stream and (2) combination variable fog and straight stream (CVFSS), or fog nozzles. The **smooth-bore nozzle** is only capable of producing a solid stream of water. Because the stream is solid, it generally has the advantage of being able to penetrate better than a CVFSS nozzle on straight stream. It is also less affected by wind than the straight stream of a CVFSS nozzle.

A **CVFSS nozzle** is one that is capable of producing any pattern from a straight stream to 100 degree fog (see Figure 9-5). The primary advantage of the CVFSS nozzle is that, through design, the nozzle breaks down the water into small droplets that absorb heat much more readily than a solid stream, because there

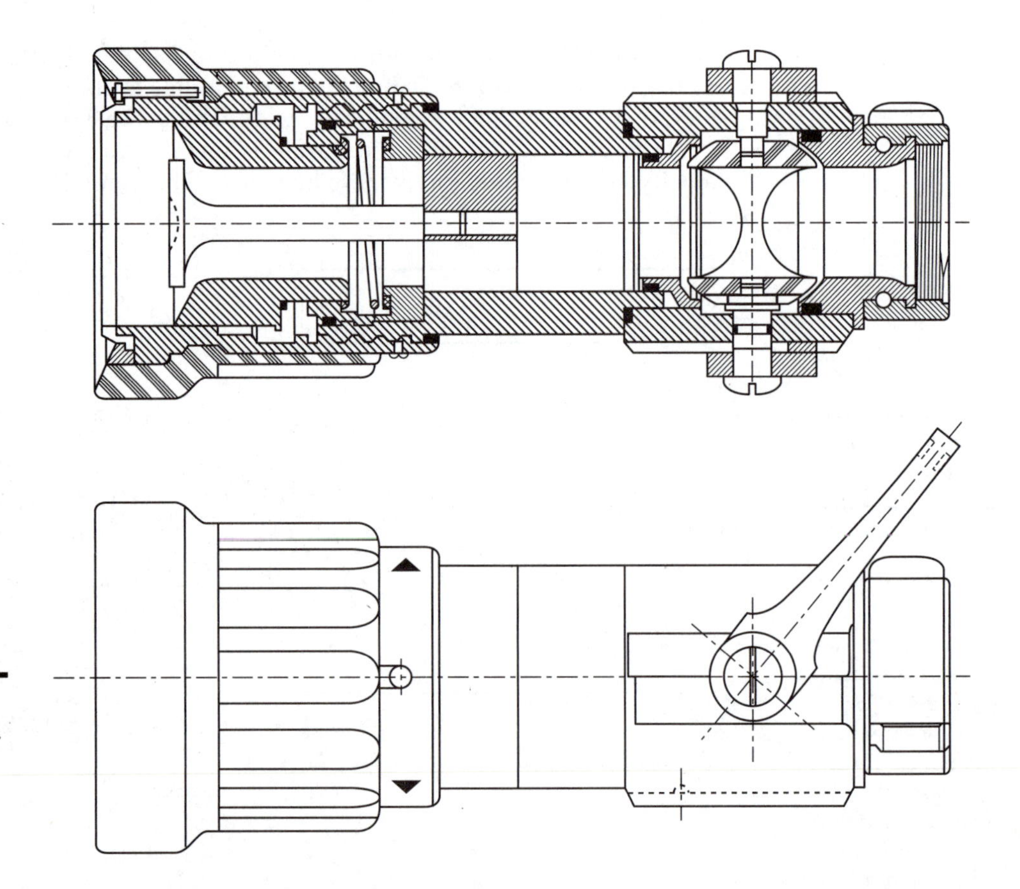

Figure 9-5 *The CVFSS (fog) nozzle. Courtesy of Elkhart Brass Manufacturing Company, Inc.*

is a much higher surface area exposed to the hot atmosphere created by the fire. The greater surface area of the droplets is the most advantageous format for water to absorb the heat of the fire and vaporize. For water to be of maximum value, it must vaporize. (Recall Chapter 1 calculations of specific heat and latent heat.)

There are fog nozzles other than the CVFSS, but CVFSS nozzles are the predominate fog nozzle today. The other types of fog nozzles usually have a fixed fog pattern and have names such as Navy All-Purpose Nozzle and Low-Velocity Fog Applicator. Except for the fact that these nozzles do not have a variable pattern, they operate in a way similar to the CVFSS nozzle.

In general, fog nozzles are designed to be a fixed gallonage. Some CVFSS nozzles can vary their flow by turning a ring that changes an aperture within the nozzle, allowing more or less water to flow. There is, however, a type of CVFSS nozzle, called the *automatic nozzle*, that allows a variable flow of water without any manual adjustment.

The automatic nozzle is designed to deliver water at any gpm within its design range and maintain a nozzle pressure of 100 psi. When pumping to an automatic nozzle, you calculate the engine pressure based on friction loss for the desired flow and a nozzle pressure of 100 psi. The advantage of the automatic nozzle is that to change the flow, you only need to recalculate the engine pressure and adjust the discharge to the new pressure. The nozzle automatically adjusts the size of the opening to accommodate either more or less water. Figure 9-6 is a cutaway of an automatic nozzle. The baffle, 5, adjusts to allow more or less water flow while maintaining the desired nozzle pressure. Of particular interest is the sliding

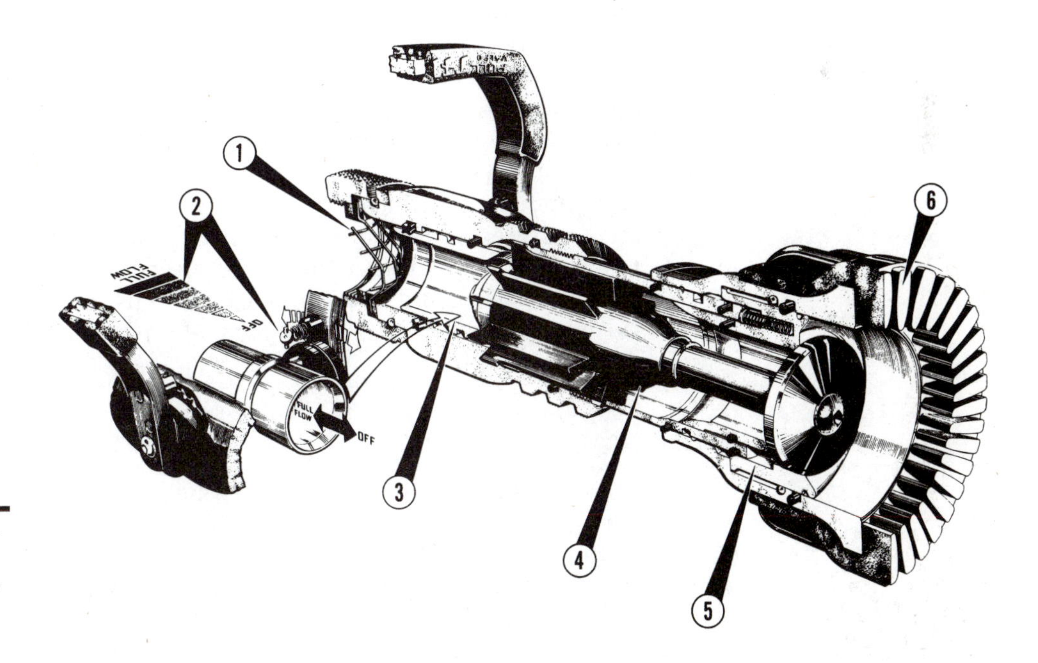

Figure 9-6 *Cutaway of an automatic nozzle. Courtesy of Task Force Tips.*

cutoff, 3, which is designed to reduce turbulence within the nozzle. Other nozzles use the more traditional ball cutoff shown in Figure 9-5.

A word of caution is in order about automatic nozzles. With a standard nozzle, if you have a kink in your line, a poor pattern will indicate inadequate flow. With the automatic nozzle, the nozzle will adjust to the lower flow and maintain the nozzle pressure, disguising the less-than-expected flow. To avoid this situation, you must be extra alert for kinks in hose.

We have just defined smooth-bore and CVFSS nozzles. In reality, the terms *nozzle* and *tip* are often used interchangeably. In general, a cutoff with a tip is called a nozzle and is defined by the type tip it has, for example, a cutoff with a smooth-bore tip is called a smooth-bore nozzle.

Nozzle Pressure

nozzle pressure the pressure required at the nozzle to allow it to deliver the designed gpm and pattern

Nozzle pressure is the pressure required at the nozzle to allow it to deliver the designed gpm and pattern. For the nozzle to operate properly and deliver the correct amount of water, it is critical that the nozzle have the correct pressure. Nozzle pressures do not vary much and are usually specific to a particular type. For instance, hand line smooth-bore nozzles usually have a 50-psi nozzle pressure while master stream smooth-bore nozzles have an 80-psi nozzle pressure. These pressures are not absolutes and can be varied to some extent. If the nozzle pressures for smooth-bore tips are varied, use caution not to overpump the nozzle or a poor, ineffective stream will result. If too low a pressure is used, insufficient water will flow and extinguishing capacity will be lost.

Fog nozzles require a pressure of 100 psi, 75 psi, or 50 psi, depending on manufacture and use. In general, 100 psi is the standard pressure for a fog nozzle, regardless of whether it is a hand line nozzle or master stream tip. The 75- and 50-psi nozzles are hand line nozzles specifically designed to provide required flows at reduced nozzle reaction. The primary difference between the high-pressure fog nozzle (100 psi) and a low-pressure fog nozzle (75 or 50 psi) is water droplet size. As the nozzle pressure is reduced, the water droplet size increases.

Nozzle pressures for fog nozzles are specific and cannot be varied from the manufacturer's design pressure. A 100-psi nozzle cannot be pumped at 75 psi and get an acceptable pattern and flow. Similarly, a 50 psi nozzle pumped at 100 psi will be overpumped and produce a poor fog pattern. In short, give a fog nozzle the pressure it is designed for.

Nozzle Reaction

As the water exits the nozzle, regardless of whether it is a straight stream or fog stream, there is an opposing reaction called nozzle reaction (see Figure 9-7). This condition is an example of Newton's famous third law of motion: "Whenever one object exerts a force on a second object, the second exerts an equal and opposite

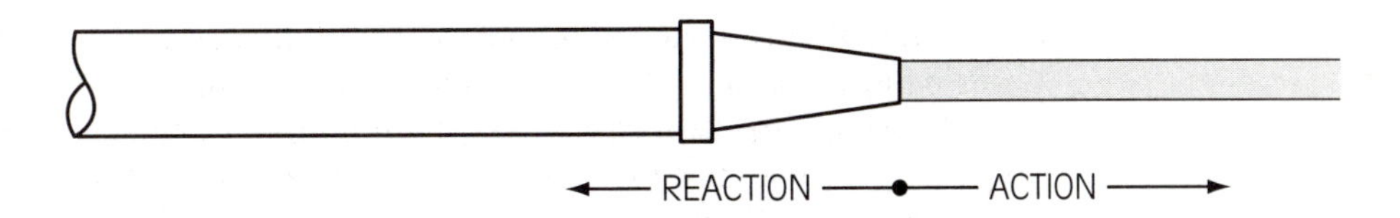

Figure 9-7 *Newton's third law is responsible for nozzle reaction.*

force on the first." In a hose line or fire stream frame of reference, the first object would be the hose line. Technically the hose line is not creating any force; however, it is directing the force of the water created by velocity and the weight of the water, much like a rocket engine directs exhaust to create thrust. The second object would be the water as it exits the nozzle. Because the hose line and nozzle direct the force created by the velocity and weight of the water, the water exiting the nozzle creates an equal and opposite reaction. In practical terms, **nozzle reaction** is the opposing force created by water exiting the nozzle as predicted by Newton's third law.

nozzle reaction
the opposing force created by water exiting the nozzle

Because nozzle reaction is a factor of both the velocity of the water and the weight of the water as it leaves the nozzle, a change in either factor can affect the nozzle reaction. For example, a flow of *x* gpm at *y* nozzle pressure produces a certain nozzle reaction we will call z. If we were to increase the nozzle diameter, thus requiring a lower nozzle pressure to get the same gpm flow, we would also get a lower nozzle reaction. If we were to reduce the nozzle diameter, requiring a higher nozzle pressure to get the same flow, the nozzle reaction would also be higher.

The principle just outlined is important to understand, but it is best understood when we can see the proof of the concept. To prove the concept it is necessary to be able to find the nozzle reaction at various pressures. The formula for nozzle reaction is $NR = 1.5 \times D^2 \times P$, in which *NR* is nozzle reaction, *D* is the diameter of the nozzle, and *P* is the nozzle pressure. To use the formula, we only need to know the diameter of the nozzle and the nozzle pressure. The answer is given in pounds.

$NR = 1.5 \times D^2 \times P$

Example 9-4 What is the nozzle reaction for a 1-inch tip operating at 50 psi nozzle pressure?

ANSWER
$$\begin{aligned} NR &= 1.5 \times D^2 \times P \\ &= 1.5 \times 1^2 \times 50 \\ &= 1.5 \times 1 \times 50 \\ &= 75 \text{ pounds} \end{aligned}$$

Example 9-5 What would be the nozzle reaction for 210 gpm if the tip size were $1\frac{1}{8}$ inch?

ANSWER The 1-inch tip in Example 9-4 flows 210 gpm at 50 psi nozzle pressure. Here we are going to find the nozzle reaction for the same flow through a larger tip, thus requiring a lower nozzle pressure, and the nozzle reaction should also be lower. We need to start by determining the nozzle pressure for the 1⅛-inch tip to deliver 210 gpm.

$$P = (gpm/29.77 \times D^2)^2$$
$$= (210/29.77 \times 1\tfrac{1}{8}^2)^2$$
$$= (210/29.77 \times 1.266)^2$$
$$= (210/37.69)^2$$
$$= 5.57^2$$
$$= 31.04 \text{ psi nozzle pressure}$$

Now find the nozzle reaction for 210 gpm from a 1⅛-inch tip at 31.04 psi nozzle pressure.

$$NR = 1.5 \times D^2 \times P$$
$$= 1.5 \times 1\tfrac{1}{8}^2 \times 31.04$$
$$= 1.5 \times 1.266 \times 31.04$$
$$= 58.94 \text{ pounds}$$

In these examples, we have lowered the nozzle reaction by 16.06 pounds while maintaining the same flow, proving that for the same flow, by reducing the nozzle pressure, the nozzle reaction will decrease.

A note of caution is in order here. As the nozzle pressure is reduced, the reach of the stream will also be reduced. However, where reach is not a primary factor, this trade-off may be acceptable.

The foregoing formula for nozzle reaction is for smooth-bore nozzles only. However, fog nozzles also have nozzle reaction and follow the same laws regarding flow, nozzle pressure, and nozzle reaction as smooth-bore nozzles. To find the nozzle reaction for fog nozzles, it is necessary to find the equivalent diameter for the nozzle pressure and gpm. This number is then inserted in the formula for nozzle reaction in place of D. Fortunately, in Chapter 5 we developed the formula $D = \sqrt{gpm/(29.77 \times C \times \sqrt{P})}$ to solve for D when we know gpm and nozzle pressure. Since we are looking for D^2 in the formula for nozzle reaction, we need to square $\sqrt{gpm/(29.77 \times C \times \sqrt{P})}$ to get $gpm/29.77 \times \sqrt{P}$. We can drop the correction factor for calculating nozzle reaction. When $gpm/29.77 \times \sqrt{P}$ is inserted directly into the nozzle reaction formula in place of D^2 we get a new formula for nozzle reaction: $NR = 1.5 \times (gpm/[29.77 \times \sqrt{P}]) \times P$. Although this formula is workable, by performing the appropriate mathematical operations on it we arrive at the simpler formula $NR = .0504 \times gpm \times \sqrt{P}$ for finding nozzle reaction for a fog nozzle. This formula also works for smooth-bore tips if the gpm is known.

$NR = .0504 \times gpm \times \sqrt{P}$

Example 9-6 What is the nozzle reaction for a fog nozzle on a 1¾-inch line if the nozzle is flowing 150 gpm at 100 psi nozzle pressure?

ANSWER $NR = .0504 \times gpm \times \sqrt{P}$

$= .0504 \times 150 \times \sqrt{100}$

$= .0504 \times 150 \times 10$

$= 75.6$ pounds

The nozzle reaction calculated for fog nozzles by the formula $NR = .0504 \times gpm \times \sqrt{P}$ gives the nozzle reaction for a straight stream pattern. However, fog nozzles can be adjusted to any pattern from straight stream to 110-degree fog. As the pattern shifts away from straight stream, the angle of the water leaving the nozzle changes, causing the nozzle reaction to change.

Because nozzle reaction is an opposite reaction, we need to find a way to determine the nozzle reaction for patterns other than straight stream. To do so, we need to analyze the forces exerted by the water as it leaves the nozzle by use of vectors. Vectors are the measure of strength and direction of forces. Figure 9-8 is an example of how the forces for a 60-degree fog pattern would look if analyzed

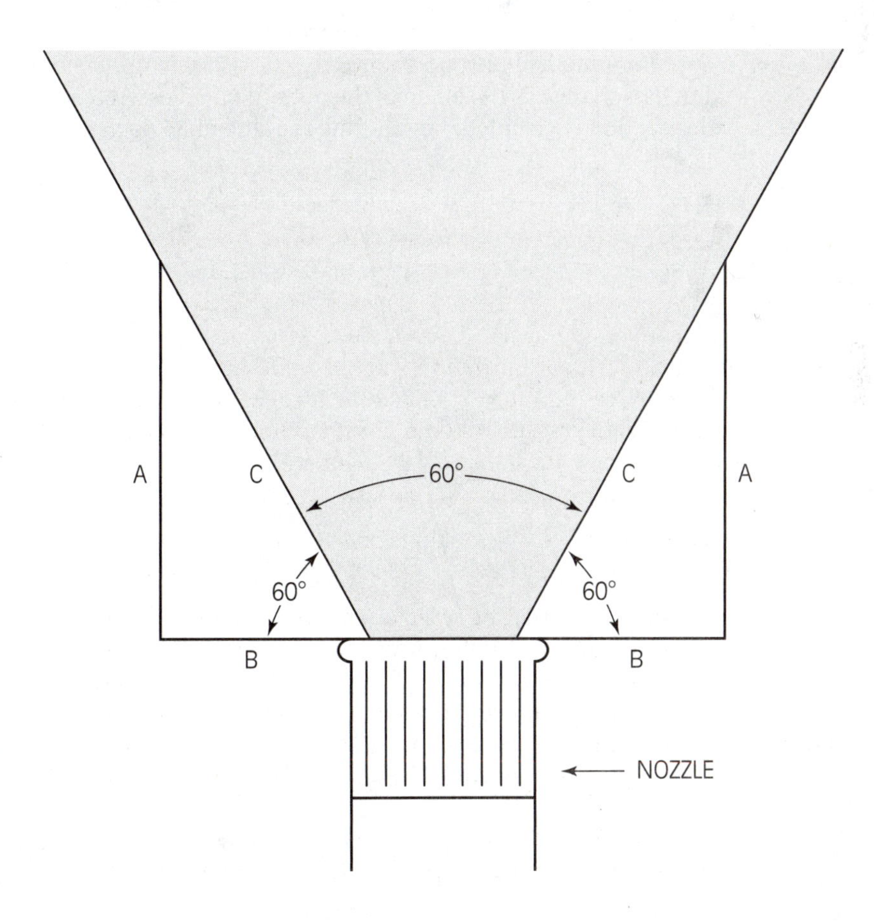

Figure 9-8 *Vector force A is responsible for the nozzle reaction you feel.*

by their component vectors. Vector B is perpendicular to the nozzle reaction and is in equilibrium because it exists equally all around the nozzle. Because this force is in equilibrium, it produces no net force. Vector C of the stream is at an angle, and because it is produced equally around the periphery of the nozzle, it reacts against itself and is also in equilibrium. Vector C is the force calculated by the foregoing nozzle reaction formula. Vector A of the vector forces is the only force not in equilibrium and is responsible for the nozzle reaction for the 60-degree fog pattern.

The strength of vector A is a function of the strength of vector C and angle BC. More specifically, the nozzle reaction will be the nozzle reaction calculated by the formula above, multiplied by the sine of angle BC. Sine is a trigonometric function of an angle that when multiplied by the length of the hypotenuse of a right triangle will find the length of the side opposite the angle. By breaking the forces of a fog nozzle into vectors, a right triangle is formed. In Figure 9-8, the calculated nozzle reaction is used in place of the length of the hypotenuse, and the opposing force (nozzle reaction) is represented by vector A. The sine of an angle can be looked up in any good math book that has trigonometric functions and most scientific calculators can calculate sine.

For practical purposes it is only necessary to understand how the nozzle reaction can change as the angle of the pattern changes. At a 30-degree pattern, the nozzle reaction is calculated by finding the sine of 75 degrees (see Figure 9-9). The sine

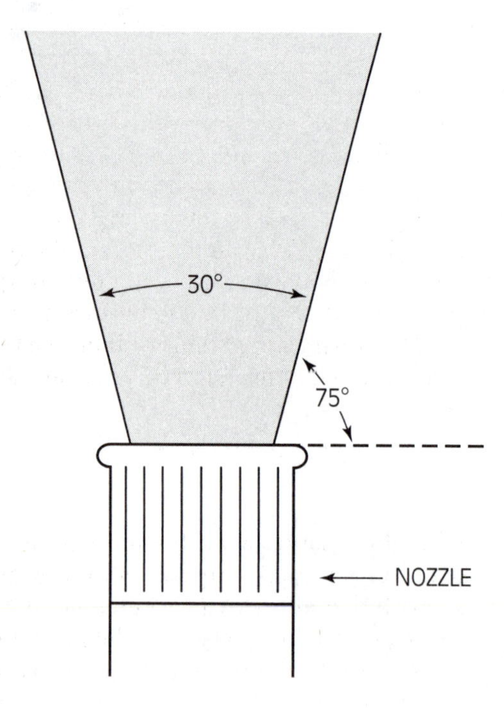

Figure 9-9
Nozzle reaction = calculated nozzle reaction × sine 75°.

of 75 degrees is .97. At a 60-degree pattern, the nozzle reaction will be .87 times the nozzle reaction of a straight stream. A 90-degree pattern will have a nozzle reaction of .7 times the straight stream reaction. Finally, at 100 degrees, the widest fog pattern, the nozzle reaction will be .64 times that of a straight stream.

Example 9-7 What is the nozzle reaction of a fog nozzle flowing 225 gpm at 100 psi nozzle pressure if it is on a 60-degree pattern?

ANSWER First, calculate the nozzle reaction.

$$NR = .0504 \times gpm \times \sqrt{P}$$
$$= .0504 \times 225 \times \sqrt{100}$$
$$= .0504 \times 225 \times 10$$
$$= 113.4 \text{ pounds}$$

Now multiply the answer by .87 to get the actual nozzle reaction.

$$\text{Actual nozzle reaction} = 113.4 \times .87$$
$$= 98.66 \text{ pounds}$$

Stream Straighteners

In order to get the best possible stream from a smooth-bore nozzle, it is important that the water flow as straight as possible as it leaves the nozzle. The problem is that fluids have a natural tendency to rotate in a counterclockwise direction in the Northern Hemisphere. This action is a result of the rotation of the Earth and is called the *Coriolis effect*. (In the Southern Hemisphere, the Coriolis effect causes water to rotate clockwise. There is no rotation at the equator.)

In order that the Coriolis effect not be a factor trying to degrade stream integrity, and to reduce the effects of turbulence in the water, stream straighteners are installed in master stream devices. Figure 9-10 is an illustration of what a stream straightener looks like. It is simply a pattern, similar to the one shown, that is made of thin metal or plastic, and extends several inches up the barrel of a master stream device. The barrel is then attached to master stream device and the tip is attached to the barrel. The stream straightener will negate the Coriolis Effect and allow for a smoother flow of water. Some hand line nozzles used with smooth bore tips also have stream straighteners.

BACK PRESSURE

Back pressure was defined in Chapter 3 as "the pressure exerted by a column of water against the discharge of a pump." This concept is essentially the same as the one introduced in Chapter 2 that allowed us to calculate the pressure at the base of a column of water. As calculated in Chapter 2, the pressure exerted by the column of water is useful because it can create hydrant pressure, but when a pump

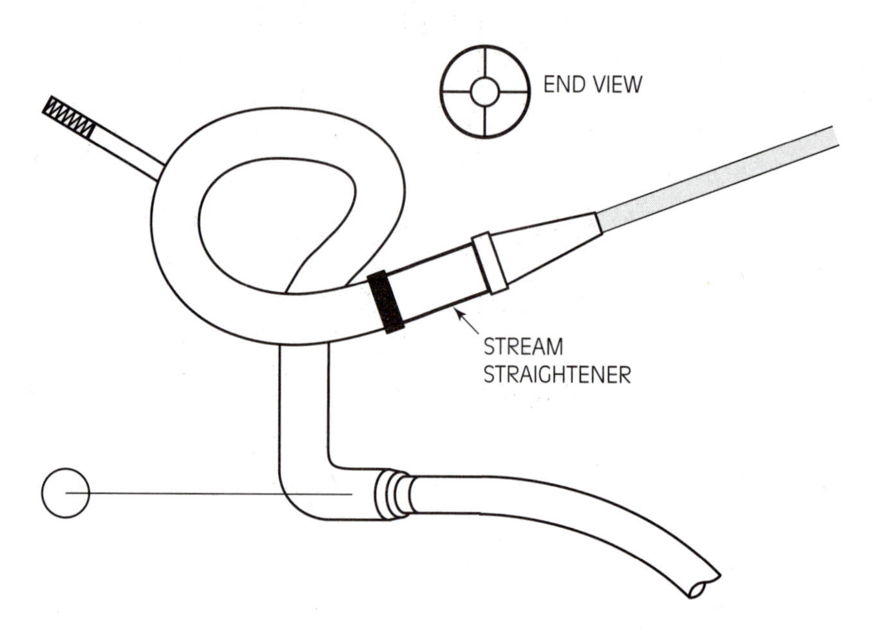

Figure 9-10 *Stream straightener for monitor nozzle.*

has to pump water to an elevated position, this pressure works against us. Look at the illustration in Figure 9-11. The pumper at ground level is pumping to a line on the third floor of a building. Assuming each floor is 8 feet, and another foot for the thickness of each floor, the nozzle on the third floor would be about 18 feet above the pumper. A column of water this tall would exert ($P = .433 \times H$, where $H = 18$ feet) 7.79 psi. The problem now is that when we begin to calculate the amount of pressure needed to deliver the specified gpm, we must also over-

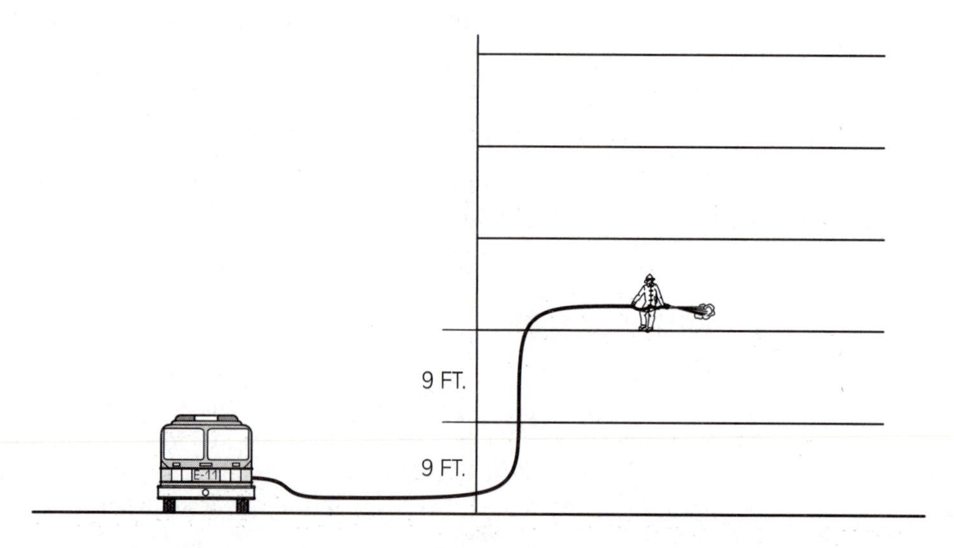

Figure 9-11 *The nozzle is operating 18 feet above the pumper. You need to compensate for back pressure.*

come back pressure. In the example where the nozzle is operating on the third floor, in addition to pumping the correct nozzle pressure and friction loss, we also need to add 7.79 pounds of pressure to overcome back pressure.

Back pressure must be accounted for in all instances where the nozzle is operating higher than the level of the pumper. It does not matter that the nozzle is attached to a preconnect attack line or on a 100 foot line hooked up to a standpipe system, back pressure will be the same in either instance. Even in instances where a fire department pumper is supplementing the water supply for a sprinkler system on an upper floor or pumping to an aerial device, back pressure must be taken into account.

Example 9-8 How much back pressure must be compensated for on a ladder pipe if the stream is at 80 feet elevation?

ANSWER Even though the angle of the ladder can actually change the actual height the stream is off the ground, use the extension of the ladder to calculate elevation.

$$P = .433 \times H$$
$$= .433 \times 80$$
$$= 34.64 \text{ psi backpressure}$$

There are also times when the weight of the water can work for us, such as when a hose line is 18 feet below the pumper. To compensate for the weight of the column of water when the nozzle is operating 18 feet below the pumper, we need to subtract 7.79 psi from the required friction loss and nozzle pressure.

The amount of pressure to add to compensate for back pressure depends on where the nozzle is operating. In Figure 9-12 we have a hose line operating off

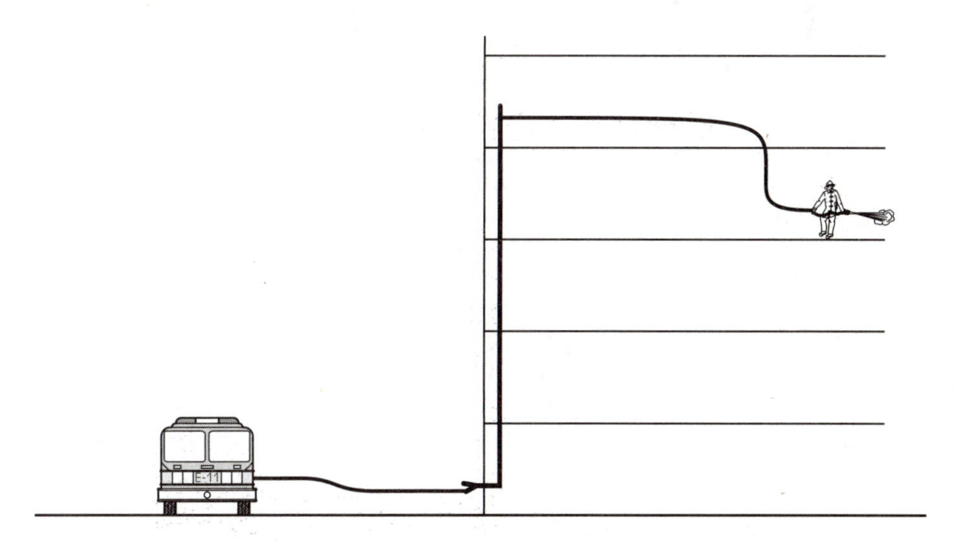

Figure 9-12 *Calculate back pressure for the nozzle on the fourth floor.*

a standpipe on the fifth floor while the nozzle is on the fourth floor. The issue is how much back pressure is added? Since the hose is being taken off the standpipe system on the fifth floor, do we need to overcome the back pressure to the fifth floor? If we do calculate back pressure to the fifth floor, how do we account for the pressure gained by having the line descend one story? The short explanation to this problem is that we pump for the location of the nozzle and do not worry about how it got there. In the example, if we pump for the fourth floor, we will be a few pounds short of the correct pressure at the fifth floor due to failure to compensate for one story worth of back pressure. But, by the time the water comes back down one story we gain what we originally lost, and we have the correct pressure at the nozzle.

Elevation

■ Note
When calculating elevation, simply add pressure when the nozzle is above the pumper, and subtract pressure when the nozzle is below the pumper.

The need to compensate for pressure, either added or subtracted, due to the vertical position of the nozzle in reference to the pump, can be simplified into one concept: elevation. Elevation still encompasses the issue of back pressure, but it also accounts for pressure gain as a line descends below the level of the pumper.

When calculating elevation, simply add pressure when the nozzle is above the pumper, and subtract pressure when the nozzle is below the pumper. Keep in mind that the elevation is calculated for the position of the nozzle regardless of how it got there.

SPECIAL APPLIANCES

When calculating the required pressure to provide the best stream, we often have to include allowances for special appliances. Special appliances are devices such as wagon pipes, monitor nozzles, and ladder pipes. These appliances need special consideration because of friction loss considerations in the device. For instance, friction loss in a monitor nozzle may be as high as 20 psi for the device itself. This 20 psi must be included in any calculations to obtain the correct engine pressure to deliver the correct gpm.

Sprinkler systems and standpipe systems are also included in the category of special appliances. It is critical that sprinkler systems are not overpumped, so particular attention must be given to sprinkler systems when supplementing the water supply. If the pressure to a sprinkler system is too high, the water flowing from each head will exit with such force that it will be atomized. The water droplets will then be too small and light to penetrate the fire plume to the seat of the fire and extinguish it. In short, if a sprinkler system is overpumped, the water spray will become so fine as to be ineffective and the fire will claim the building.

Standpipe systems are less critical. The standpipe system in a building can be thought of as simply an extension of the pump discharge. It does have friction loss that must be accounted for, but that is easy enough to do. We usually account for the friction loss in a standpipe system by assigning a fixed pressure for the system. Then, when we calculate the pressure for the line in use, we simply add the fixed system pressure as a special appliance requirement.

Today some sprinkler and standpipe systems in the same building use a common riser from the fire department siamese. Any time a combination system is encountered, it is critical that the pressure to the system be limited. Because it is possible that hose lines taken off the riser may easily require a pressure higher than the system maximum, the sprinkler system must be given priority. From the account at the beginning of the chapter about One Meridian Plaza, it is easy to understand why the sprinkler system must be a top priority. Once the fire is declared under control and hose lines are sent in to overhaul, the system can be left to serve solely as a standpipe. In order to comply with this it is important to know the maximum pressure for the system, and each system may be different. If the system maximum pressure is not known, a pressure of 175 psi should not be exceeded.

Foam Systems

A unique category of special appliances is foam systems. At one time when the word foam was mentioned around firefighters, the natural inclination was to think in terms of Class B fires or high expansion foam. Today that has changed. In addition to Class B and high expansion foam, there are now Class A foam systems and compressed air foam systems (CAFS) for Class A fires.

If any kind of built-in foam system is used, the manufacturer's specifications should be followed. This caveat includes Class A foam, CAFS, or foam proportion systems for Class B foam.

Fortunately, the most common foam-producing device is the simple foam eductor similar to the one illustrated in Figure 3-7. Using a foam eductor successfully requires attention to a few simple rules. First, limit the amount of hose that comes off the eductor. Usually the limit is 150 feet. Second, have a flow of water that at least meets the minimum required by the eductor. For example, the eductor may need a flow of 95+ gpm in order to work. If you try to flow only 80 gpm through this eductor, there will not be enough velocity to create the necessary vacuum in the eductor. Without enough vacuum, you will be unable to draft the foam solution out of the container. Third, pump 200 psi to the eductor and do not be concerned about friction loss in the hose coming off the eductor or the nozzle pressure. Finally, calculate friction loss only for the hose supplying the eductor.

Standard CVFSS nozzles can be used with many types of foam, although some foams require special foam nozzles.

APPLICATION ACTIVITIES

Apply your knowledge of fire streams to solve the following questions.

Problem 1 The sector chief orders you to have your company direct a stream into the third floor of a building. If the window is approximately 20 feet off the ground, how far from the building will you have to position the nozzle?

ANSWER To achieve the desired 45 degree angle, the nozzle is placed a distance from the building equal to the desired reach. In this case the nozzle should be about 20 feet from the building.

Problem 2 What is the maximum horizontal range for a $^{15}/_{16}$-inch tip at 50 psi?

ANSWER Add 5 for ⅛ over ¾ inch. (If it is not a full ⅛ inch, do not add 5.)

$$HR = \tfrac{1}{2}\, NP + 26^*$$
$$= \tfrac{1}{2}\, 50 + 31$$
$$= 25 + 31$$
$$= 56 \text{ feet}$$

Problem 3 What is the maximum vertical range for the line in Problem 2?

ANSWER Add 5 for ⅛ over ¾ inch.

$$VR = \tfrac{5}{8}\, NP + 26^*$$
$$= \tfrac{5}{8}\, 50 + 31$$
$$= 31.25 + 31$$
$$= 62.25 \text{ feet rounded to 62 feet.}$$

Problem 4 What is the nozzle reaction on a ladder pipe at 80-foot extension if it is using a 1½-inch tip at 80 psi?

ANSWER

$$NR = 1.5 \times D^2 \times P$$
$$= 1.5 \times 1.5^2 \times 80$$
$$= 1.5 \times 2.25 \times 80$$
$$= 270 \text{ pounds}$$

Problem 5 What is the nozzle reaction on a 2½-inch hand line with a fog tip flowing 225 gpm at 100 psi?

ANSWER

$$NR = .0504 \times gpm \times \sqrt{P}$$
$$NR = .0504 \times 225 \times \sqrt{100}$$
$$= .0504 \times 225 \times 10$$
$$= 113.4 \text{ pounds}$$

Problem 6 How much back pressure would be created in a standpipe system in a seven-story building if each story were 12 feet and we are pumping to a line on the fifth floor?

ANSWER We only need to overcome enough back pressure to get our water up to the fifth floor. Because the line will be operating at or close to the floor on the fifth floor, calculate for a full 4 floors of height, or 48 feet.

$$
\begin{aligned}
P &= .433 \times H \\
&= .433 \times 48 \\
&= 20.78 \text{ psi}
\end{aligned}
$$

Summary

Knowing how to calculate the correct friction loss, or knowing how to determine the exact gpm flow is not enough to define an effective fire stream. Although proper line placement is a subject for another text, an effective fire stream must be able to place the correct amount of water at the right point in the proper form to absorb the BTUs being generated.

For the stream to be effective, it must be able to reach the fire. This is best accomplished at a 45-degree angle. For penetration and reach, a smooth-bore or straight stream is needed. Where reach is not a major concern, fog streams are superior due to their heat absorbing ability.

As seen in prior chapters of this book, the laws of physics are evident in the principles that govern the behavior of a charged hose line. As water exits the nozzle, it exerts an equal and opposite reaction (Newton's third law) on the firefighter holding the nozzle. The strength of that reaction can be calculated as long as we know either the nozzle diameter and nozzle pressure or the gpm and the nozzle pressure.

When using stream patterns other than straight stream, the nozzle reaction can vary. The strength of the reaction the firefighter will have to hold gets less as the stream is opened up into an ever-wider fog pattern. At 100 degrees the nozzle reaction is almost half that of a straight stream at the same pressure and gpm.

Other factors that have to be taken into consideration when attempting to provide an acceptable stream include the rotation of the earth, friction loss in special appliances, and backpressure. Only by considering all these factors will an effective fire stream be possible.

Review Questions

1. Why is it so important that water be applied at precisely the correct area of a fire?
2. Why is it so important that the water be applied in the correct form?
3. In the end, how much water actually stopped the spread of fire at One Meridian Plaza?
4. Up to the limit named, how much water should be inside a 15-inch circle?
5. Why is it considered necessary for a fire stream to penetrate an opening and be deflected off the ceiling?
6. With all else being equal, how does nozzle diameter affect the range of a stream?
7. What is the correct angle for the maximum horizontal range?
8. What is the best angle for an effective fire-fighting stream?
9. What is the maximum horizontal range for a 1-inch tip at 50 psi nozzle pressure?
10. What is the maximum vertical range for a 1-inch tip at 50 psi nozzle pressure?

11. Which of Newton's three laws explains nozzle reaction?
12. Calculate nozzle reaction for a ⅞-inch tip at 40 psi nozzle pressure.
13. How is the formula $NR = 1.5 \times D^2 \times P$ converted to find nozzle reaction for a fog nozzle?
14. Which has the greatest nozzle reaction, a 30-degree pattern or a 60-degree pattern?
15. How is back pressure different from friction loss?
16. Is compensation for elevation added or subtracted from the required engine pressure?
17. What are special appliances?

List of Formulas

Horizontal range of streams:

$$HR = \frac{1}{2}\, NP + 26^*$$

Vertical range of streams:

$$VR = \frac{5}{8}\, NP + 26^*$$

Nozzle reaction:

$$NR = 1.5 \times D^2 \times P$$

Nozzle reaction when gpm and pressure are known:

$$NR = .0504 \times gpm \times \sqrt{P}$$

---

* Add 5 for each ⅛ inch of nozzle diameter more than ¾ inch.

Calculating Engine Pressure

Learning Objectives

Upon completion of this chapter, you should be able to:

- Understand the engine pressure formula.
- Select and insert the correct components into the engine pressure formula.
- Be able to calculate the correct engine pressure for a variety of basic situations.
- Be able to calculate the correct engine pressure using the relay formula.

INTRODUCTION

Up to this point, this book has been concerned with a specific aspect of hydraulics. Now we put this knowledge to work to find the correct pressure for various hose layouts.

The point of learning hydraulics is to be able to accurately calculate the pressure needed in hose lines on the fireground. It is not that complicated, but a thorough understanding of hydraulics helps us be more exact. In fact, for the purposes of this chapter and the rest of this book, we rely on exact friction loss rounded to two decimal places and gpm to the closest whole number. Only after the answer has been found do we round engine pressure to the closest whole number. If we practice being exact when we do calculations on paper, when we transfer the calculations to real world applications we will be well within practical and acceptable parameters. If, however, we practice with friction loss charts that have been rounded off, we can never expect a high degree of accuracy. Aristotle said it best when he said, "We are what we repeatedly do. Excellence, then, is not an act, but a habit."

Several formulas are used to calculate engine pressure. Some require knowledge of special factors or are specific to limited situations. The best formula, however, is one that is accurate, easy to remember without knowledge of special factors, and is as useful in the classroom as on the fireground. Just such a formula exists and is used to calculate engine pressure in this chapter and in Chapter 11.

THE ENGINE PRESSURE FORMULA

To correctly calculate the engine pressure it is necessary to add or subtract several integers, called addends and subtrahends in the language of mathematics, to arrive at an engine pressure or sum. The formula used here is the most universally used formula because of its potential for accuracy and versatility. By using this formula both in the classroom and the fireground, it will become second nature and only one formula will need to be learned.

EP = NP + FL 1 + FL 2 ± E + SA

The engine pressure formula is *EP* = *NP* + *FL* 1 + *FL* 2 ± *E* + *SA*, where *EP* = engine pressure, *NP* = nozzle pressure, *FL* = friction loss, *E* = elevation, and *SA* = special appliance. Friction loss appears twice because we often have to deal with situations where there is more than one size hose involved or one size hose may have more than one friction loss. Elevation is either plus or minus (±) because it can be added or subtracted depending on the location of the nozzle in relation to the pumper. In this chapter we solve basic engine pressure calculations using progressively more elements of the engine pressure formula. Each element is explained as it is used. Some jurisdictions might refer to engine pressure as pump discharge pressure. Both terms mean the same thing, but engine pressure is used here.

EP = NP + FL 1 + FL 2 ± E + SA

The most fundamental engine pressure calculation involves simply adding the nozzle pressure and the required friction loss. In Chapter 6 we learned how to

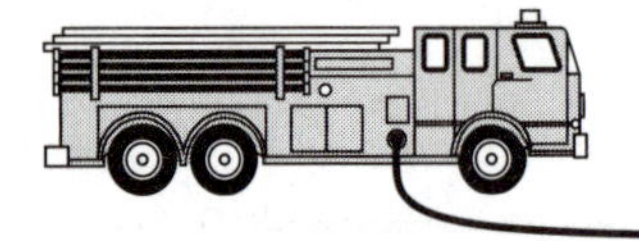

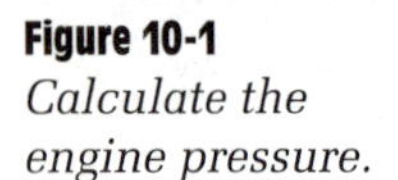
Figure 10-1 *Calculate the engine pressure.*

calculate friction loss. The friction loss we calculated was for 100 feet of hose. When calculating friction loss to put into our engine pressure formula, we need to determine the amount of friction loss for the total hose lay. For this we use the formula $FL = FL\ 100 \times L$, where $FL\ 100$ is the friction loss per 100 feet, and L is the length of hose in hundreds. Remember, to determine the length of line in hundreds simply move the decimal point two places to the left. A line 650 feet in length is an L of 6.5.

$FL = FL\ 100 \times L$

Example 10-1 Find the engine pressure needed for a 350 foot 1½-inch hose line if the nozzle pressure is 100 psi and it is flowing 100 gpm (see Figure 10-1).

ANSWER Calculate $FL\ 100$ by the formula $FL\ 100 = Cf \times 2Q^2$ from Chapter 6.

$FL\ 100$ for 100 gpm in 1½-inch hose is 24 psi

$$FL = FL\ 100 \times L$$
$$= 24 \times 3.5$$
$$= 84 \text{ psi}$$

Now find the engine pressure:

$$EP = NP + FL\ 1 + FL\ 2 \pm E + SA$$
$$= 100 + 84 + 0 \pm 0 + 0$$
$$= 184 \text{ psi}$$

The correct engine pressure for a 1½-inch hose line flowing 100 gpm at 100 psi nozzle pressure is 184 psi. This is the most basic engine pressure calculation, yet it is the one that is used for the majority of all fireground situations.

$EP = NP + FL\ 1 + FL\ 2 \pm E + SA$

It may be necessary at times to calculate engine pressure when a hose line is made up of more than one size hose. A practical fireground scenario might involve making a knock down with a 2½-inch hand line and then extending the line with 1½-inch hose. The flow in each size hose will remain the same, but friction loss in each size hose will have to be calculated independently.

Example 10-2 What is the engine pressure for a line that consists of 150 feet of 2½-inch hose that has been extended with 100 feet of 1½-inch hose? The 1½-inch nozzle is operating at 100 psi nozzle pressure and 100 gpm (see Figure 10-2).

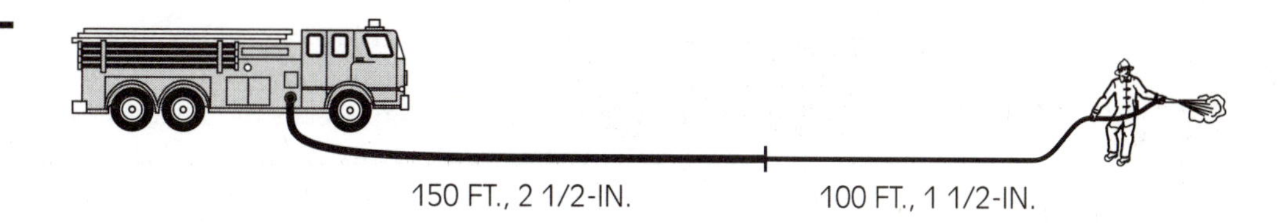

Figure 10-2 *Calculate engine pressure with two different sizes of hose.*

ANSWER From Example 10-1 we already know the *FL* 100 for $1\frac{1}{2}$-inch is 24 psi.

FL 100 for $2\frac{1}{2}$-inch hose flowing 100 gpm is 2 psi

$$FL = FL\ 100 \times L$$

$$FL\ 1\ (2\tfrac{1}{2}) = 2 \times 1.5$$

$$FL = 3 \text{ psi}$$

$$FL\ 2\ (1\tfrac{1}{2}) = 24 \times 1$$

$$FL = 24 \text{ psi}$$

Now find the engine pressure:

$$EP = NP + FL\ 1 + FL\ 2 \pm E + SA$$

$$= 100 + 3 + 24 \pm 0 + 0$$

$$= 127 \text{ psi}$$

In this example, the larger diameter hose was laid first and then the smaller diameter hose was used to extend it. If, for some reason, the hose had been laid in with the smaller diameter hose first and then the larger diameter hose, the problem would be solved in the exact same way. The order of the different diameter hose is irrelevant. It is not likely that hand lines will be extended with larger diameter hose, but it is very possible that supply lines can have two different diameter hose in a single line in any order.

$EP = NP + FL\ 1 + FL\ 2 \pm E + SA$

In Chapter 9, the concept of back pressure was introduced. Here is where back pressure is applied. In order to arrive at the correct engine pressure, elevation above or below the pumper must sometimes be taken into consideration.

Where the exact elevation is known we can use the formula $P = .433 \times H$ from Chapter 2 to determine the exact pressure adjustment required. In most cases we do not know the exact height but only know how many stories the nozzle is above or below the pumper. If we assume the average story of a building to be 12 feet from floor to floor, using the formula $P = .433 \times H$ gives us a standard pressure per floor of $P = .433 \times 12$ or 5.196 psi. This amount is usually rounded off to a constant of 5 psi for each floor above the first. The actual distance between floors of a building can vary. Some texts use a figure of 10 feet, and the author is aware of buildings where the distance between floors is actually 14 feet. If you have only one- and two-family dwellings in your jurisdiction, you may want to use a constant of 4 psi per story above the first because the floor-to-floor height of the average house is only about 9 feet. For the purposes of this text, when calculating elevation 5 psi per story is used.

There will also be situations where elevation is estimated. Where possible, the formula $P = .433 \times H$ should be used to calculate the exact pressure adjustment needed. However, there are situations on the fireground where quick calculations are necessary. In such cases it is common practice to use ½ pound per foot of elevation. For example, where an aerial ladder is extended with the ladder pipe in operation, if the ladder is extended 80 feet, the required compensation for elevation would be 40 psi.

One final reminder on elevation: remember from Chapter 9 that elevation is calculated for the position of the nozzle. If the line goes to the fourth floor and then back down to the second floor, you calculate elevation for the second floor.

Example 10-3 What engine pressure is required at the pumper if it is pumping to a 300 foot 1¾-inch line, flowing 150 gpm at 100 psi nozzle pressure, operating on the fourth floor (see Figure 10-3)?

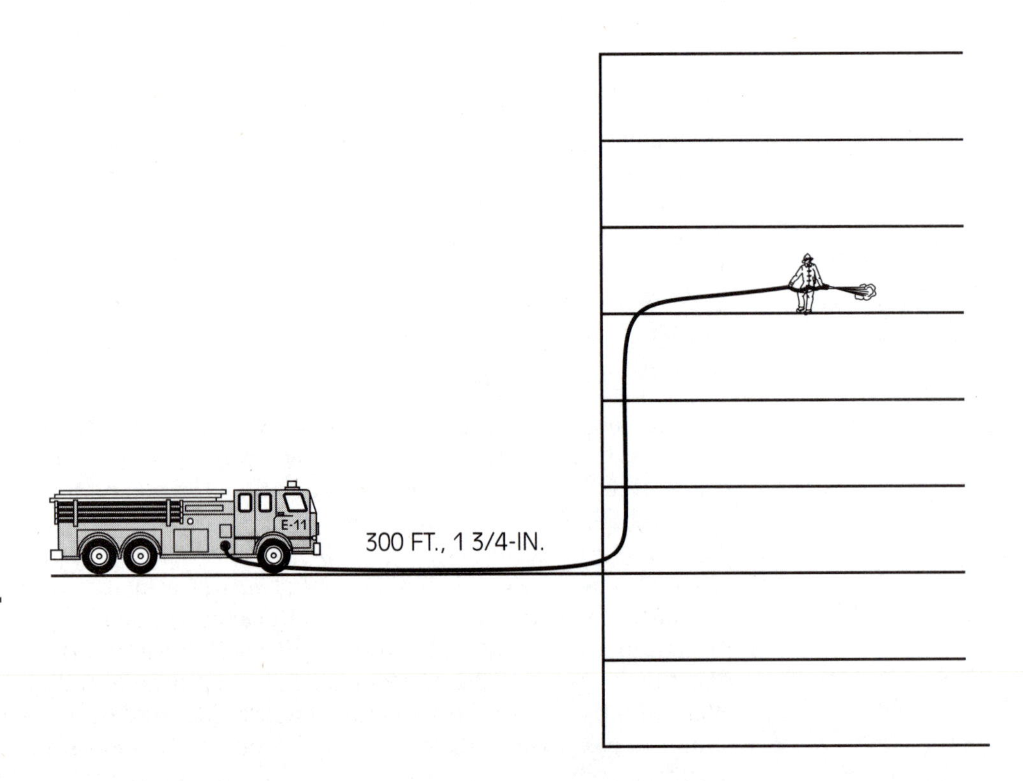

Figure 10-3 *Calculate the engine pressure with the nozzle on the fourth floor.*

ANSWER The *Cf* for 1¾-inch hose is 7.76.

FL 100 for 150 gpm in 1¾-inch hose is 35 psi

$$FL = FL\ 100 \times L$$
$$= 35 \times 3$$
$$= 105 \text{ psi}$$

Now find the engine pressure. Remember, the nozzle is on the fourth floor, or three floors above the first. Elevation will be 5 pounds for each floor above the first, or 15 psi.

$$EP = NP + FL\ 1 + FL\ 2 \pm E + SA$$
$$= 100 + 105 + 0 + 15 + 0$$
$$= 220 \text{ psi}$$

It is important to remember that elevation can just as easily be a subtrahend (a number to be subtracted). Again, do not count the first floor, but count every floor below the first. A basement requires a minus 5 psi of elevation while a sub-basement requires a minus 10 psi of elevation.

Example 10-4 An engine company has advanced its 250-foot preconnect 1¾-inch line to a second level underground parking garage for an auto fire. What is the required engine pressure if the line is flowing 150 gpm at 100 psi nozzle pressure (see Figure 10-4)?

ANSWER We already know from Example 10-3 that *FL* 100 for the 1¾-inch line is 35 psi.

$$FL = FL\ 100 \times L$$
$$= 35 \times 2.5$$
$$= 87.5$$

Now find the engine pressure. Elevation is –10 psi.

$$EP = NP + FL\ 1 + FL\ 2 \pm E + SA$$
$$= 100 + 87.5 + 0 - 10 + 0$$
$$= 177.5 \text{ or } 178 \text{ psi}$$

EP = *NP* + *FL* 1 + *FL* 2 ± *E* + *SA*

In addition to nozzles and hose, we often lose friction in other devices, such as monitor nozzle or ladder pipes. In order to deliver the correct gpm at a pressure that will produce the desired pattern, friction loss in these special appliances must be accounted for. Table 10-1 shows friction loss allowances for selected devices. Keep in mind that the figures given in this table are not absolute figures because

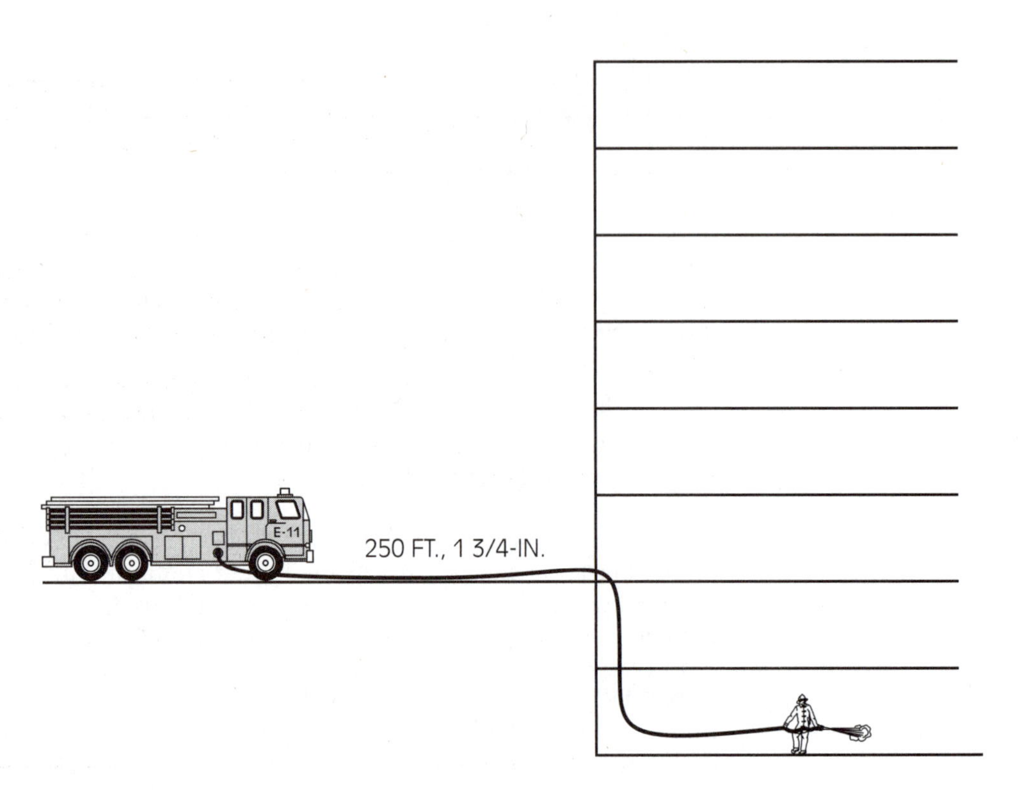

Figure 10-4 *Calculate engine pressure for a line in the parking garage.*

they can vary from one manufacturer to another. Instead they are provided for illustrative purpose and are used through the remainder of this text for the purpose of standardization.

For water tower operations on aerial devices that have piping installed to deliver water to the nozzle, the manufacturer usually stipulates a pressure. That pressure needs to be made available to pump operators supplying the aerial device.

Table 10-1 *Special appliance allowances.*

Ladder pipe (includes siamese, 100 feet of 3-inch hose, and the ladder pipe)	50 psi
Standpipe	25 psi
Monitor nozzle/wagon pipe	20 psi
Sprinkler systems	150 psi
Combination sprinkler/standpipe system	175 psi
Foam eductor	200 psi

When calculating engine pressure when a special appliance is used, all the elements of the engine pressure formula are used. Even special appliances usually have a nozzle pressure, there is friction loss in the hose to the appliance, and it can be located above or below the pumper.

Example 10-5 What is the engine pressure where a pumper is pumping to a 1¾-inch hand line on the third floor of a building? The line is 100 feet long, flowing 150 gpm at 100 psi nozzle pressure and is connected to a standpipe system. The standpipe system is supplied by a 50-foot section of 2½-inch hose (see Figure 10-5).

ANSWER A standpipe system should be thought of as a water distribution system. While they are usually vertical, horizontal standpipe systems are also used in special applications. Regardless, the engine pressure will be calculated exactly the same way.

We already know the friction loss for 150 gpm in 1¾-inch hose is 35 psi per 100 feet. From Table 10-1, the friction loss allowed for a standpipe system is 25 psi.

First calculate the friction loss for 100 feet of 2½-inch hose flowing 150 gpm.

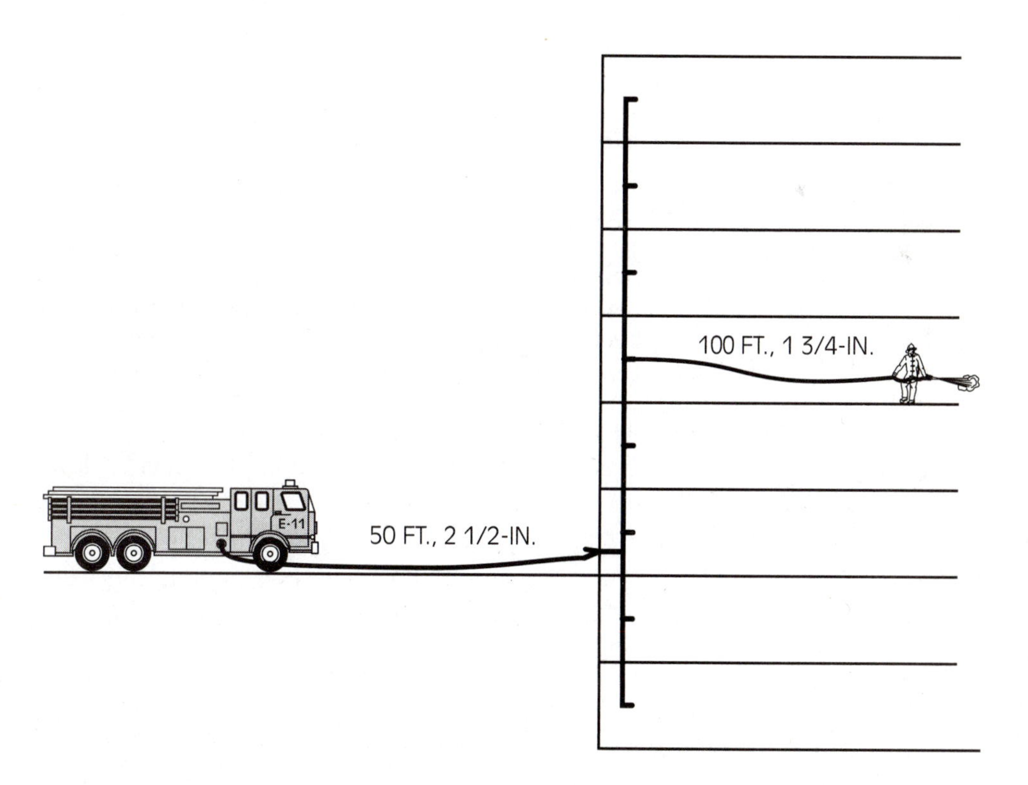

Figure 10-5
Operating from a standpipe with 100 feet of 1¾-inch hose.

$$\begin{aligned} FL\ 100 &= Cf \times 2Q^2 \\ &= 1 \times 2 \times 1.5^2 \\ &= 1 \times 2 \times 2.25 \\ &= 4.5 \text{ psi} \end{aligned}$$

Remember, you only have 50 feet of 2½-inch hose.

$$\begin{aligned} FL &= FL\ 100 \times .5 \\ &= 4.5 \times .5 \\ &= 2.25 \text{ psi} \end{aligned}$$

Now solve for engine pressure.

$$\begin{aligned} EP &= NP + FL\ 1 + FL\ 2 \pm E + SA \\ &= 100 \times 2.25 + 35 + 10 + 25 \\ &= 172.25 \text{ or } 172 \text{ psi} \end{aligned}$$

The one instance when a special appliance will be encountered where it will be the only element of the engine pressure formula used is when a sprinkler system is used. The recommended course of action for a sprinkler system is to pump a set pressure, usually 150 psi is recommended. Because it is unknown how many heads may have opened, where they are located, and how much water is flowing, the set pressure is used. A note of caution is in order here: Make certain the system is not overpumped. If too much pressure is put into the system, water will come out of the sprinkler head at too high a velocity and the water will be atomized. When water is atomized the droplets are too small and light to penetrate the fire plume and fall to the seat of the fire. Without getting to the seat of the fire, extinguishment or even control is unlikely. If pumping to a sprinkler system with a two-stage pump, the transfer valve should be in the parallel position.

One final special case under special appliances is the foam eductor. As mentioned in Chapter 9, foam eductors are designed to work at a specified pressure and flow. When calculating the engine pressure for an eductor you only need to add friction loss, elevation, and the special appliance allowance.

Example 10-6 Calculate the engine pressure for a foam line that has 200 feet of 2½-inch hose supplying a foam eductor designed to flow a minimum of 95 gpm at 200 psi.

ANSWER Remember to limit the hose coming off the eductor to 150 feet and use a nozzle that will flow at least 95 gpm. In this example the line is flowing 100 gpm. *FL* 100 for 100 gpm through the 2½-inch hose is 2.

$$\begin{aligned} FL &= FL\ 100 \times 2 \\ &= 2 \times 2 \\ &= 4 \end{aligned}$$

$$EP = NP + FL\,1 + FL\,2 \pm E + SA$$
$$= 0 + 4 + 0 + 0 + 200$$
$$= 204 \text{ psi}$$

With 200 feet of 2½-inch hose supplying the foam eductor, the eductor will work properly at an engine pressure of 204 psi. This problem may seem to be missing the size of hose coming off the eductor, but it is irrelevant because the only two things that count are the gpm and length of the line.

ENGINE PRESSURE AND PARALLEL LINES

Real life fireground evolutions often involve multiple hose evolutions. Where one pumper is responsible for pumping multiple lines to another pumper or device, the evolution is referred to as parallel line. Parallel line lays can be divided into two different categories: (1) where all lines are of equal diameter and (2) where the lines are of unequal diameter.

Parallel Lines of Equal Diameter Supplied by One Pumper

The evolution where one pumper will be supplying multiple lines to a master stream device is common. However, three rules governing these situations must be taken into consideration in order to arrive at the correct engine pressure. These rules apply to evolutions where all lines are the same diameter.

■ Note Each line into the device is assumed to be carrying an equal share of water.

Rule 1. *Each line into the device is assumed to be carrying an equal share of the water.* For instance, if two lines are supplying a monitor nozzle that is flowing 600 gpm, the gpm are divided by two to get the quantity of water assumed to be flowing through each hose or 300 gpm. If three lines are supplying the monitor nozzle, the 600 gpm are divided by 3 to get a flow through each hose of 200 gpm.

■ Note When the length of the parallel lines is unequal, the average length is used.

Rule 2. *When the length of the parallel lines is unequal, the average length is used.* In Figure 10-6 a pumper is supplying 600 gpm to a monitor nozzle. It is supplying the water through one 3-inch line 400 feet long and another 3-inch line that is only 300 feet long. To find the

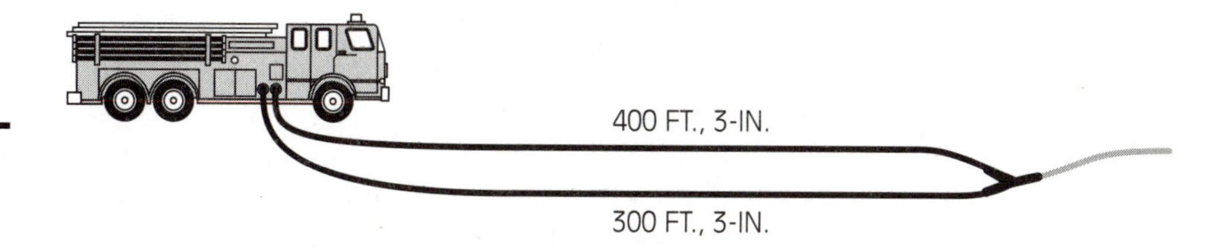

Figure 10-6 *Average the length of the hose lines.*

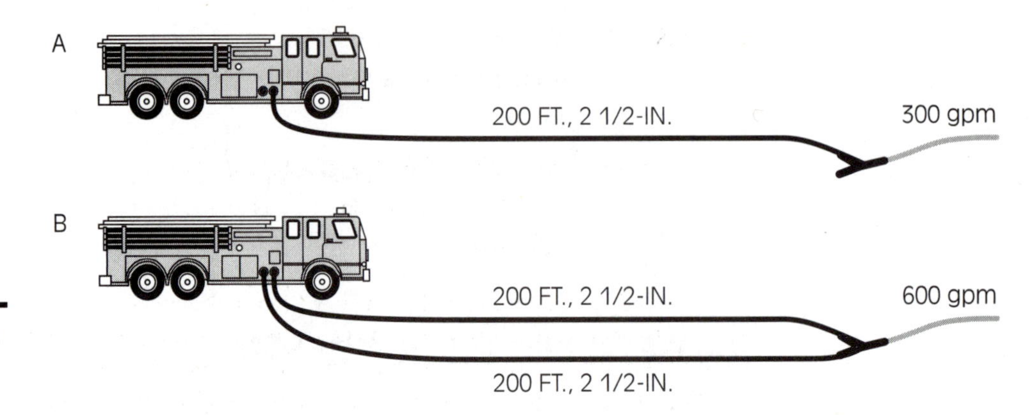

Figure 10-7 *A and B require the same engine pressure.*

average, add the length of the two lines together and divide by two, that is, 400 + 300 = 700 ÷ 2 = 350 feet. This procedure works regardless of the number of lines coming into the device, as long as they are from the same pumper.

■ Note
When calculating engine pressure where rule 1 and rule 2 apply, friction loss through only one line is used.

Rule 3. *When calculating engine pressure where rule 1 and rule 2 apply, friction loss through only one line is used.* Where lines are parallel, friction loss is not cumulative. The best way to explain this concept is with an illustration as shown in Figure 10-7.

Example 10-7 In Figure 10-7A, the pumper is supplying a monitor nozzle with a flow of 300 gpm at 80 psi nozzle pressure through a 200-foot-long, 2½-inch hose line. If the friction loss in 2½-inch hose at 300 gpm is 18 psi, what will the engine pressure be?

ANSWER For 10-7A only one line is being used and it is carrying all the water.

$$FL = FL\,100 \times L$$
$$= 18 \times 2$$
$$= 36 \text{ psi}$$

$$EP = NP + FL\,1 + FL\,2 \pm E + SA$$
$$= 80 + 36 + 0 \pm 0 + 20$$
$$= 136 \text{ psi}$$

Example 10-8 In Figure 10-7B the pumper is supplying a monitor nozzle with a flow of 600 gpm at 80 psi nozzle pressure through two 200-foot-long, 2½-inch hose lines. Each line is flowing only 300 gpm so the friction loss is 18 psi. What is the required engine pressure?

ANSWER $FL = FL\ 100 \times L$

$= 18 \times 2$

$= 36$ psi

$EP = NP + FL\ 1 + FL\ 2 \pm E + SA$

$= 80 + 36 + 0 \pm 0 + 20$

$= 136$ psi

Notice that in both Example 10-7 and Example 10-8, the nozzle pressure is the same and the allowance for the special appliance is the same. The only possible variable is the friction loss, but because each line is flowing the same gpm, even friction loss is the same, making the engine pressure in each example exactly the same, even though in Example 10-8 twice as much water is being pumped. Rule 3 applies for any number of lines as long as the lines are parallel and of the same diameter. The friction loss in only one line is used because they all have the same gpm flow.

Parallel Lines of Unequal Diameter Supplied by One Pumper

When a pumper is supplying parallel lines of unequal diameter only rule 2 applies. Each line does not handle an equal amount of water because, as already mentioned in Chapter 6, the pressures in parallel lines will equalize at the point where both lines have the same friction loss. When the diameter of the hose is the same, it means the gpm flow in each line is the same, but when the diameter of the hose is different, the gpm flow in each line is different. From Example 6-12 in Chapter 6 we found that at 11.25 psi friction loss per 100 feet in 2½-inch hose we could get a flow of 237 gpm. But for the same friction loss in 3-inch hose the flow is 375 gpm.

To find the friction loss for a given flow through parallel lines of unequal diameter, it is absolutely necessary that we know the conversion factor for the combination of hose sizes. Recall from Chapter 6 the conversion factor for one 2½-inch and one 3-inch hose is .15. Now to find the friction loss per 100 feet for any flow through this combination of hose sizes we can just insert .15 in the *FL* 100 formula in place of *Cf*.

Example 10-9 What is the engine pressure for a pumper that is supplying 500 gpm to a monitor nozzle at 80 psi nozzle pressure, if it is pumping through 300 feet of parallel line, one 2½-inch and one 3-inch (see Figure 10-8)?

ANSWER First find the friction loss for 500 gpm through this combination of hose.

$FL\ 100 = Cf \times 2Q^2$

$= .15 \times 2 \times 5^2$

$= .15 \times 2 \times 25$

$= 7.5$ psi

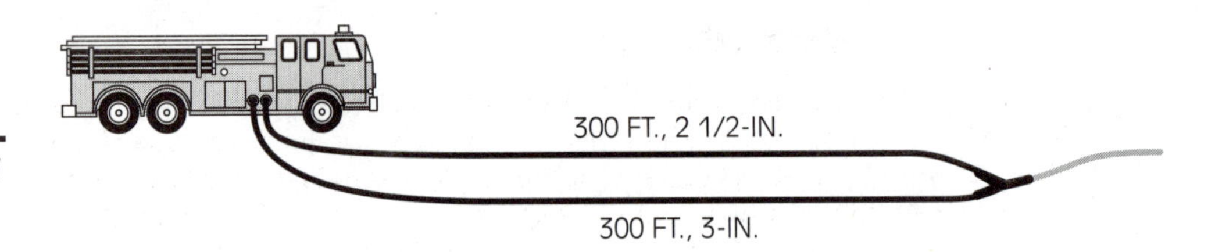

Figure 10-8 *Unequal diameter parallel lines.*

Now find the total friction loss.

$FL = FL\ 100 \times L$
$= 7.5 \times 3$
$= 22.5$

Finally, find the engine pressure.

$EP = NP + FL\ 1 + FL\ 2 \pm E + SA$
$= 80 + 22.5 + 0 \pm 0 + 20$
$= 122.5$ or 122 psi

Even when parallel lines are unequal diameter, rule 2 still applies. Add up the lengths of all the lines then divide by the number of lines to get the average length.

Example 10-10 In Example 10-9, if the 2½-inch line had only been 250 feet, what would the engine pressure be?

ANSWER We already know the *FL* 100 for a 2½-inch and 3-inch hose, so now the first step is to determine the average length of lines. The 2½-inch hose is 250 feet long and the 3-inch hose is 300 feet long.

$250 + 300 = 550/2 = 275$ feet

Now find the total friction loss.

$FL = FL\ 100 \times L$
$= 7.5 \times 2.75$
$= 20.6$

Now find the engine pressure.

$EP = NP + FL\ 1 + FL\ 2 \pm E + SA$
$= 80 + 20.6 + 0 \pm 0 + 20$
$= 120.6$ or 121 psi

RELAY FORMULA

relay
any situation in which one pumper is supplying water to another pumper

A **relay** is any situation in which one pumper is supplying water to another pumper. Relays often involve only two pumpers, one at the hydrant and the second one at the fire such as illustrated in Figure 10-9. This situation is often referred to as a *two-pump operation* and is the most common relay evolution. The more common idea of a relay is several pumpers in line, such as shown in Figure 10-10. What makes either evolution a relay is the fact that water is being relayed, in turn, from the pumper at the water source to the pumper at the fire with any number of pumpers in between.

$EP = 20 + FL \pm E$

The formula for use in any relay application is $EP = 20 + FL \pm E$. EP, FL, and E have the same meaning as in the engine pressure formula, however, the 20 is a new element. The 20 is actually 20 psi and is the minimum amount of pressure we want at the intake of the next pumper in line. (Some jurisdictions may use a pressure higher than 20 psi, but in no case should a pressure less than 20 psi be used.) In the relay formula friction loss is calculated just as it would be in the engine pressure formula.

Example 10-11 What is the engine pressure for pumper 2 in Figure 10-11, pumping through two 500-foot, 3-inch supply lines delivering 800 gpm?

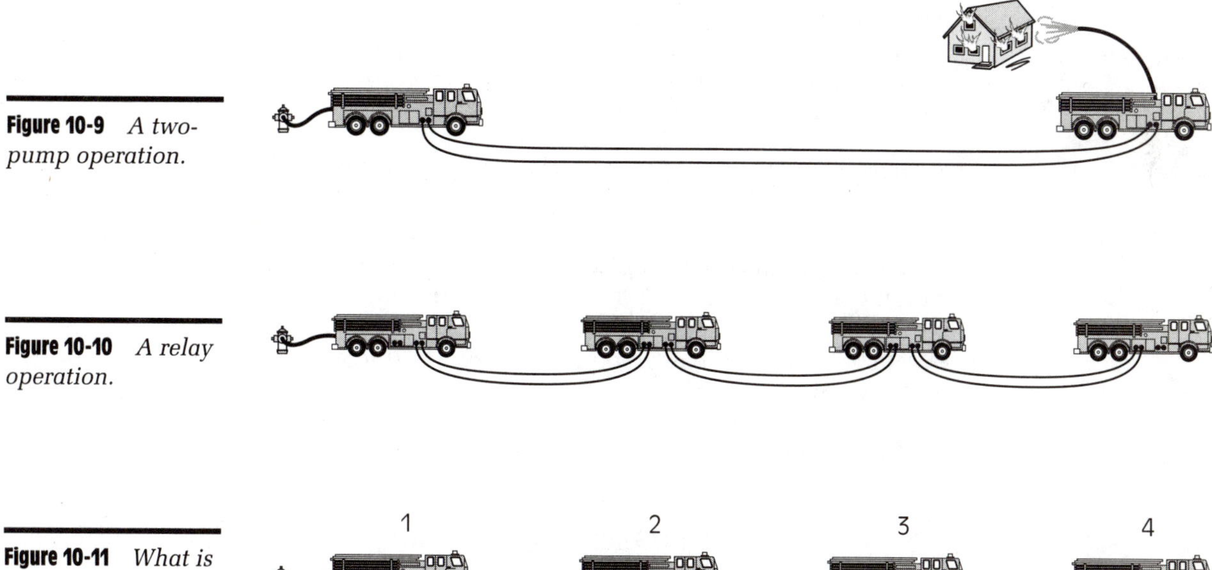

Figure 10-9 *A two-pump operation.*

Figure 10-10 *A relay operation.*

Figure 10-11 *What is the engine pressure for pumper 2?*

ANSWER First, we need to find the friction loss for 800 gpm through two 3-inch supply lines. Remember, each line is only flowing 400 gpm, so we only need to find friction loss for 400 gpm through 3-inch hose.

$$
\begin{aligned}
FL\ 100 &= Cf \times 2Q^2 \\
&= .4 \times 2 \times 4^2 \\
&= .4 \times 2 \times 16 \\
&= 12.8 \text{ psi}
\end{aligned}
$$

Now find the total friction loss.

$$
\begin{aligned}
FL &= FL\ 100 \times L \\
&= 12.8 \times 5 \\
&= 64 \text{ psi}
\end{aligned}
$$

Now solve for engine pressure using the relay formula.

$$
\begin{aligned}
EP &= 20 + FL \pm E \\
&= 20 + 64 \pm 0 \\
&= 84 \text{ psi}
\end{aligned}
$$

It is not unusual for fire departments to establish a minimum pump pressure when pumping in a relay. For instance, a standard operating procedure may require a minimum discharge pressure of 125 psi. This would mean that the driver of the pumper in Example 10-10 would actually pump at 125 psi instead of the calculated pressure of 84 psi.

APPLICATION ACTIVITIES

You are now ready to begin solving problems using the engine pressure and relay formula. In these problems you will be required to solve each element of the problem, including gpm and *FL* 100, as well as the engine pressure.

Problem 1 What is the engine pressure for a 250 foot 2½-inch line with a 1-inch tip at 45 psi nozzle pressure? (C = .98)

ANSWER First, find the gpm for a 1-inch tip at 45 psi.

$$
\begin{aligned}
gpm &= 29.77 \times D^2 \times C \times \sqrt{P} \\
&= 29.77 \times 1^2 \times .98 \times \sqrt{45} \\
&= 29.77 \times 1 \times .98 \times 6.7 \\
&= 195.4 \text{ or } 195 \text{ gpm}
\end{aligned}
$$

Now, find the friction loss 100 for 195 gpm in 2½-inch hose.

$$FL\ 100 = Cf \times 2Q^2$$
$$= 1 \times 2 \times 1.95^2$$
$$= 1 \times 2 \times 3.8$$
$$= 7.6 \text{ psi}$$

The total friction loss is:

$$FL = FL\ 100 \times L$$
$$= 7.6 \times 2.5$$
$$= 19 \text{ psi}$$

Now, solve for engine pressure.

$$EP = NP + FL\ 1 + FL\ 2 \pm E + SA$$
$$= 45 + 19 + 0 \pm 0 + 0$$
$$= 64 \text{ psi}$$

Problem 2 What is the engine pressure for 150 feet of 1¾-inch hose being supplied by 200 feet of 2½-inch hose? The nozzle pressure is 75 psi for a fog nozzle that is flowing 180 gpm (see Figure 10-12).

ANSWER Because it is a fog nozzle, the flow is already known. The first step is to calculate the friction loss in both 1¾-inch hose and 2½-inch hose.

$$FL\ 100 = Cf \times 2Q^2 \text{ (for the 2½-inch hose)}$$
$$= 1 \times 2 \times 1.8^2$$
$$= 1 \times 2 \times 3.24$$
$$= 6.48 \text{ psi}$$

Total friction loss for the 2½-inch hose is:

$$FL = FL\ 100 \times L$$
$$= 6.48 \times 2$$
$$= 12.96$$

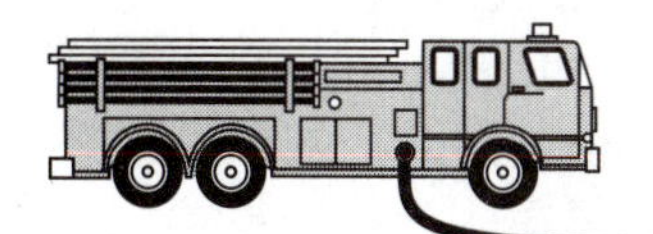

Figure 10-12 *Calculate the engine pressure.*

Now find friction loss in the 1¾-inch hose.

$$FL\ 100 = Cf \times 2Q^2$$
$$= 7.76 \times 2 \times 1.8^2$$
$$= 7.76 \times 2 \times 3.24$$
$$= 50.28$$

The total friction loss for the 1¾-inch hose is:

$$FL = FL\ 100 \times L$$
$$= 50.28 \times 1.5$$
$$= 75.42 \text{ psi}$$

Now solve for engine pressure.

$$EP = NP + FL\ 1 + FL\ 2 \pm E + SA$$
$$= 75 + 12.96 + 75.42 \pm 0 + 0$$
$$= 163.38 \text{ or } 163 \text{ psi}$$

Problem 3 Calculate the engine pressure for both the pumper at the hydrant and the pumper at the fire in the following evolution. Pumper 1 is hooked up to a hydrant and is pumping through two supply lines, one 300 feet of 3-inch hose and the other 400-feet of 3½-inch hose, to pumper 2. Pumper 2 is supplying a monitor nozzle with a 1¾-inch tip at 80 psi nozzle pressure through two 250-foot, 3-inch lines. Pumper 2 is also supplying a 150-foot, 2½-inch hand line with a 1⅛ inch tip at 50 psi tip pressure.

ANSWER First, we need to calculate the gpm being used.

$$gpm = 29.77 \times D^2 \times C \times \sqrt{NP}$$
$$= 29.77 \times 1.125^2 \times .98 \times \sqrt{50}$$
$$= 29.77 \times 1.27 \times .98 \times 7.07$$
$$= 261.95 \text{ or } 262 \text{ gpm for the } 2\tfrac{1}{2}\text{-inch hand line}$$

$$gpm = 29.77 \times D^2 \times C \times \sqrt{NP}$$
$$= 29.77 \times 1.75^2 \times .98 \times \sqrt{80}$$
$$= 29.77 \times 3.06 \times .98 \times 8.94$$
$$= 798.1 \text{ or } 798 \text{ gpm for the monitor nozzle}$$

Total gpm is 262 + 798 = 1,060 gpm

Then we calculate the engine pressure for pumper 1. Pumper 1 is supplying all the water being used by pumper 2.

What is the average length of supply line?

300 feet (3-inch hose) + 400 feet (3½-inch hose) =
700 feet/2 = 350 foot average

What is the friction loss 100 in parallel lines of one 3-inch hose and one 3½-inch hose? In Chapter 6, Problem 8 we already calculated the conversion factor for one 3-inch hose and one 3½-inch hose as .062.

$$\begin{aligned} FL\ 100 &= Cf \times 2Q^2 \\ &= .062 \times 2 \times 10.6^2 \\ &= .062 \times 2 \times 112.36 \\ &= 13.93 \text{ psi} \end{aligned}$$

Now calculate the total friction loss.

$$\begin{aligned} FL &= FL\ 100 \times L \\ &= 13.93 \times 3.5 \\ &= 48.76 \text{ psi} \end{aligned}$$

Use the relay formula to calculate engine pressure for pumper 1.

$$\begin{aligned} EP &= 20 + FL \pm E \\ &= 20 + 48.76 \pm 0 \\ &= 68.76 \text{ or } 69 \text{ psi} \end{aligned}$$

To determine the engine pressure for pumper 2 it is necessary to calculate the engine pressure for the hand line and the monitor nozzle separately. Let us calculate the engine pressure for the hand line first.

$$\begin{aligned} FL\ 100 &= Cf \times 2Q^2 \\ &= 1 \times 2 \times 2.62^2 \\ &= 1 \times 2 \times 6.86 \\ &= 13.72 \text{ psi} \end{aligned}$$

The total friction loss for the 2½-inch hose is:

$$\begin{aligned} FL &= FL\ 100 \times L \\ &= 13.72 \times 1.5 \\ &= 20.58 \text{ psi} \end{aligned}$$

Now calculate the engine pressure for this line.

$$\begin{aligned} EP &= NP + FL\ 1 + FL\ 2 \pm E + SA \\ &= 50 + 20.58 + 0 \pm 0 + 0 \\ &= 70.58 \text{ or } 71 \text{ psi} \end{aligned}$$

Now calculate the engine pressure for the monitor nozzle. First, we need to calculate the friction loss. Remember that because there are two hoses of the same diameter, we only need to find friction loss through one 3-inch hose for half the total gpm.

$FL\ 100 = Cf \times 2Q^2$

$= .4 \times 2 \times 3.99^2$

$= .4 \times 2 \times 15.92$

$= 12.74$ psi

Now find the total friction loss.

$FL = FL\ 100 \times L$

$= 12.74 \times 2.5$

$= 31.85$ psi

Finally, the engine pressure for the monitor nozzle is:

$EP = NP + FL\ 1 + FL\ 2 \pm E + SA$

$= 80 + 31.85 + 0 \pm 0 + 20$

$= 131.85$ or 132 psi

gate back
partially close off the discharge until the correct pressure is obtained

gate up
partially opening the discharge until the correct pressure is obtained

In the foregoing problem, the engine pressure for pumper 1 is clear. However, the pressure for pumper 2 is not so clear. The monitor nozzle requires nearly twice the pressure as the hand line. In such instances the higher pressure is always the correct pressure. Discharges to lines/nozzles requiring lower pressures will have to be gated back. To **gate back** means to partially close off the discharge until the correct pressure is obtained. Some instructors prefer to teach the **gate up** technique, which involves starting with a closed discharge and partially opening the discharge until the correct pressure is obtained. In reality, the nozzle crew will probably not know which line is overpumped until the lines are in operation.

Problem 4 What is the engine pressure for a ladder pipe being operated off an aerial ladder if the ladder is extended to 70 feet at a 70-degree angle, operating a fog nozzle flowing 500 gpm and being supplied by two 100-foot 2½-inch lines.

ANSWER We begin by calculating the friction loss 100 for the 2½-inch hose. By now it should be second nature to understand that we will be calculating the friction loss in the 2½-inch hose for only 250 gpm.

$FL\ 100 = Cf \times 2Q^2$

$= 1 \times 2 \times 2.5^2$

$= 1 \times 2 \times 6.25$

$= 12.5$ psi. This amount is also the total friction loss because the lines are only 100 feet long.

Note: The nozzle pressure will be 100 psi because it is a fog nozzle.

Special appliance allowance is 50 psi and includes the siamese, 100 feet of hose from siamese to the ladder pipe, and the ladder pipe itself. Calculate elevation at ½ pound per foot of extension or 35 psi.

$$EP = NP + FL\,1 + FL\,2 \pm E + SA$$
$$= 100 + 12.5 + 0 + 35 + 50$$
$$= 197.5 \text{ or } 198 \text{ psi}$$

Problem 5 Calculate the engine pressure for pumper 2 in Figure 10-13. Pumper 2 is supplying 1,000 gpm through parallel lines. One line is 600 feet of 3-inch hose, and the second line is 400 feet of 3-inch hose and 200 feet of 2½-inch hose.

ANSWER This problem is really fairly simple. We do, however, need to break it down into two parts. The first part will be finding the friction loss for 400 feet of parallel 3-inch lines and the second part will be finding the friction loss for 200 feet of parallel lines, one 3-inch and the other 2½-inch.

The friction loss for 400 feet of parallel 3-inch lines is: each line is flowing 500 gpm.

$$FL\,100 = Cf \times 2Q^2$$
$$= .4 \times 2 \times 5^2$$
$$= .4 \times 2 \times 25$$
$$= 20 \text{ psi}$$

$$FL = FL\,100 \times L$$
$$= 20 \times 4$$
$$= 80 \text{ psi}$$

Now find the friction loss for 2½-inch and 3-inch parallel lines flowing 1,000 gpm.

$$FL\,100 = Cf \times 2Q^2$$
$$= .15 \times 2 \times 10^2$$
$$= .15 \times 2 \times 100$$
$$= 30 \text{ psi}$$

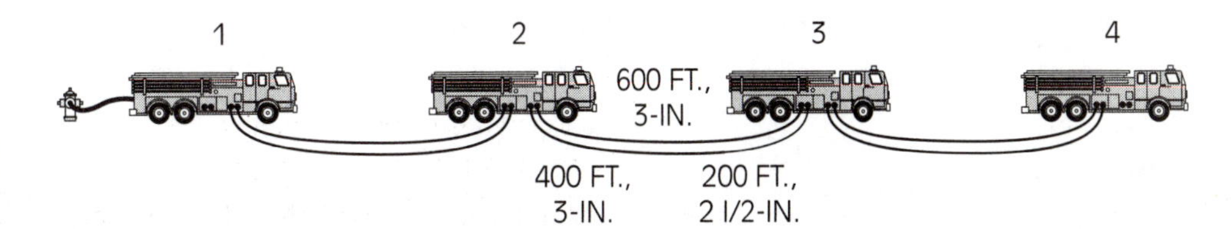

Figure 10-13 *Calculate the engine pressure for pumper 2. Note that the hose lines are of two diameters.*

$$FL = FL\,100 \times L$$
$$= 30 \times 2$$
$$= 60$$

The total friction loss is 80 + 60 = 140 psi

$$EP = 20 + FL \pm E$$
$$= 20 + 140 \pm 0$$
$$= 160 \text{ psi}$$

Summary

Calculating engine pressure is the process of putting together all the theories, formulas, and principles studied in chapters 1 through 9. To calculate engine pressure we need to first know the nozzle pressure, then add the friction loss, elevation gain or loss, and allowance for special appliance. Together these comprise the engine pressure formula $EP = NP + FL\,1 + FL\,2 \pm E + SA$.

By using the engine pressure formula and substituting the proper pressure for the appropriate term, any engine pressure can be easily calculated. To be accurate, it is important that shortcuts are not taken and rounding off to whole numbers should be left until the problem is complete.

Review Questions

1. Calculate the engine pressure for a preconnect 200-foot, 1¾-inch line with a CVFSS nozzle rated at 150 gpm at 75 psi nozzle pressure.
2. You are operating a 300-foot, 2½-inch hand line on the fourth floor of a building; if it has a 1-inch tip and is operating at 50 psi nozzle pressure, what is the engine pressure? ($C = .98$)
3. You are operating a pumper at a multiple alarm fire. After the fire has been knocked down and the decision is made to begin overhaul, your officer orders 100 feet of 1½-inch hose be used to extend a 2½-inch hand line that is 200 feet long. What will the new engine pressure be if the line is operating on the roof of a warehouse building approximately 25 feet tall? The 1½-inch line is using a CVFSS nozzle designed to flow 125 gpm at 100 psi nozzle pressure.
4. If you are operating a pumper pumping through two 3-inch lines, one 300 feet and the other 350 feet long, to a monitor nozzle on the roof of a three-story building and using a 1⅝-inch tip at 80 psi, what is your engine pressure? ($C =. 99$)
5. You are the pump operator from Review Question 4 and you were just told that the 350-foot line to the monitor is not 3-inch hose but 2½-inch hose. What is the correct engine pressure?
6. A second pumper is supplying the pumper in Review Question 5. What is the correct engine pressure if it is pumping through 600 feet of 4-inch hose?
7. An aerial ladder is in position to climb and is placed to the roof of a six-story building. There is heavy fire on the fourth floor and a hand line has been taken up the aerial ladder and tied in at the fourth floor level to protect the ladder and aid in extinguishing the fire. If the hand line is a 200 foot of 1¾-inch and is using a CVFSS nozzle operating at 180 gpm and 100 psi nozzle pressure, what is the correct engine pressure?
8. You are pumping to a 15⁄16 smooth-bore nozzle ($C = .98$) on a 1¾-inch line. The line has been taken to the roof of a five-story building and then taken down a set of stairs to the fourth floor. If the line is 400 feet long and the nozzle pressure is 45 psi, what is the correct engine pressure?

9. Pumper number 1 is at a fire supplying a 2½-inch hand line operating with a 1-inch tip ($C = .98$) at 50 psi nozzle pressure, a 1¾-inch hand line with a CVFSS nozzle flowing 150 gpm, and a wagon pipe with a 1½-inch tip ($C = .99$) at 80 psi nozzle pressure. This pumper is being supplied by pumper 2 connected to a hydrant and supplying the first pumper through parallel lines of one 3-inch and one 3½-inch hose, each 750 feet long. What is the engine pressure of pumper 2?
10. You are operating the second pumper in a three pumper relay. If pumper 3 is on a hydrant and is pumping the correct pressure, what should your intake pressure be?
11. What engine pressure is needed for a foam eductor to operate in the parking garage of an apartment building? The eductor is supplied by a 250-foot, 2½-inch line and is 15 feet below the level of the pumper. The line off the eductor is flowing 125 gpm.
12. You are pumping to a 150-foot, 2½-inch line, flowing 225 gpm through a CVFSS nozzle at 100 psi nozzle pressure. What is your engine pressure if the line is operating off a standpipe supplied by a single 2½-inch hose 100 feet long and the nozzle is on the eleventh floor?

List of Formulas

Finding total friction loss:

$$FL = FL\ 100 \times L$$

Engine pressure formula:

$$EP = NP + FL\ 1 + FL\ 2 \pm E + SA$$

Relay formula:

$$EP = 20 + FL \pm E$$

Advanced Problems in Hydraulics

Learning Objectives

Upon completion of this chapter, you should be able to:

- Apply the engine pressure formula to evolutions involving siamesed lines.
- Apply the engine pressure formula to evolutions involving wyed lines.
- Solve maximum lay problems.
- Solve maximum flow problems.
- Calculate actual nozzle pressure, flow, and friction loss when lines are overpumped or underpumped.

INTRODUCTION

In Chapter 10 we learned how to use the engine pressure formula for basic calculations. Each element of the engine pressure formula was explained, in turn, as the full formula was developed. These examples represent the most common evolutions for the engine pressure formula.

In this chapter we explore evolutions of the engine pressure formula that are a bit more complicated or not very common. Some of the evolutions in this chapter have more practical application for preplanning than actual fireground operations, but all the evolutions discussed have important application to the study of hydraulics.

As in the previous chapter, when solving for engine pressure in this chapter, use the entire engine pressure formula. Some elements will not be needed but they should never be left out of the formula.

SIAMESED LINES

Siamesed line evolutions are very similar to parallel line evolutions. The difference is that when we define an evolution as siamesed lines, we have two or more pumpers pumping into the same device. That device can be anything from an actual siamese to another pumper.

Rule 1 for parallel lines of equal diameter applies to siamesed lines as well. "Each line into the device will be assumed to be carrying an equal share of the water." The size of the line being used is irrelevant. For example, if two pumpers are pumping into a monitor nozzle flowing 600 gpm and one is using 2½-inch line and the other is using 3-inch line, they will both calculate engine pressure for 300 gpm.

Example 11-1 Calculate the engine pressure for pumper 1 and pumper 2 in Figure 11-1. The monitor nozzle is flowing 600 gpm at 80 psi (1½-inch tip). Pumper 1 is pumping through 200 feet of 3-inch hose and pumper 2 is pumping through 250 feet of 2½-inch hose.

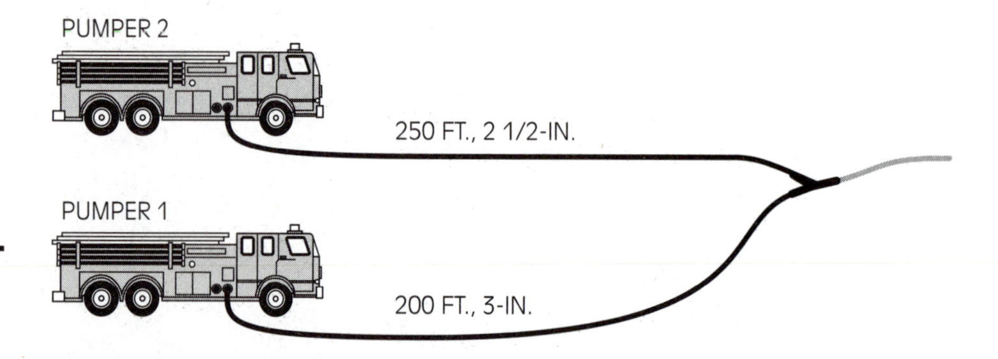

Figure 11-1 *Unequal diameter siamesed lines.*

ANSWER First, find the friction loss for 300 gpm in 3-inch hose.

$$FL\,100 = Cf \times 2Q^2$$
$$= .4 \times 2 \times 3^2$$
$$= .4 \times 2 \times 9$$
$$= 7.2 \text{ psi}$$

Total friction loss is then

$$FL = FL\,100 \times L$$
$$= 7.2 \times 2$$
$$= 15 \text{ psi}$$

Now find the engine pressure for pumper 1.

$$EP = NP + FL\,1 + FL\,2 \pm E + SA$$
$$= 80 + 15 + 0 \pm 0 + 20$$
$$= 115 \text{ psi}$$

We already know from Example 10-7 in Chapter 10 that a flow of 300 gpm in $2\frac{1}{2}$-inch hose has a friction loss per 100 feet of 18 psi.

Total friction loss for the $2\frac{1}{2}$-inch hose is:

$$FL = FL\,100 \times L$$
$$= 18 \times 2.5$$
$$= 45$$

Now find the engine pressure for pumper 2.

$$EP = NP + FL\,1 + FL\,2 \pm E + SA$$
$$= 80 + 45 + 0 \pm 0 + 20$$
$$= 145 \text{ psi}$$

In Example 11-1, both pumpers are pumping the exact same amount of water to the same device, a monitor nozzle. However, because they are each using a different size hose, their engine pressure is different.

This same principle can be carried further. If three pumpers were each pumping to another pumper or a three-way siamese, each would calculate engine pressure for one-third of the total water required.

Example 11-2 In Figure 11-2, pumper 4 is pumping to a combination of devices/lines that are using a total of 900 gpm. Calculate the engine pressure for pumper 1 if it has a 300-foot, $2\frac{1}{2}$-inch line supplying pumper 4. Calculate the engine pressure for pumper 2 if it has a 350-foot, 3-inch line supplying pumper 4 and is located 20 feet above pumper 4. Finally calculate the engine pressure for pumper 3 if it has a 300-foot, 3-inch line supplying pumper 4.

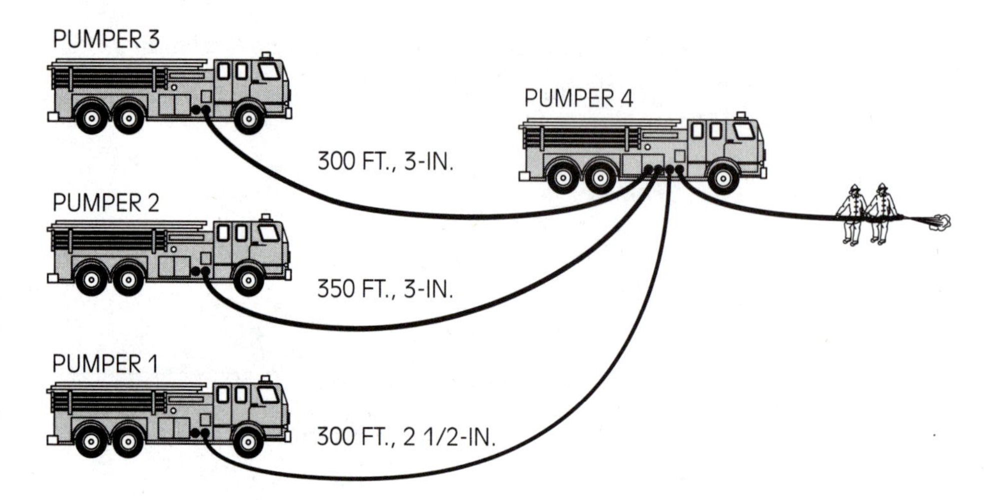

Figure 11-2
Calculate engine pressure for pumper 1, pumper 2, and pumper 3.

ANSWER Because there are three lines coming into pumper 4, each line must flow one-third of the total water, or 300 gpm.

First, find the engine pressure for pumper 1. We already know from our previous example that the friction loss for 300 gpm in $2\frac{1}{2}$-inch hose is 18 psi per 100 feet.

$$FL = FL\,100 \times L$$
$$= 18 \times 3$$
$$= 54$$

Engine pressure for pumper 1 is:

$$EP = 20 + FL \pm E$$
$$= 20 + 54 \pm 0$$
$$= 74 \text{ psi}$$

Did you remember to use the relay formula because each of these pumpers is pumping to another pumper?

Now find the friction loss and then engine pressure for pumper 2.

The friction loss for 300 gpm in 3-inch hose is 7.2 from Example 11-1.

$$FL = FL\,100 \times L$$
$$= 7.2 \times 3.5$$
$$= 25.2 \text{ psi}$$

Do not forget to account for the elevation.

$$P = .433 \times H$$
$$= .433 \times 20$$
$$= 8.66 \text{ psi}$$

$$EP = 20 + FL \pm E$$
$$= 20 + 25.2 - 8.66$$ (Remember, because pumper 2 is higher than pumper 4, you must subtract the elevation gain.)
$$= 36.54 \text{ or } 37 \text{ psi}$$

Finally, calculate the engine pressure for pumper 3. We already know the friction loss per 100 feet.

$$FL = FL\,100 \times L$$
$$= 7.2 \times 3$$
$$= 21.6 \text{ psi}$$

$$EP = 20 + FL \pm E$$
$$= 20 + 21.6 \pm 0$$
$$= 41.6 \text{ or } 42 \text{ psi}$$

One last evolution that should be considered where siamesed lines are used is where a single pumper is pumping more than one line to the device. Such an evolution is illustrated in Figure 11-3. In Figure 11-3, pumper 1 is pumping to a monitor nozzle through 350 feet of 3-inch hose, and pumper 2 is pumping through 450 foot parallel lines of one 2½-inch hose and one 3-inch hose.

Example 11-3 Calculate the engine pressure for both pumpers assuming the monitor nozzle is flowing 900 gpm with a fog nozzle.

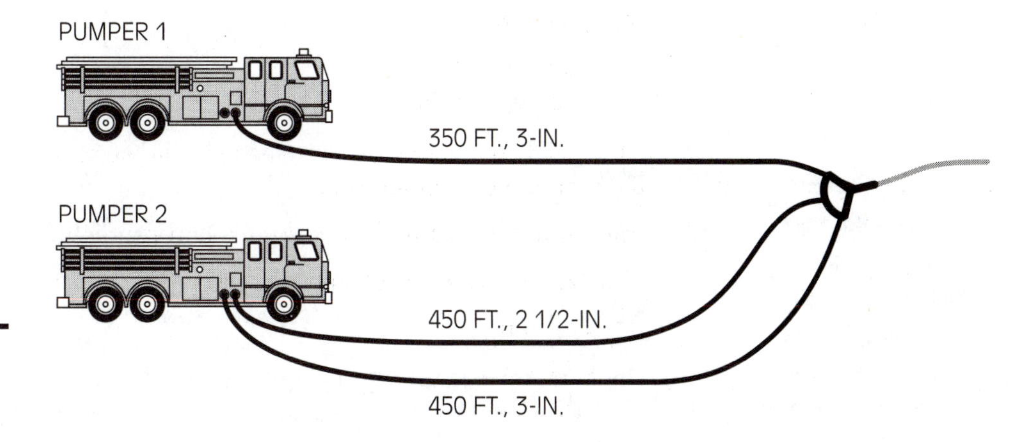

Figure 11-3 *Calculate engine pressure.*

ANSWER Recall the rule that says that each line must handle a proportional amount of water. In this instance the rule still applies. Each line is expected to carry 300 gpm, so pumper 1 will be supplying 300 gpm and pumper 2 will be supplying 600 gpm.

We already know the friction loss for 300 gpm in 3-inch hose is 7.2 psi per 100 feet.

$$FL = FL\ 100 \times L$$
$$= 7.2 \times 3.5$$
$$= 25.2$$

Now calculate the engine pressure for pumper 1.

$$EP = NP + FL\ 1 + FL\ 2 \pm E + SA$$
$$= 100 + 25.2 + 0 \pm 0 + 20$$
$$= 145.2 \text{ or } 145 \text{ psi}$$

Now solve for the engine pressure of pumper 2 just like any parallel lay problem. First we need to find the friction loss for 100 feet of 2½- and 3-inch parallel lines.

$$FL\ 100 = Cf \times 2Q^2$$
$$= .15 \times 2 \times 6^2$$
$$= .15 \times 2 \times 36$$
$$= 10.8 \text{ psi}$$

$$FL = FL\ 100 \times L$$
$$= 10.8 \times 4.5$$
$$= 48.6 \text{ psi}$$

The engine pressure is:

$$EP = NP + FL\ 1 + FL\ 2 \pm E + SA$$
$$= 100 + 48.6 + 0 \pm 0 + 20$$
$$= 168.6 \text{ or } 169 \text{ psi}$$

■ Note

When calculating engine pressure for wyed lines, include the total gpm when calculating the friction loss to the wye.

WYED LINES

Calculating engine pressure for wyed lines is fairly straightforward, however, two rules simplify the process.

Rule 1. *When calculating engine pressure for wyed lines, include the total gpm when calculating the friction loss to the wye.* In Figure 11-4 a single pumper is supplying both a 1¾-inch line flowing 150 gpm and a 2½-inch line flowing 211 gpm, through 200 feet of 3-inch hose. When

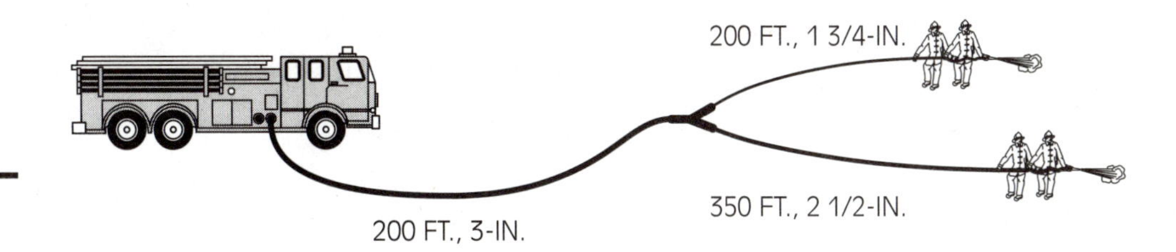

Figure 11-4 *Wyed line evolution.*

calculating the friction loss for the 200 feet of 3-inch hose, the total 361 gpm must be used.

■ Note
When calculating the engine pressure for evolutions with wyed lines, if one line requires a higher pressure, the higher pressure will be the engine pressure.

Rule 2. *When calculating the engine pressure for evolutions with wyed lines, if one line requires a higher pressure, the higher pressure will be the engine pressure.* When calculating engine pressure for wyed lines, separate calculations must be performed for each line that extends from the wye. A line that requires a higher pressure determines the correct engine pressure.

Example 11-4 Calculate the correct engine pressure for the evolution in Figure 11-4 if the 1¾-inch line is 200 feet long using a CVFSS at 75 psi nozzle pressure, and the 2½-inch line is 350 feet long with a 50 psi nozzle pressure.

ANSWER As already stated, we need to calculate friction loss in the 3-inch hose for a total of 361 gpm. The friction loss for the 3-inch hose will be *FL* 1.

$$FL\ 100 = Cf \times 2Q^2$$
$$= .4 \times 2 \times 3.61^2$$
$$= .4 \times 2 \times 13.03$$
$$= 10.42 \text{ psi}$$

$$FL = FL\ 100 \times L$$
$$= 10.42 \times 2$$
$= 20.84$ psi. This is *FL* 1.

We have previously calculated the friction loss for 150 gpm in 1¾-inch hose to be 34.92 psi per 100 feet.

$$FL = FL\ 100 \times L$$
$$= 34.92 \times 2$$
$= 69.84$ psi. This is *FL* 2 for the 1¾-inch hose side of the wye.

$$EP = NP + FL\ 1 + FL\ 2 \pm E + SA$$
$$= 75 + 20.84 + 69.84 \pm 0 + 0$$
$= 165.68$ or 166 psi. This is the correct engine pressure for the 1¾-inch portion of the hose line.

FL 1 is the same as just given for calculating the needed engine pressure for the 2½-inch hose. Our next step is to calculate the friction loss for the 2½-inch hose.

$$FL\ 100 = Cf \times 2Q^2$$
$$= 1 \times 2 \times 2.11^2$$
$$= 1 \times 2 \times 4.45$$
$$= 8.9 \text{ psi}$$

$$FL = FL\ 100 \times L$$
$$= 8.9 \times 3.5$$

= 31.15 psi. This is *FL* 2 for the 2½-inch hose side of the wye.

$$EP = NP + FL\ 1 + FL\ 2 \pm E + SA$$
$$= 50 + 20.84 + 31.15 \pm 0 + 0$$

= 101.99 or 102 psi. This is the correct engine pressure needed for the 2½-inch portion of the hose line.

Since the 1¾-inch line requires the highest pressure, 166 psi is the pump pressure.

In situations where the pressures vary as illustrated here, the higher pressure determines the actual engine pressure. In these situations, it is common practice to gate back at the wye to the line requiring the lower pressure. Without gauges, gating back is a hit-and-miss proposition at best, but with a little practice it can be done sufficiently well to prevent excessive overpressurization that causes poor stream pattern.

Example 11-5 What is the correct engine pressure for the evolution in Figure 11-5? The 1½-inch line is 150 feet long, flowing 100 gpm at 100 psi nozzle pressure. The 1¾-inch is 200 feet long, flowing 150 gpm at 75 psi nozzle pressure. There is 400 feet of 2½-inch hose between the pumper and the wye.

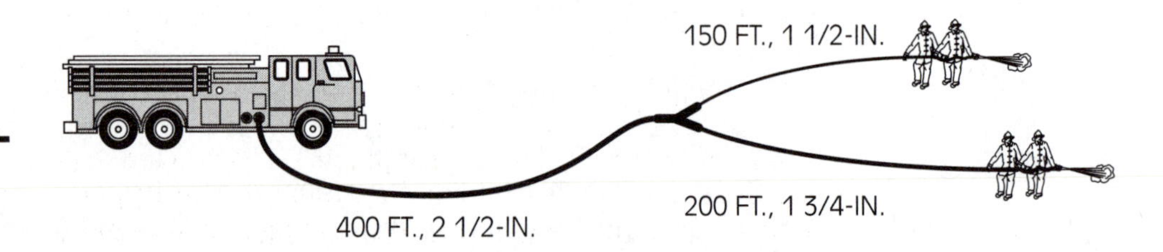

Figure 11-5 *Calculate the engine pressure.*

ANSWER We already know that 1½-inch hose flowing 100 gpm has a friction loss of 24 psi per 100 feet and 1¾-inch hose flowing 150 gpm has a friction loss of 34.92 psi per 100 feet. Begin by finding the friction loss per 100 feet in 2½-inch hose flowing 250 gpm.

$$FL\ 100 = Cf \times 2Q^2$$
$$= 1 \times 2 \times 2.5^2$$
$$= 1 \times 2 \times 6.25$$
$$= 12.5 \text{ psi}$$

$$FL = FL\ 100 \times L$$
$$= 12.5 \times 4$$
$= 50$ psi. This is *FL* 1.

Now calculate the total friction loss and the correct engine pressure for the 1½-inch line.

$$FL = FL\ 100 \times L$$
$$= 24 \times 1.5$$
$= 36$ psi. This is *FL* 2 for the 1½-inch hose side of the wye.

$$EP = NP + FL\ 1 + FL\ 2 \pm E + SA$$
$$= 100 + 50 + 36 \pm 0 + 0$$
$= 186$ psi. This is the engine pressure for the 1½-inch line.

Now calculate *FL* 2 and the engine pressure for the 1¾-inch line.

$$FL = FL\ 100 \times L$$
$$= 34.92 \times 2$$
$= 69.84$ psi. This is *FL* 2 for the 1¾-inch hose side of the wye.

$$EP = NP + FL\ 1 + FL\ 2 \pm E + SA$$
$$= 75 + 50 + 69.84 \pm 0 + 0$$
$$= 194.84 \text{ or } 195 \text{ psi}$$

The correct engine pressure for Example 11-5 is 195 psi. In actual practice, with pressures this close it would be impossible to attempt to gate back the pressure to the 1½-inch line.

A practical application of this principle would be what some departments refer to as a *skid load*. A skid load is two 1½-inch or 1¾-inch hand lines attached to a wye attached to 2½-inch supply line. The skid load can be advanced up a long drive to a house or left in front of a building while the pumper goes to the hydrant. When the supply line is charged one or both of the hand lines can be used.

MAXIMUM LAY CALCULATIONS

■ Note
The purpose of maximum lay problems is to determine how long a hose line we can lay and still give us a specified amount of water.

The purpose of maximum lay problems is to determine how long a hose line we can lay and still give us a specified amount of water. This information can be used for preplanning specific target hazards, water supply training, or calculating the benefits of large diameter hose (LDH), which is hose with a diameter of 4 inches or more.

To do maximum lay problems we need to know three pieces of information: (1) the desired quantity of water, (2) the hose size available, and (3) the pressure available to supply the water in the first place. The pressure can be either hydrant pressure or pressure from a pumper attached to the hydrant.

To determine the maximum lay, follow the following procedure:

1. Determine the gpm needed.
2. Determine the size of hose available to get the water from the source to the scene.
3. Calculate the friction loss per 100 feet in the available hose for the gpm needed.
4. Subtract 20 psi from the maximum pressure available. This amount is intake pressure for the pumper at the scene.
5. Divide the friction loss from step 3 into the pressure remaining after 20 psi has been subtracted. The answer is the maximum lay in hundreds.
6. Finally, convert the maximum lay, in hundreds, into feet by multiplying by 100 and rounding off to the last 50 increment. For example, if the lay works out to 3.75 in hundreds, it is 375 feet. However, there are no 25-foot sections of hose, so the answer would be rounded down to the next 50-foot increment, or 350 feet. Some LDH only comes in 100-foot sections, so you would have to round it down to the next 100 foot increment.

Example 11-6 If it is necessary to deliver a minimum flow of 700 gpm, what is the maximum lay of 4-inch hose that can do this? The hose is directly connected to a hydrant that has a static pressure of 60 psi.

ANSWER We know the gpm and the size hose, so the next step is to determine the friction loss per 100 feet. The conversion factor for 4-inch hose is .1.

$$
\begin{aligned}
FL\,100 &= Cf \times 2Q^2 \\
&= .1 \times 2 \times 7^2 \\
&= .1 \times 2 \times 49 \\
&= 9.8 \text{ psi}
\end{aligned}
$$

Now subtract 20 psi from the hydrant pressure to determine how much pressure is available for friction loss, then divide by 9.8 psi.

$60 - 20 = 40 \div 9.8 = 4.08$ or 408 feet. This would be rounded down to a maximum lay of 400 feet.

In Example 11-6 we have found that even with 4-inch hose, if we need to deliver 700 gpm we can only lay hose a maximum of 400 feet. If we had to go further than that, we would have to do it in a relay. If a relay were to be set up from the first pumper, the pumpers forming the relay would have a maximum pressure much higher than the hydrant pressure. Two things would determine their top pressure: (1) the capacity of the pump and (2) some LDH has maximum pressure limits.

The capacity of the pump is important because it determines the maximum discharge pressure. Recall from Table 7-1, as pump capacity increases the maximum pressure possible decreases. For instance, if we were to use a 1,250 gpm pump in Example 11-6 and it were necessary to relay water to a second pumper, the 1,250 gpm pump would be limited to a maximum discharge pressure of 200 psi when delivering 700 gpm. We determine this by calculating that 700 gpm is 56 percent of the capacity of a 1,250-gpm pump. If we do not exceed 50 percent of the capacity of the pump, we can pump at a maximum of 250 psi pressure. But because we need to pump at 56 percent of the capacity of the pump, we have to lower our maximum pressure. Because we can pump up to 70 percent of the capacity of the pump at 200 psi, this becomes the maximum discharge pressure.

Another restriction we need to be aware of is that LDH often has pressure limitations. For instance, it is not uncommon to find that LDH is limited to a maximum pressure of 185 psi. In fact, unless otherwise known, it should always be assumed that a limitation exists. Before doing maximum lay problems in real life situations, determine if a maximum pressure exists and what it is.

Example 11-7 If we were to establish a relay off the hydrant in Example 11-6, and the first pumper had a maximum capacity of 1,250 gpm, how far could the second pumper be from the first (see Figure 11-6)?

ANSWER All the important facts are already known. The maximum pressure as previously explained is 185 psi because the hose is LDH, and the friction loss for the 700 gpm in 4-inch hose is 9.8 psi.

Since this hose is limited to 185 psi, pressure available for friction loss is:

$185 - 20 = 165$

The maximum lay is:

$165 \div 9.8 = 16.84$ or 1,684 feet. This would be rounded down to a maximum of 1,600 feet. Remember LDH comes in 100 foot sections. (You may carry a single 50-foot section of LDH on each pumper to better complete evolutions such as this one)

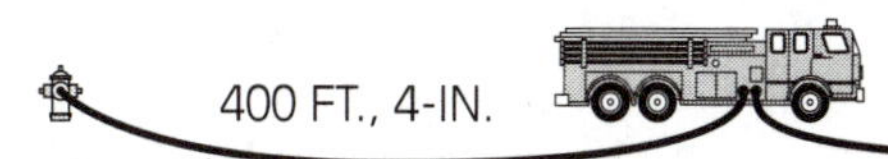

Figure 11-6 *Find the length of hose between the two pumpers.*

MAXIMUM FLOW CALCULATIONS

■ Note
Maximum flow calculations allow us to calculate the maximum amount of water we can deliver under specified hose layouts.

Maximum flow calculations allow us to calculate the maximum amount of water we can deliver under specified hose layouts. By using maximum flow calculations, we can calculate the advantage of using larger diameter supply line, or how to get the most out of a water supply by placing a pumper on the hydrant.

To do the maximum flow calculations we need to know the following:

- What size supply line is being used?
- How long is the hose lay?
- What is the maximum pressure available at the source?
- If there is a pumper at the source, what is its capacity?
- If the source is a hydrant, what is its capacity?

With these five pieces of information, we can calculate the maximum amount of water we can deliver with any specified hose lay. Just as with the maximum lay calculations, maximum flow calculations can be useful for preplanning of target hazards and can be used to calculate maximum flows at fire scenes.

In order to calculate the maximum flow, follow the following procedure:

1. Determine the length and size of hose.
2. Determine the maximum discharge pressure at the source. With a line directly connected to a hydrant this is the hydrant static pressure. If a pumper is at the source, either at draft or connected to a hydrant, it may require some trial-and-error calculations to determine the maximum pressure. With practice, you will be able to get a feel for the correct pressure without so much trial and error.
3. Subtract 20 psi from the maximum pressure to find the amount of pressure available for friction loss.
4. Divide the amount of pressure available for friction loss by the amount of hose in hundreds, giving the friction loss available per 100 feet.
5. From the available friction loss, calculate the gpm.

Example 11-8 In Figure 11-7, Pumper 1 (rated at 1,000 gpm) is attached to a hydrant pumping through two 3-inch lines to pumper 2. If the 3-inch lines are 600 feet long, how much water can pumper 1 supply to pumper 2?

ANSWER We were given both the diameter and length of the lines. The first step is to determine the maximum discharge pressure of pumper 1. If pumper 1 is rated at 1,000 gpm at 150 psi, then at 200 psi it can only pump 700 gpm, and at 250 psi it can only pump 500 gpm. It may be necessary to start at 150 psi and calculate the flow at each pressure/flow range until you find one that works. In this particular problem, unless you have a conversion factor for dual 3-inch hose lines, you will actually calculate the gpm for one line. To determine the total gpm you must double the quantity.

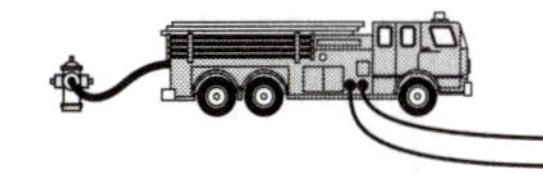
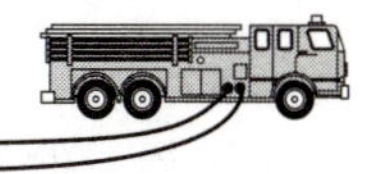

Figure 11-7 *Calculate maximum flow.*

If we assume a maximum pressure of 150 psi the problem would be:

150 – 20 = 130 psi maximum pressure available for friction loss.

130 ÷ 6 = 21.66 psi is the *FL* 100

To find the gpm from the *FL* 100:

$Q = \sqrt{FL\ 100/Cf \times 2}$

$= \sqrt{21.66/.4 \times 2}$

$= \sqrt{21.66/.8}$

$= \sqrt{27.08}$

= 5.20 or 520 gpm. For both lines the total capacity will be 1,040 gpm.

Even though the pumper is only rated at 1,000 gpm, it is so close we will accept this answer. If we were to go to a maximum pressure of 200 psi, the friction loss available per 100 feet would be much higher. This would be associated with a flow much higher than the 700 gpm possible at 200 psi.

In Example 11-8 the correct answer is 1,040 gpm at a pressure of 150 psi. At 150 psi the pumper can pump its maximum capacity. Even with a maximum rated capacity of 1,000 gpm, the extra 40 gpm is not a concern for two reasons. First, it is possible to simply calculate friction loss for a maximum of only 1,000 gpm and adjust our discharge pressure accordingly. Second, from a hydrant with a residual pressure above 20 psi it is probable that the 1,000-gpm-rated pumper can actually pump the 1,040 gpm. Remember the rating of a pumper indicates its performance at draft.

Example 11-9 Calculate the maximum gpm possible if pumper 1 is attached to a hydrant and is pumping through 1,200 feet of 4-inch LDH to pumper 2. Pumper 1 is rated at 1,250 gpm. Remember, unless otherwise stated, LDH is limited to a maximum pressure of 185 psi.

ANSWER We begin by calculating the maximum flow at a discharge pressure of 150 psi.

150 – 20 = 130 psi

130 ÷ 12 = 10.83 psi is *FL* 100

$$Q = \sqrt{FL\ 100/Cf \times 2}$$
$$= \sqrt{10.83/.1 \times 2}$$
$$= \sqrt{10.83/.2}$$
$$= \sqrt{54.15}$$
$$= 7.36 \text{ or } 736 \text{ gpm}$$

This amount is only 58 percent of the capacity of the pump. We can obviously pump more water than this. We need to recalculate using a maximum pressure of 185 psi.

$185 - 20 = 165$ psi

$165 \div 12 = 13.75$ psi is *FL* 100

$$Q = \sqrt{FL\ 100/Cf \times 2}$$
$$= \sqrt{13.75/.1 \times 2}$$
$$= \sqrt{13.75/.2}$$
$$= \sqrt{68.75}$$

$Q = 8.29$ or 829 gpm. Since 829 gpm is not higher than 70 percent of the rated capacity of a 1,250 pumper (875 gpm), this answer is the maximum flow we can get.

In Example 11-9 we needed to calculate the maximum flow twice to find the correct answer. The first time the gpm was way too low. By recalculating at the higher discharge pressure of 185 psi we were able to use a higher *FL* 100, which had a flow that is almost exactly 70 percent of the pump capacity. This amount fits perfectly with the guidelines in Table 7-1. If we were now to try the problem again using 250 psi, assuming no pressure restriction, we would find the *FL* 100 is 19.17 psi, which works out to a gpm of 992 gpm. Since at 250 psi the pump can only flow 625 gpm, 992 gpm obviously will not work.

FINDING CORRECT NOZZLE PRESSURE, GPM, AND FRICTION LOSS

There are times when we discover we have been pumping an incorrect engine pressure to a hand line for any number of reasons. Since we already know how important it is to use the correct gpm to calculate friction loss, it is easy to understand how an incorrect engine pressure can change the gpm. As part of a thorough understanding of hydraulics, we should be able to determine just how much water is flowing when an incorrect engine pressure is used. From the gpm we can then calculate what the friction loss should have been.

To calculate the actual flow, we first need to subtract for both elevation and special appliances. The excess or shortage of pressure is distributed proportionally between nozzle pressure and friction loss. Since allowance for both elevation and special appliances are constants, they cannot change and should not be a part of the calculation.

The first step in calculating the correct flow is finding out how much nozzle pressure we actually had. Begin by establishing a ratio of the correct nozzle pressure and the correct engine pressure, which is also the ratio of the incorrect nozzle pressure and the incorrect engine pressure. Mathematically this ratio is written *CNP*/*CEP* = *INP*/*IEP*, where *CNP* is the correct nozzle pressure, *CEP* is the correct engine pressure, *INP* is the incorrect nozzle pressure, and *IEP* is the incorrect engine pressure. Since this formula represents a proportion, values *CEP* and *INP* are referred to as *means* and values *CNP* and *IEP* are referred to as *extremes*.

$$\frac{CNP}{CEP} = \frac{INP}{IEP}$$

To work this problem we need to know *any* three of the values. Generally, we will know the correct nozzle pressure and can then determine the friction loss for the calculated flow, from which we can calculate the correct engine pressure. The next step is to determine the incorrect engine pressure. This is easy; it will be what the discharge gauge indicated. Now fill in the formula, *CNP*/*CEP* = *INP*/*IEP* with the appropriate values for *CNP*, *CEP*, and *IEP* and solve for *INP*. We solve for *INP* by multiplying the extremes and then dividing by the known means.

Example 11-10 In Problem 1 in Chapter 10 it was determined that a 250 foot, 2½-inch line with a 1-inch tip at 45 psi will flow 195 gpm, giving us a friction loss of 7.6 psi per hundred feet, or 19 psi for the 250 feet. The engine pressure was 64 psi. In the heat of the moment the pump operator allowed the engine pressure to get up to 100 psi. What was the actual nozzle pressure at this engine pressure?

ANSWER $CNP = 45$, $CEP = 64$, $INP = ?$, $IEP = 100$

$$CNP/CEP = INP/IEP$$

$$45/64 = INP/100$$

Multiply the extremes and then divide by the known means.

$45 \times 100/64 = 70.31$ psi. This is the actual nozzle pressure, or *INP*, when the line has the incorrect engine pressure of 100 psi.

Now let us prove that 70.31 psi is the correct nozzle pressure.

The gpm for a 1-inch tip at 70.31 psi is:

$$\begin{aligned} GPM &= 29.77 \times D^2 \times C \times \sqrt{P} \\ &= 29.77 \times 1^2 \times .98 \times \sqrt{70.31} \\ &= 29.77 \times 1 \times .98 \times 8.39 \\ &= 244.78 \text{ or } 245 \end{aligned}$$

Now calculate the friction loss for 250 feet of 2½-inch hose flowing 250 gpm.

$FL\ 100 = Cf \times 2Q^2$

$= 1 \times 2 \times 2.45^2$

$= 1 \times 2 \times 6$

$= 12$ psi

$FL = FL\ 100 \times L$

$= 12 \times 2.5$

$= 30$

Finally we need to verify the engine pressure.

$EP = NP + FL\ 1 + FL\ 2 \pm E + SA$

$= 70.31 + 30 + 0 \pm 0 + 0$

$= 100.31$ or 100 psi

We can accept that the answer is off by .31 psi. Considering that the incorrect engine pressure of 100 psi is 56 percent higher than the correct engine pressure of 64 psi, this is remarkably accurate. We have just verified the process.

Example 11-11 In Problem 2 of Chapter 10, the engine pressure was calculated to be 163 psi. Find out what the new nozzle pressure would be and verify the final engine pressure if the line were to be underpumped at 150 psi.

ANSWER This problem presents a twist because it has a fog nozzle. We have to start by finding an equivalent diameter for the nozzle. We do this by using the formula

$D = \sqrt{gpm/(29.77 \times C \times \sqrt{P})}$ from Chapter 5.

$= \sqrt{180/(29.77 \times .98 \times \sqrt{75})}$

$= \sqrt{180/(29.77 \times .98 \times 8.66)}$

$= \sqrt{180/252.65}$

$= \sqrt{.712}$

$= .844$ inches

Next calculate the incorrect nozzle pressure.

$CNP/CEP = INP/IEP$

$75/163 = INP/150$

$INP = 75 \times 150/163 = 69$ psi

We can verify the process; first find the gpm at 69 psi nozzle pressure.

$$gpm = 29.77 \times D^2 \times C \times \sqrt{P}$$
$$= 29.77 \times .844^2 \times .98 \times \sqrt{69}$$
$$= 29.77 \times .712 \times .98 \times 8.3$$
$$= 172.41 \text{ or } 172$$

Calculate the friction loss for the 1¾-inch hose.

$$FL\ 100 = Cf \times 2Q^2$$
$$= 7.76 \times 2 \times 1.72^2$$
$$= 7.76 \times 2 \times 2.96$$
$$= 45.94 \text{ psi}$$

$$FL = FL\ 100 \times L$$
$$= 45.94 \times 1.5$$
$$= 68.91 \text{ psi}$$

Now calculate the friction loss for the 2½-inch hose.

$$FL\ 100 = Cf \times 2Q^2$$
$$= 1 \times 2 \times 1.72^2$$
$$= 1 \times 2 \times 2.96$$
$$= 5.92 \text{ psi}$$

$$FL = FL\ 100 \times L$$
$$= 5.92 \times 2$$
$$= 11.84 \text{ psi}$$

Finally, verify the engine pressure.

$$EP = NP + FL\ 1 + FL\ 2 \pm E + SA$$
$$= 69 + 68.91 + 11.84 \pm 0 + 0$$

$= 149.75$ or 150 psi. Once again the process works.

APPLICATION ACTIVITIES

Put your knowledge of hydraulics to work by solving the following problems.

Problem 1 In Figure 11-8, the pumper is pumping through 350 feet of 2½-inch hose to a wye. From one discharge of the wye there is a 150-foot, 1½-inch hose with a CVFSS nozzle flowing 125 gpm at 100 psi nozzle pressure. From the other discharge is a 200-foot 1¾-inch hose line using a 15/16-inch tip at 50 psi nozzle pressure. Calculate the correct engine pressure for the pumper.

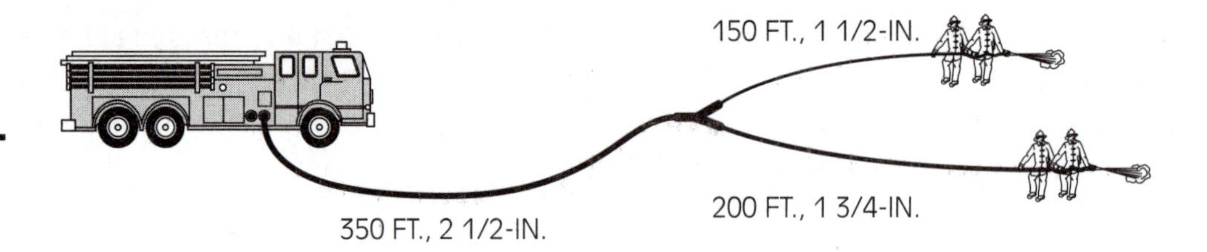

Figure 11-8 *What is the correct engine pressure?*

ANSWER We first need to determine how much total water is flowing. We are given the gpm of the 1½-inch hose line as 125 gpm and from Chapter 5, Problem 2 we calculated the gpm for the 15⁄16-inch tip at 50 psi nozzle pressure to be 181 gpm.

$125 + 181 = 306$ gpm total

Calculate the friction loss for the 2½-inch hose.

$$
\begin{aligned}
FL\,100 &= Cf \times 2Q^2 \\
&= 1 \times 2 \times 3.06^2 \\
&= 1 \times 2 \times 9.36 \\
&= 18.72 \text{ psi}
\end{aligned}
$$

$$
\begin{aligned}
FL &= FL\,100 \times L \\
&= 18.72 \times 3.5 \\
&= 65.52 \text{ psi}
\end{aligned}
$$

The friction loss for the 1½-inch hose is:

$$
\begin{aligned}
FL\,100 &= Cf \times 2Q^2 \\
&= 12 \times 2 \times 1.25^2 \\
&= 12 \times 2 \times 1.56 \\
&= 37.44 \text{ psi}
\end{aligned}
$$

$$
\begin{aligned}
FL &= FL\,100 \times L \\
&= 37.44 \times 1.5 \\
&= 56.16 \text{ psi}
\end{aligned}
$$

The friction loss for the 1¾-inch hose is:

$$
\begin{aligned}
FL\,100 &= Cf \times 2Q^2 \\
&= 7.76 \times 2 \times 1.81^2 \\
&= 7.76 \times 2 \times 3.28 \\
&= 50.91 \text{ psi}
\end{aligned}
$$

$FL = FL\ 100 \times L$
$= 50.91 \times 2$
$= 101.82$ psi

Now calculate the engine pressure.

For the 1½-inch hose:

$EP = NP + FL\ 1 + FL\ 2 \pm E + SA$
$= 100 + 65.52 + 56.16 \pm 0 + 0$
$= 221.68$ or 222 psi

For the 1¾-inch hose:

$EP = NP + FL\ 1 + FL\ 2 \pm E + SA$
$= 50 + 65.52 + 101.82 \pm 0 + 0$
$= 217.34$ or 217 psi

You would pump at 222 psi

Problem 2 You are the driver on a pumper that responds to a second-alarm fire. When you arrive on the fireground you are ordered to pump to a ladder pipe that is attempting to put into service a CVFSS nozzle that will flow 1,000 gpm at 100 psi nozzle pressure. The ladder is extended to 70 feet. You lay a 250 foot, 3-inch line into the siamese at the base of the ladder. If there is another engine pumping into the siamese with a 2½-inch line, what is your engine pressure?

ANSWER The size of line the other pumper is using is of no concern to you. Because you are supplying one of two lines into the siamese, you will supply half of the water needed for the nozzle. Calculate engine pressure accordingly.

First calculate the friction loss for the 3-inch hose.

$FL\ 100 = Cf \times 2Q^2$
$= .4 \times 2 \times 5^2$
$= .4 \times 2 \times 25$
$= 20$ psi

$FL = FL\ 100 \times L$
$= 20 \times 2.5$
$= 50$ psi

Find the elevation.

$P = .433 \times H$
$= .433 \times 70$
$= 30.31$ psi. This is the elevation.

Now find the engine pressure.

$$EP = NP + FL\,1 + FL\,2 \pm E + SA$$
$$= 100 + 50 + 0 + 30.31 + 50$$
$$= 230.31 \text{ or } 230 \text{ psi}$$

Problem 3 You are laying out from a pumper on a hydrant. If you want to be able to deliver 700 gpm through a parallel hose lay, how far can you lay a line? One of the hoses is 2½-inch and the other hose is 3-inch. The pumper on the hydrant is rated at 1,000 gpm.

ANSWER In Chapter 6 we calculated the *Cf* for one 2½-inch and one 3-inch hose as .15.

$$FL\,100 = Cf \times 2Q^2$$
$$= .15 \times 2 \times 7^2$$
$$= .15 \times 2 \times 49$$
$$= 14.7 \text{ psi}$$

To deliver 70 percent of the capacity of a 1,000 gpm capacity pump, the maximum pressure is limited to 200 psi.

200 – 20/14.7 = 12.24 or 1,224 feet. This must be rounded down to 1,200 feet.

Problem 4 You have just laid out 500 feet of 4-inch hose from a hydrant. If the static pressure on the hydrant is 75 psi, what is the maximum amount of water that you can flow?

ANSWER Begin by subtracting the intake pressure needed for the pumper from the static pressure of the hydrant. 75 – 20 = 55 psi, the maximum amount of pressure you have available for friction loss.

55 ÷ 5 = 11. This is the allowable friction loss per 100 feet.

$$Q = \sqrt{FL\,100/Cf \times 2}$$
$$= \sqrt{11/.1 \times 2}$$
$$= \sqrt{11/.2}$$
$$= \sqrt{55}$$
$$= 7.42 \text{ or } 742 \text{ gpm}$$

Problem 5 You are pumping to a 2½-inch hand line with a 1⅛-inch tip. You are told that the line is 350 feet long. After you calculate the correct engine pressure and have been pumping to this line for several minutes, you are informed that the line is actually 500 feet long. During the time that you were pumping the pressure required for a 350 foot line, how much water was actually flowing?

ANSWER You must begin by calculating the engine pressure for the 350 foot line. This answer is the incorrect engine pressure. First, calculate the gpm at 50 psi nozzle pressure.

$$
\begin{aligned}
gpm &= 29.77 \times D^2 \times \sqrt{P} \\
&= 29.77 \times 1\tfrac{1}{8}^2 \times \sqrt{50} \\
&= 29.77 \times 1.27 \times 7.07 \\
&= 267.3 \text{ or } 267
\end{aligned}
$$

Now find the friction loss for this flow.

$$
\begin{aligned}
FL\ 100 &= Cf \times 2Q^2 \\
&= 1 \times 2 \times 2.67^2 \\
&= 1 \times 2 \times 7.13 \\
&= 14.26 \text{ psi}
\end{aligned}
$$

$$
\begin{aligned}
FL &= FL\ 100 \times L \\
&= 14.26 \times 3.5 \\
&= 49.91 \text{ psi}
\end{aligned}
$$

The engine pressure is then:

$$
\begin{aligned}
EP &= NP + FL\ 1 + FL\ 2 \pm E + SA \\
&= 50 + 49.91 + 0 \pm 0 + 0 \\
&= 99.91 \text{ or } 100 \text{ psi}
\end{aligned}
$$

Now calculate what the correct engine pressure should have been.

$$
\begin{aligned}
FL &= FL\ 100 \times L \\
&= 14.26 \times 5 \\
&= 71.3 \text{ psi}
\end{aligned}
$$

$$
\begin{aligned}
EP &= NP + FL\ 1 + FL\ 2 \pm E + SA \\
&= 50 + 71.3 + 0 \pm 0 + 0 \\
&= 121.2 \text{ or } 121 \text{ psi}
\end{aligned}
$$

Find the incorrect nozzle pressure.

$$
\begin{aligned}
CNP/CEP &= INP/IEP \\
50/121 &= INP/100 \\
50 \times 100/121 &= 41.32 \text{ psi}
\end{aligned}
$$

Now calculate the gpm for a 1⅛-inch tip with 41.32 psi nozzle pressure.

$$gpm = 29.77 \times D^2 \times \sqrt{P}$$
$$= 29.77 \times 1\tfrac{1}{8}^2 \times \sqrt{41.32}$$
$$= 29.77 \times 1.27 \times 6.43$$
$$= 243.11 \text{ or } 243$$

Summary

By now it should be evident that hydraulics is more than just calculating engine pressure. Some of the more unusual applications in need of hydraulic calculations have been presented in this chapter. Calculating engine pressure when pumping to siamesed lines and wyed lines is not that unusual, but to get it right you have to understand the rules that apply in each case. If you do not know the rules, you can never get the correct engine pressure.

Maximum lay and maximum flow problems represent a different kind of calculation. They can be used to determine how much water, or how far you can deliver it, with current hose, or to assist in making a decision to change the amount or size of hose you carry. Maximum flow calculation can also be applied to postfire critiques to determine if you used the pumping apparatus on the fireground to its best advantage.

The final calculation, determining the actual nozzle pressure, flow, and friction loss, while a necessary part of understanding hydraulics, also has a practical application. As illustrated in Problem 5, a simple error can throw off your engine pressure significantly, which illustrates the need to be as accurate as possible in order to flow the correct amount of water.

Review Questions

1. If pumper 1 is being supplied by two different pumpers, pumper 2 using a 4-inch supply line and pumper 3 using a 3-inch supply line, how much of the water is pumper 1 getting from pumper 2?
2. Two pumpers are supplying a monitor nozzle. Calculate the engine pressure for each pumper. Pumper 1 is supplying the monitor nozzle through one line of 3-inch hose 350 feet long. Pumper 2 is supplying the monitor nozzle through two lines, one 2½-inch hose 300 feet long and a second line of 3-inch hose 350 feet long. The monitor nozzle is using a 2-inch tip at 80 psi nozzle pressure. ($C = .99$)
3. You are pumping through 300 feet of 2½-inch line to a wye. From the wye there is one 1¾-inch line, 150 feet long with a ⅞-inch tip at 45 psi and a 2½-inch line 100 feet long with a 1-inch tip at 50 psi. What is the correct engine pressure?
4. A 250 foot 2½-inch line was operating off a ladder into the third floor. After the fire was knocked down, the nozzle was removed and a wye attached. The wye has two 1½-inch, 100-foot lines with CVFSS nozzles that flow 100 gpm at 100 psi nozzle pressure. What is the correct engine pressure?
5. You must deliver 1,300 gpm to a target hazard 1,500 feet from the closest hydrant. If you place a pumper on the hydrant and all the pumpers are rated at 1,500 gpm, how many pumpers will it take to relay the water? You have parallel supply lines of one 3-inch and one 3½-inch hose. Remember to include one pumper at the scene.

6. If you were to lay a 4-inch supply line from a hydrant that has a static pressure of 70 psi, what is the maximum distance you could lay and still deliver 1,200 gpm?
7. You have laid 500 feet of 4-inch supply line from a hydrant with a 60 psi static pressure. What is the maximum amount of water that you can get from the hydrant?
8. You have laid 500 feet of 4-inch supply line from a hydrant. Pumper 2, a 2,000 gpm capacity pumper, has hooked up to the hydrant and will pump to you. What is the maximum gpm that pumper 2 can supply to you?
9. As the pump operator you are pumping to a 200-foot preconnect 1¾-inch hand line with a CVFSS nozzle. You pump the correct pressure for a flow of 150 gpm at 100 psi nozzle pressure. Later you discover that a new nozzle was in use that only required 75 psi nozzle pressure. How much water were you actually flowing?
10. You are pumping to a monitor nozzle, ($C = .98$) with a 1¾-inch tip, through what you believe are parallel 2½-inch lines, 350 feet long. Later you discover that one of the lines only has 50 feet of 2½-inch hose and the other 300 feet is 3-inch. When you were pumping at the correct pressure for the parallel 2½-inch lines and assuming the correct nozzle pressure should have been 80 psi, what was the actual gpm and total friction loss?
11. For this question assume a pumper laying out from you has the same size supply line you have. How much hose can another pumper lay out from you and still count on you to be able to supply your pump's capacity? (Use knowledge of your department to solve this problem.)

List of Formulas

For finding the actual nozzle pressure (INP) when an incorrect engine pressure is used:

$$\frac{CNP}{CEP} = \frac{INP}{IEP}$$

Water Supply

Learning Objectives

Upon completion of this chapter, you should be able to:

- Identify the sources of water for firefighting.
- Identify the parts of a municipal water supply system.
- Test a fire hydrant to determine the available gpm.
- Calculate the efficiency gained by connecting a pumper to a hydrant.
- Identify the needs of a rural water supply.

INTRODUCTION

On a planet that is two-thirds covered with water, we are fortunate that water turned out to be such an ideal extinguishing agent. You would think that with such an abundance of water it would be easy enough to get water to any fire. Unfortunately, getting water to the fire is often as challenging as extinguishing the fire itself.

Where we get the water can be divided into four categories: (1) onboard water, (2) municipal water supply systems, (3) ponds, rivers, and lakes, and (4) private water supplies. Every fire department in the world must employ one, or a combination, of these sources.

As you read this chapter, remember that the water supply is the most important element in fire suppression. If the water is not there, the fire cannot be extinguished. Or if the water is there but not properly used, the fire still may not go out. Even with a limited water supply, fires can be extinguished successfully, if the water can be delivered from the source of supply to the fire in the most efficient way possible.

WATER SUPPLY REQUIREMENTS

How much water does it take to extinguish a given fire? To answer this question requires knowledge of several factors, including type of construction, distance between buildings, occupancy, and a factor for the probability of fire spread within the building. When placed in a single formula, these factors determine the NFF.

There are actually several formulas for calculating NFF depending on who is doing the calculations. For preplanning purposes almost any formula will work to give us a ballpark figure of how much water is needed to extinguish a given fire, including those given in Chapter 1. For determining how much water is needed to extinguish a specific fire for insurance rating purposes, only one method, the ISO formula, counts. The ISO is an independent organization that sets base insurance classifications for fire departments in most of the United States.

It is not the intent of this text to require the calculation of NFF beyond the formulas given in Chapter 1. Those calculations are more properly left to texts on firefighting operations. It is, however, necessary to understand the overall concept that governs how much water a municipality should have to meet its needs, including fighting fires.

MDC
the amount of water the system demands, per minute, at its peak demand time

In order to determine how much water a municipality needs to be able to meet ISO requirements, we need to know two factors: (1) the maximum daily consumption (**MDC**), that is, the amount of water the system demands, per minute, at its peak demand time, and (2) the maximum NFF. Together these tell us how many gpm a municipal system needs to be able to deliver to meet its domestic needs and fight a major fire at the same time. To determine how much water must be in storage, we need to determine how long we need the flow. The ISO wants NFFs of up to 2,500 gpm for two hours, NFFs from 2,501 up to 3,500 gpm for three hours, and

NFFs in excess of 3,500 for four hours. ISO requires a minimum NFF of 500 gpm and a maximum of 12,000 gpm. Therefore at an NFF of 500 gpm, the system should have a minimum of 60,000 gallons in storage in addition to domestic water needs.

$MDC = \frac{Pop. \times 214.5}{1440}$

You can estimate the amount of water needed to meet MDC by using 214.5* gallons per person. Simply multiply this figure by the population of the jurisdictions that are served by the water system to find the total gallons of water needed to meet the MDC. Then divide your answer by 1,440 (the number of minutes in a day) to find the flow rate. This calculation is represented by the formula $MDC = Pop. \times 214.5/1{,}440$, where *MDC* is the maximum daily consumption rate in gpm, *Pop.* is the population of the jurisdiction served by the water supply system, 214.5 is the MDC per person, and 1,440 converts the MDC into the MDC needed per minute, or the MDC rate.

$SC = (MDC + NFF) \times T$

Now we have gpm needed at the peak demand time (MDC) to meet the domestic needs of the community. Add this to the largest NFF to find the amount of water the water supply system must have in storage. This translates very neatly into the formula $SC = (MDC + NFF) \times T$, where *SC* is the storage capacity, and *MDC* and *NFF* are as previously defined and *T* is the amount of time, in minutes, that the flow is needed.

Example 12-1 A small town of 15,000 residents has a target hazard with a NFF of 2,200 gpm. What storage capacity must the system meet in order to satisfy domestic water needs and fight a fire?

ANSWER Begin by calculating the MDC.

$$MDC = Pop. \times 214.5/1440$$
$$= 15{,}000 \times 214.5/1440$$
$$= 3{,}217{,}500/1440$$
$$= 2{,}234 \text{ gpm}$$

Add this to the NFF, then multiply by the time, in minutes, that the flow is needed.

Since the NFF is under 2,500, in this case the flow must be maintained for 2 hours.

$$SC = (MDCr + NFF) \times T$$
$$= (2{,}234 + 2{,}200) \times 120$$
$$= 532{,}080 \text{ gallons}$$

---

* In his book, *Fire Suppression Rating Schedule Handbook* (Professional Loss Control Education Foundation, U.S.A., 1993), Dr. Harry Hickey states that where adequate records are unavailable, the MDC can be calculated as 150% of (or 1.5 times) the average daily consumption (ADC). The American Water Works Association (AWWA) has calculated the ADC to be 143 gallons per day per person.

If you just wanted to know the flow needed in the system to meet the minimum flow rate, all you would have to do is add the MDC and NFF. Doing so gives us the formula $MFr = MDC + NFF$, where MFr is the minimum flow rate and MDC and NFF are as defined previously. In this formula all the figures will be in gpm.

MFr = MDC + NFF

Example 12-2 Calculate the minimum flow rate to the target hazard for the hypothetical town in Example 12-1.

ANSWER

$$MFr = MDC + NFF$$
$$= 2{,}234 + 2{,}200$$
$$= 4{,}434 \text{ gpm}$$

During a major fire this system will need to be able to deliver a minimum of 4,434 gpm.

SOURCES OF WATER SUPPLY

Water supply sources can be divided into four categories: onboard water, rural water sources, private water supplies, and municipal water systems. On most fires no single source is used exclusively. As an example, in many jurisdictions with municipal water systems, the initial attack is begun with water carried on board the pumper. Once the driver has made the necessary connections with the supply line, the supply line is charged and the driver changes over. Changing over is the process of shutting off water from the onboard tank and opening the intake and admitting water from the charged supply line.

In order to utilize water supplies to their maximum advantage, it is necessary to understand their particulars. As you study the characteristics of each water source, remember that these sources are often used in conjunction with each other.

Onboard Water

Any firefighting apparatus that has an onboard pump is required to have an onboard water supply. How much water depends on the specific kind of apparatus. According to NFPA 1901, *Automotive Fire Apparatus*, pumper fire apparatus is required to be equipped with a tank that has a minimum capacity of 300 gallons. This same standard requires initial attack fire apparatus, or minipumpers, to have a minimum tank capacity of 200 gallons.

NFPA 1901

Automotive Fire Apparatus

Because only small fires can be extinguished with this amount of water, many fire departments use much larger capacity tanks. Tanks of 500 to 1,000 gallons are very common. Even these tanks have a limited extinguishing capacity.

The advantage of onboard water is that it is instantly available. The driver only needs to engage the pumps and open the valve to the tank, if it is not kept open, and water is available. Onboard water is convenient for extinguishing small fires, such as trash fires, small brush fires, even automobile fires and small storage sheds. In rural operations, if used sparingly and efficiently, it can allow a primary search or rescue while a water supply is being established. Beyond these examples, extinguishment should generally not be attempted without a secondary water source.

Where a secondary source of water in the form of a municipal water system is not present, fire departments must rely on trucking in their water supply. To do this requires water tankers. NFPA 1901 defines a water tanker as any apparatus that is *intended* to function as a mobile water supply apparatus. It further requires that it have an onboard tank of at least 1,000 gallon capacity. Considering that we commonly use water at a minimum rate of 150 gpm, 1,000 gallons will only last 6 minutes and 40 seconds. This amount is nowhere near adequate. A water tanker should be required to carry a minimum of 2,500 gallons. At 150 gpm, 2,500 gallons will last for 16 minutes and 40 seconds.

Rural Water Sources

When operating beyond the reach of a municipal water system, local fire departments need to develop alternate sources of water supply. Water tankers, mentioned previously, are the primary source of water. Because structure fires can often require large quantities of water, it may be necessary to employ several tankers to fight a single fire.

When a single tanker is not sufficient to fight a fire, multiple tankers are organized to shuttle water to the fire scene. The first tanker in often erects a portable drafting basin, a large cistern capable of holding a few thousand gallons of water, and then off-loads its water into the drafting basin. The pumper(s) on the scene being used to attack the fire then draft water from the basin. The rate of water usage on the fireground determines the frequency at which the basin must be refilled to ensure an uninterrupted water supply. Remember, at just 150 gpm flow a 2,500 gallon basin will need refilling in 16 minutes, at 300 gpm it will need refilling in 8 minutes, and so on.

Where a tanker shuttle is used to deliver water to the fireground, there must also be a water source to fill the tankers. Sources of water can be a nearby hydrant from a municipal water supply system, assuming they will allow you to take their treated water, ponds, lakes, and rivers. Regardless of source, a pumper of at least 1,000 gpm capacity must be stationed at the source to fill the tankers as they arrive. The idea is to have enough tankers of sufficient capacity that a constant flow can be maintained at the fireground even though the tankers may have to travel several miles to the water source. The further the source of water from the fire the more tankers needed to maintain a continuous flow.

Some departments employ very large tankers, some with capacities of as much as 5,000 or 6,000 gallons. These tankers are referred to as *nurse* tankers. Rural departments often employ nurse tankers by parking them at a convenient spot on the fireground and then keeping them full. The nurse tanker then pumps directly to the pumper(s) on the fireground used to attack the fire so a portable

NFPA 1231

Water Supplies for Suburban and Rural Fire Fighting

drafting basin does not need to be used. NFPA 1231, *Water Supplies for Suburban and Rural Fire Fighting* contains additional information on rural water supplies and getting the most out of the water source.

In the previous section, Water Supply, a method to determine the maximum needed water supply was demonstrated. In rural parts of the country there is no water supply; this would normally result in an ISO classification of semiprotected ISO rating of 9 if within 5 miles of the local fire station or of an unprotected ISO rating of 10 if beyond 5 miles. The ISO recognizes that many parts of this country do not have any local water supply for fighting fires and has made allowances for the local fire department to supply water to raise the insurance classification. (Normally the water supply is evaluated independent of the fire department for grading purposes.) To meet a recognized minimum water supply in areas without an adequate municipal supply system, the local fire department must be able to supply 250 gpm for a total of 2 hours, which is a minimum of 30,000 gallons of water over the 2-hour period without interruption, and it must be started within 5 minutes of the arrival of the first pumper on the scene.

Municipal Water Supply Systems

A municipal water supply system provides water for firefighting throughout the most densely populated areas of the United States. Understandably, it is the water system most often envisioned when water supply is mentioned.

A municipal water supply system, as it concerns the fire service, is composed of three elements: (1) a storage system, (2) a distribution system, and (3) fire hydrants. The storage system can be in the form of reservoirs, water tanks, or towers. Water to fill these storage elements comes from springs, rivers, lakes, and wells. Exactly which is used to fill the reservoirs and tanks depends upon availability and need. Large towns and cities often have several sources of water to meet their needs.

Pressure for the water supply system is provided by one of two methods: (1) gravity or (2) pumps. Gravity has distinct advantages over pumps for supplying pressure to the water system. Gravity never breaks down or needs routine maintenance, it never needs to be replaced, and it provides a constant and reliable pressure. Simply by placing reservoirs or water tanks on high ground, gravity can be used to provide the pressure needed for the everyday needs of residents and industry and to provide hydrant pressure. Often, where elevated water towers are used (see Figure 12-1), gravity feeds water to the residents during the day at a constant pressure and flow. At night, when demand is down, a nearby reservoir (higher than the tank) or pump is used to fill the tank again and make it ready for the next day's needs.

For some jurisdictions, gravity feed is not possible. In these jurisdictions, pumps are used to supply the pressure needed for the water supply system, which means that pumps must be running constantly, day and night, in order to maintain the needed flow and pressure around the clock. The disadvantages of

Figure 12-1 *Typical elevated water storage tank/water tower.*

pumps are that they can break down, they need routine maintenance, and can need replacing.

Some municipal water supply systems actually employ both gravity and pumps to maintain pressure in their systems. It may seem that the pumps are fighting gravity to provide pressure, but such is not the case. Large systems are often broken down into many smaller systems. Some of the smaller subsystems can be gravity fed while others, where gravity feed is not feasible, are pumped. These subsystems, regardless of how pressurized, are isolated from each other. Physically there are connections that have valves in them that are kept closed. If an unexpected high demand hits one of the subsystems, such as a major fire, the valves can be opened and water admitted from other subsystems.

Water gets from the point of storage to the point of use by a distribution system. There are three elements of a distribution system: (1) feeder mains, (2) secondary feeder mains, and (3) distribution mains. The terms used here to describe these elements of the water supply system are not universal; different systems may call them different names but they do illustrate the principle. To best describe the role of each element of the distribution system, it is easier to understand if we describe it in reverse order.

The distribution main is the part of the supply system that has the hookup to each individual home, apartment, school, business, factory, or other building. Distribution mains can be whatever size is required to meet the domestic needs of the particular area of the system. This means pipe smaller than 6 inches can be used. However, fire hydrants should never be attached to pipe less than 6 inches in diameter in residential areas and 8 inches in diameter in commercial and industrial areas.

Secondary feeder mains supply water from the feeder main to the distribution mains. They are used to effectively break down large areas of the jurisdiction into smaller grids. One secondary feeder may supply water to a residential area while another secondary feeder runs through the commercial area, feeding the distribution mains that run up and down the blocks feeding the various businesses.

Finally, the secondary feeder main is supplied by the feeder main. The feeder main carries water from the point of storage to the secondary feeders.

The secondary feeder mains and the distribution mains are designed to form a grid or loop (see Figure 12-2), allowing water to feed each section of the grid from more than one direction. If a break occurs, it can be isolated and a minimum of customers will need to be shut down. This configuration also ensures that a minimum number of fire hydrants are placed out of service if a break occurs.

Private Water Supply

Private water supplies can range from a tank of water to handle the needs of a sprinkler system until the fire department arrives (see Figure 12-3), to a small version of a municipal system. In general, they are used where municipal water systems are either not present or are not capable of handling the needs of a particular occupancy.

Just as with a municipal water supply system, it is critical that all elements of the private water supply be maintained. Where the responsibility for maintenance of a municipal water supply system rests on the local water authority, the owner of a private system is responsible for the maintenance of that system. The big disadvantage here is we have no way to verify that these systems are being properly maintained. We can test municipal hydrants, but unless owners of private systems allow us, we cannot test their hydrants.

Water storage for a private water supply system can be in the same forms as a municipal system. Some private systems are simply private extensions of the municipal system where occupants install and maintain their own piping, hydrants, and valves on private property.

Figure 12-2 *Water supply grid.*

Figure 12-3 *The water tank and pump house provide water and pressure for fire protection in this shopping center.*

FIRE HYDRANTS

A key element of any water supply system used for fighting fires is the fire hydrant. The fire hydrant allows the fire department to gain access to the water system quickly and efficiently. By hooking up to a fire hydrant, a pumper gains access to water in the municipal (private) water system at whatever pressure is in the system. Fire hydrants come in two main types: dry barrel and wet barrel.

In order to remain reliable, hydrants should be inspected on a regular basis. The inspection should include operation of moving parts, inspection of threads for possible damage, and assurance that caps on each discharge are in place. One final inspection needed is to ensure that the threads at each discharge are in place and secure. The threads of the hydrant are not integral and can loosen. If not secure, they can leak and even separate from the hydrant under pressure.

Dry Barrel Fire Hydrants

The dry barrel fire hydrant is the hydrant most often used in this country. It is intended to be used in climates where freezing weather is possible. The most distinguishing feature of the dry barrel fire hydrant is that it has an operating stem that is operated from the very top of the fire hydrant and that goes down through the fire hydrant to operate an admission valve at the bottom (see Figure 12-4). The admission valve is at the bottom of the hydrant, below the frost line, to prevent the valve from freezing. When not in operation, a drain valve, also in the bottom of the hydrant, allows any water in the hydrant to drain out so it cannot freeze. The drain valve closes when the hydrant is open.

When using a dry barrel fire hydrant, all the outlets receive water any time the hydrant is turned on, which means that any hose to be connected to the fire hydrant must be connected before the hydrant is turned on or "charged." After the hydrant is charged, if we want to connect another hose to the fire hydrant we must first shut it down. However, if we first place a hydrant valve or some sort of gate valve on the hydrant before we charge it, we can charge additional lines as needed.

steamer connection comes from the fact that the first mechanically driven fire engines, called steamers, hooked up to the large discharge of hydrants

The exact style and number of outlets of a hydrant can vary depending on need and manufacturer. A typical fire hydrant has two 2½-inch outlets and one 4-inch, or larger, steamer connection. The term **steamer connection** comes from the fact that the first mechanically driven fire engines, called steamers, hooked up to the large discharge of hydrants.

Wet Barrel Fire Hydrants

The wet barrel fire hydrant is used only in areas where there is no fear of freezing at any time of the year. Water is constantly in the barrel of the wet barrel fire hydrant and is controlled at each individual discharge (see Figure 12-5). Like the dry barrel fire hydrant, the wet barrel fire hydrant can has 2½-inch connections

MUELLER® SUPER CENTURION® FIRE HYDRANT PARTS 9.5

CAT. PART #	DESCRIPTION	MATERIAL	MATERIAL STANDARD
A-1	Operating nut	Bronze	ASTM B584
A-2	Weather cap (not shown; used only on pre-1988 models)	Cast iron	ASTM A126 CL B
A-3	Hold down nut O-ring	Rubber	ASTM D2000 BUNA N
A-4	Hold down nut (not shown; used only on pre-1988 models)	Bronze	ASTM B584
A-5	Bonnet O-ring	Rubber	ASTM D2000 BUNA N
A-6	Anti-friction washer	Celcon	
A-7	Oil plug	Brass	ASTM B16
A-8	Bonnet	Cast Iron	ASTM A126 CL B
A-9	Bonnet bolt and nut	Steel	ASTM A307 Plated
A-10	Bonnet O-ring (1997 and newer 3-way models; all pre-1997 models and 1-way and 2-way models have flat gasket	Rubber	ASTM D2000 BUNA N
A-11	Upper stem	Steel	ASTM A576 GR B
A-12	Stem O-ring	Rubber	ASTM D2000 BUNA N
A-13	Nozzle lock	Stainless steel	ASTM A276
A-14	Pumper nozzle	Bronze	ASTM B584
A-15	Pumper nozzle gasket	Rubber	ASTM D2000 Neoprene
A-16	Pumper nozzle O-ring	Rubber	ASTM D2000 BUNA N
A-17	Pumper nozzle cap	Cast iron	ASTM A126 CL B
A-18	Hose nozzle	Bronze	ASTM B584
A-19	Hose nozzle gasket	Rubber	ASTM D2000 Neoprene
A-20	Hose nozzle O-ring	Rubber	ASTM D2000 BUNA N
A-21	Hose nozzle cap	Cast iron	ASTM A126 CL B
A-22	Cap chain	Steel	Plated
A-23	Chain ring	Steel	Plated
A-24	Upper barrel less nozzles	Cast iron	ASTM A126 CL B
A-25	Safety coupling	Stainless steel	ASTM A890
A-26	Safety flange bolt and nut	Steel	ASTM A307 Plated
A-27	Safety flange O-ring (1997 and newer models; pre-1997 models have flat gaskets)	Rubber	Cellulose
A-28	Safety flange	Cast iron	ASTM A126 CL B
A-29	Cotter pin	Stainless steel	ASTM A276
A-30	Clevis pin	Stainless steel	ASTM A276
A-31	Lower stem	Steel	ASTM A576 GR B
A-32	Lower barrel	Cast iron	ASTM A126 CL B
A-33	Stem pin	Stainless steel	ASTM A276
A-34	Drain valve facing	Plastic	
A-35	Drain valve screw	Stainless steel	ASTM A276
A-36	Upper valve plate (includes A-34 and A-35)	Bronze ASTM	B584
A-37	Shoe bolt and nut	Steel	ASTM A307 Plated
A-38	Drain ring housing O-ring (1997 and newer models; pre-1997 models have square gasket)	Rubber	ASTM D2000 BUNA N
A-39	Seat ring top O-ring	Rubber	ASTM D2000 BUNA N
A-40	Drain ring housing	Cast iron	ASTM A126 CL B
A-41	Drain ring housing bolt and nut (not shown; used only on pre-1997 model hydrants)	Steel	ASTM A307 Plated
A-42	Drain ring	Bronze	ASTM B584
A-43	Seat ring	Bronze	ASTM B584
A-44	Seat ring bottom O-ring	Rubber	ASTM D2000 BUNA N
A-45*	Reversible main valve (1997 and newer models only; pre-1997 models use non-reversible main valve and lower valve plate—not shown)	Rubber	ASTM D2000
A-46	Lower valve plate (1997 and newer models for reversible main valve; pre-1997 models have non-reversible main valve—not shown)	Cast iron	ASTM A126 CL B
A-47	Cap nut seal	Rubber	ASTM D2000
A-48	Lock washer	Stainless steel	ASTM A276
A-49	Lower valve plate nut	Cast iron	ASTM A126 CL B
A-50	Shoe	Cast iron	ASTM A126 CL B
A-84	Hold down nut	Bronze	ASTM B584
A-85	Weather seal	Rubber	ASTM D2000
A-51	10.5 oz. hydrant lubricating oil (not shown)		

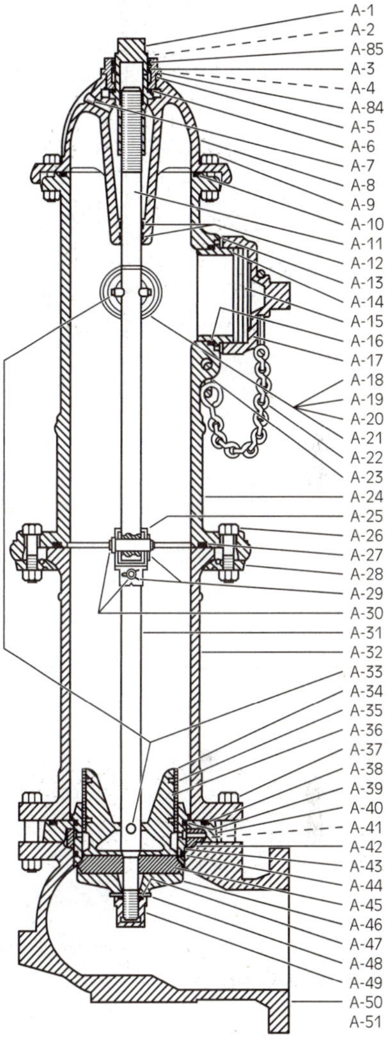

Figure 12-4 *Dry barrel fire hydrant. Courtesy of Mueller Co.*

and, usually, a steamer connection. The primary advantage of the wet barrel fire hydrant is its ability to connect additional lines without having to shut down the hydrant. Its only drawback is that it cannot be used in latitudes where freezing weather may occur.

MUELLER® HIGH FLO® WET BARREL FIRE HYDRANT PARTS **9.36**

CAT. PART #	DESCRIPTION	MATERIAL	MATERIAL STANDARD
1	Pumper nozzle cap gasket	Rubber	ASTM D2000
2	Pumper nozzle O-ring	Rubber	ASTM D2000
3	Cotter pin	Silicon bronze	
4	Anti-friction washer	Celcon	
5	Seat washer	Rubber	ASTM D2000
6	O-ring	Rubber	ASTM D2000
7	Stem	Bronze	ASTM B584
8	Stuffing box O-ring	Rubber	ASTM D2000
9	Stuffing box	Bronze	ASTB B584
10	Barrel O-ring	Rubber	ASTM D2000
11	Hose nozzle cap gasket	Rubber	ASTM D2000
12	Hose nozzle O-ring	Rubber	ASTM D2000
13	Cotter pin	Silicon bronze	Silicon bronze
14	Anti-friction washer	Celcon	
15	Seat washer	Rubber	ASTM D2000
16	O-ring	Rubber	ASTM D2000
17	Stem	Bronze	ASTM B584
18	Stuffing box O-ring	Rubber	ASTM D2000
19	Stuffing box	Bronze	ASTM B584
20	Barrel O-ring	Rubber	ASTM D2000
21	Retaining ring	Steel	
	Rivet (not illustrated)	Copper	

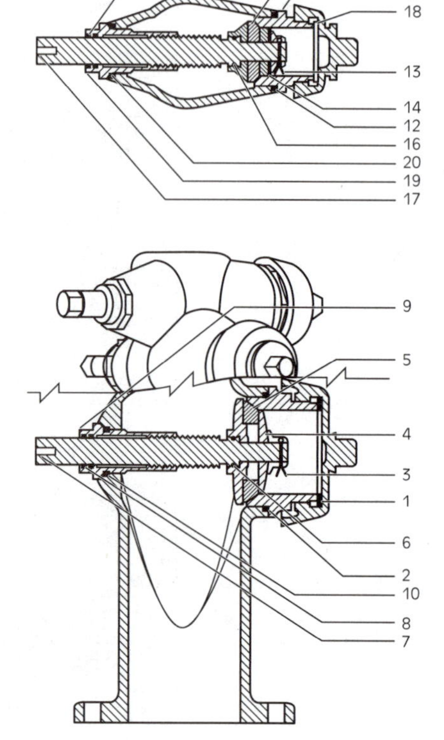

Figure 12-5 *Wet barrel fire hydrant. Courtesy of Mueller Co.*

PRESSURE LOSS IN PIPE

Hazen-Williams formula used to determine pressure (friction) loss in pipe

$$P_f = \frac{4.52 \times Q^{1.85}}{C^{1.85} \times D^{4.87}}$$

Around 1905 Allen Hazen and Gardner Williams performed flow tests on pipes used for water supply and developed the formula known today as the **Hazen-Williams formula**. The Hazen-Williams formula is used to determine pressure (friction) loss in pipe. This formula performs the same function in pipe as the formula introduced in Chapter 6 for calculating friction loss in hose. However, since the characteristics of pressure loss in pipe are significantly different than in hose, the formula is different to take these characteristics into account. The Hazen-Williams formula is: $P_f = 4.52 \times Q^{1.85}/C^{1.85} \times D^{4.87}$, where P_f is the pressure loss per foot of pipe in PSI, Q is the gpm, C is the roughness coefficient, and D is the internal diameter of the pipe. In order for this formula to be useful it is necessary to look up the appropriate values of C and D. Values of C are found in various NFPA publications, such as NFPA 13, 24, and the *Fire Protection Handbook*. Additionally,

AWWA publications contain charts with values of C. Values of D can be found in various publications from pipe manufacturers.

Different pipes have different roughness characteristics and the C factor takes that into consideration. Some of the coefficients for C for a specific kind of pipe even vary with age. It is also important to use the correct coefficient for D because the internal diameter of pipe is not what would be expected. Some pipe actually has an internal diameter larger than the stated pipe diameter, and some are smaller. For example, 8-inch class 50 ductile iron pipe has an actual internal diameter of 8.51 inches, while class 200 PVC plastic pipe has an actual internal diameter of 7.68 inches.

The Hazen-Williams formula looks a little intimidating at first glance, however it is actually a very basic algebraic expression. The exponentials in the equation are easily calculated with a scientific calculator. Be cautioned that some of the divisors and dividends can get rather large, so pay attention to keep the numbers accurate.

By presenting the Hazen-Williams formula here, it is not intended that the average firefighter should be able to calculate pressure loss in a municipal water system. Instead it is presented so that the average firefighter can understand the process of calculating pressure loss in a municipal water systems and pipe.

Example 12-3 What is the friction loss for 8-inch class 50 ductile iron pipe flowing 500 gpm?

ANSWER We already know that D for 8-inch class 50 ductile iron pipe is 8.51 and C for ductile iron is 100.

$$P_f = 4.52 \times Q^{1.85}/C^{1.85} \times D^{4.87}$$
$$= 4.52 \times 500^{1.85}/100^{1.85} \times 8.51^{4.87}$$
$$= 4.52 \times 98{,}422.526/5{,}011.872 \times 33{,}787.57$$
$$= 444{,}869.818/169{,}338{,}976.031.$$
$$= .0026 \text{ psi per foot friction loss}$$

$P_t = P_f \times L$

The pressure loss as calculated by the Hazen-Williams formula is per foot of pipe, because pipe lengths, unlike hose, can be random lengths. To find the total amount of friction loss, simply multiply by the amount of pipe by the pressure loss per foot as determined by the Hazen-Williams formula. Total pressure loss then becomes $P_t = P_f \times L$, where P_t is the total pressure loss, P_f is the pressure loss per foot as calculated above using the Hazen-Williams formula, and L is the length of pipe in feet.

Example 12-4 Calculate friction loss from the water tower to the fire hydrant in Figure 12-6. The pipe is 8-inch class 50 ductile iron.

ANSWER In Example 12-3, we calculated the friction loss for this pipe to be .0026 psi.

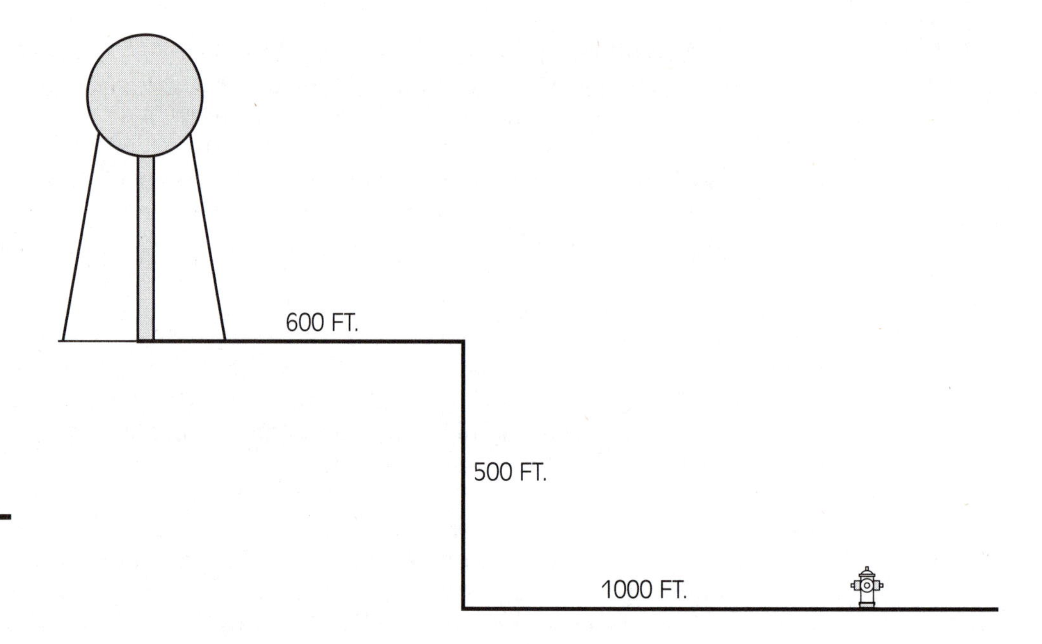

Figure 12-6 *Calculate the friction loss.*

$$P_t = P_f \times L$$
$$= .0026 \times 2{,}100$$
$$= 5.46 \text{ psi}$$

Where elbows and valves are installed in pipes, these fittings are assigned an equivalent in feet to straight pipe. For example, a 90-degree elbow for 8-inch ductile iron pipe is equivalent to 12.84 feet. If you were trying to find the friction loss for 100 feet of 8-inch ductile iron pipe with a single 90-degree elbow in the middle, you would calculate friction loss for an equivalent of 112.84 feet.

Example 12-5 Go back to Example 12-4 and recalculate the total friction loss, taking into consideration the two elbows in Figure 12-6.

ANSWER We have an equivalent length of 2,100 feet of pipe and two elbows, each equivalent to 12.84 feet, for a total of 2,125.68 feet.

$$P_t = P_f \times L$$
$$= .0026 \times 2125.68$$
$$= 5.527 \text{ psi}$$

$$P_t = \frac{4.52 \times Q^{1.85} \times L}{C^{1.85} \times D^{4.87}}$$

For those who feel comfortable with the process, a single formula that is still the Hazen-Williams formula but that calculates P_t directly can be used. That formula is $P_t = 4.52 \times Q^{1.85} \times L/C^{1.85} \times D^{4.87}$. You should note that this formula is the

Hazen-Williams formula with L included in the formula, so you are solving directly for P_t. When using this formula, remember to include equivalent length for each pipe fitting in the system.

Example 12-6 Calculate the friction loss for the system illustrated in Figure 12-6 with the same 500 gpm, only this time calculate for 6-inch pipe.

ANSWER The following are specific to 6-inch class 50 ductile iron pipe:

$C = 100$

$D = 6.4$

90-degree elbow is equivalent to 9.98 feet.

Total length is 2,100 plus 19.96 feet or 2,119.96 feet.

$P_t = 4.52 \times Q^{1.85} \times L/C^{1.85} \times D^{4.87}$

$= 4.52 \times 500^{1.85} \times 2119.96/100^{1.85} \times 6.4^{4.87}$

$= 4.52 \times 98{,}422.43 \times 2119.96/5{,}011.88 \times 8{,}435.23$

$= 943{,}105{,}290/42{,}276{,}360$

$= 22.31$ psi

WATER SUPPLY TESTING

The ability to determine the amount of water available at a particular fire hydrant is one of the most useful tools a fire department can have. It allows us to determine, before the fire occurs, if additional water sources will be necessary or if extra planning is required. It can also serve as an argument against putting in a new factory or apartment complex if water necessary to fight a fire is not there. You may think that planners and their engineers will correctly make these decisions, but it often doesn't happen that way. You need the ability to verify needed flows and intelligently challenge any discrepancies.

Note
Testing a water supply system is systematically calculating flows at selected hydrants.

Testing a water supply system is systematically calculating flows at selected hydrants in the system and is usually referred to as testing hydrants. Hydrant tests require a minimum of equipment, but it is fairly specialized equipment. At minimum you will need one blind cap with a pressure gauge and drain valve for the test hydrant and a pitot gauge for each flow hydrant.

The test procedure consists of placing the blind cap with the pressure gauge on the hydrant you are testing. This hydrant is referred to as the *test hydrant*, and is the one for which we are determining the flow. Next, flow water from adjacent hydrants, opening enough hydrants that a minimum pressure drop of 10 psi registers on the gauge at the test hydrant. These hydrants are referred to as *flow hydrants*. When opening up the flow hydrant begin with only one 2½-inch discharge open. If this does not give an adequate pressure drop at the test hydrant, open additional discharges on the flow hydrant. In order to get the required pressure drop it may be necessary to open only one hydrant in a small water system or several hydrants in a large municipal water system.

When selecting the hydrant to test, it is important that you determine the direction of flow in the water mains. If only a single hydrant flow is needed, it is preferable that the flow hydrant be downstream from the test hydrant. However, if it is necessary to flow several hydrants to get the required pressure drop, the flow hydrants should be between the test hydrant and larger mains. In short, when using multiple flow hydrants, water should flow past the flow hydrant to the test hydrant (see Figure 12-7). Determining the direction of flow is not all that difficult. Water always seeks the path of least resistance and flows from larger pipe to smaller pipe. By noting the relationship of the larger pipe and smaller pipe on a water supply map, you can determine the direction of water flow.

After we flow water from the flow hydrants it is necessary to know exactly how much water the hydrants were flowing. To learn this amount we need to know just how large the inside diameter of each discharge is, to the 1/16 inch, because these can vary. The exact size is obtained by measuring each opening

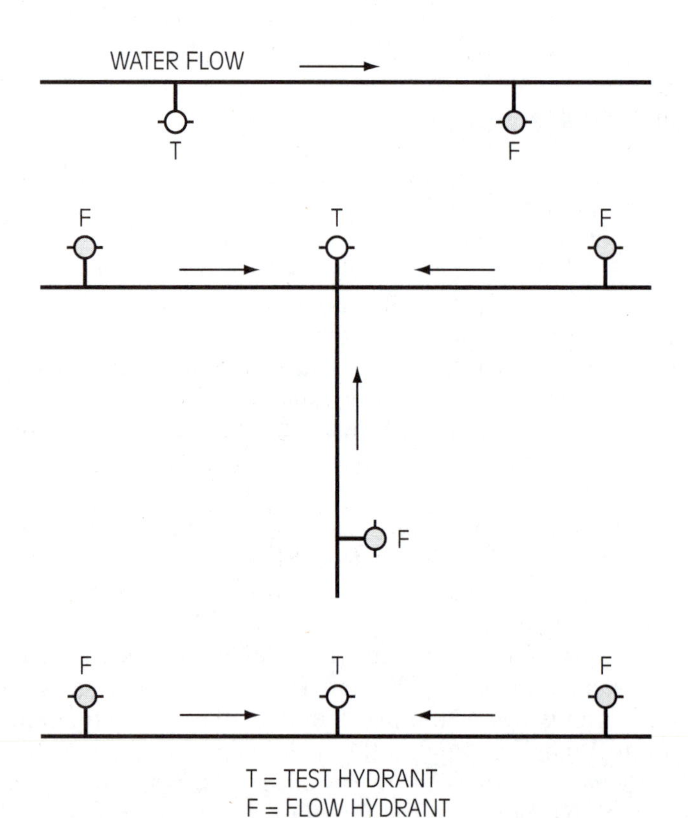

Figure 12-7
Determine the direction of water flow to the test hydrant.

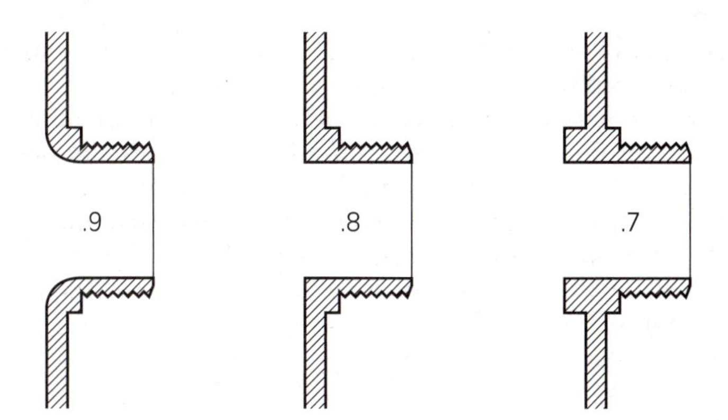

Figure 12-8 *Hydrant outlet coefficients.*

used. We also need to know the discharge coefficient of each outlet, which is obtained by feeling the inside of the hydrant where the outlet meets the hydrant barrel. The coefficients are illustrated in Figure 12-8. The pressure at the flow hydrant must be determined by use of the pitot gauge. The proper use of the pitot gauge is illustrated in Figure 12-9. The blade of the gauge should be inserted into the center of the stream, half the diameter of the opening away from the opening. To get an accurate reading, this reading should be taken from an outlet no bigger than 2½ inches.

Where pitot gauges are unavailable, blind caps with pressure gauges can be substituted. The cap with pressure gauge is put on one of the outlets of the flow

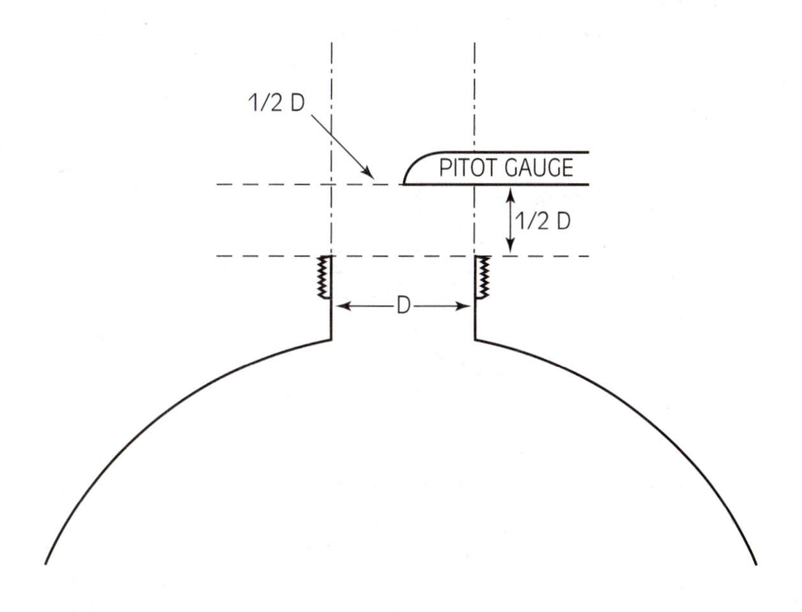

Figure 12-9 *Proper use of pitot gauge.*

velocity pressure the measure of the kinetic energy of flowing water

vena contracta the point of maximum steam contraction

hydrant and the hydrant is charged. A reading taken this way is *approximately* the same as a reading taken with a pitot gauge. The biggest disadvantage to obtaining a flow pressure in this manner is that it is actually a reading of the hydrant residual pressure and not the more accurate velocity pressure. **Velocity pressure** is the measure of the kinetic energy of flowing water. Properly done, the velocity pressure is taken at the point of vena contracta (see Figure 12-10). **Vena contracta** is the point of maximum stream contraction.

The following procedure should be followed precisely to get accurate results when testing a hydrant:

1. Place the blind cap with gauge and drain valve on the hydrant selected to be the test hydrant.
2. Open the hydrant all the way. Open the drain valve and bleed off all the air in the hydrant; then close the drain valve. Write down the static pressure.
3. Signal personnel at the flow hydrants to open their hydrant(s), in succession, until a *minimum* of 10 psi drop in pressure from static is obtained.
4. Signal personnel at the flow hydrants to write down the pressure reading of the flow from their hydrant and copy down your residual pressure.
5. Shut down the flow hydrants in succession to prevent a surge in the system. Shut down the test hydrant and remove the pressure gauge.
6. Measure the exact interior size of all outlets that discharged water during the test and determine the outlet coefficient.
7. Return all the hydrants to system ready condition.

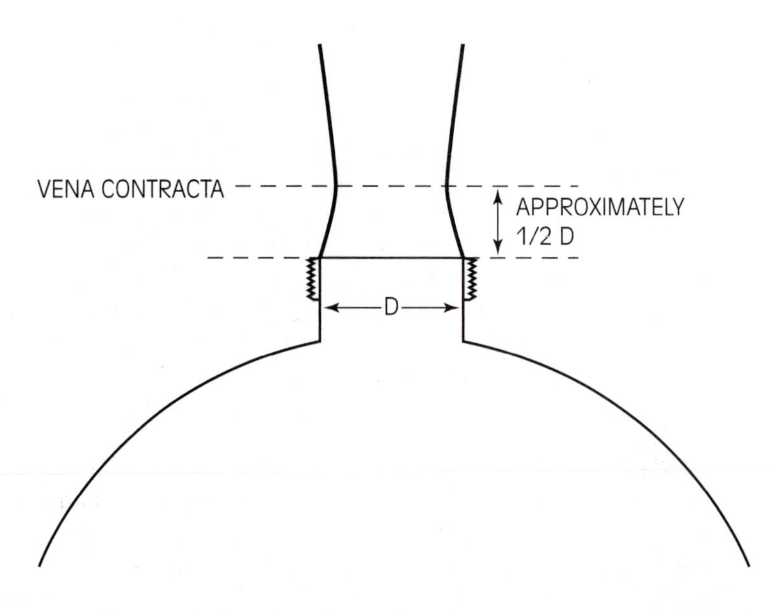

Figure 12-10 *Point of vena contracta (exaggerated).*

Calculating Hydrant Capacity

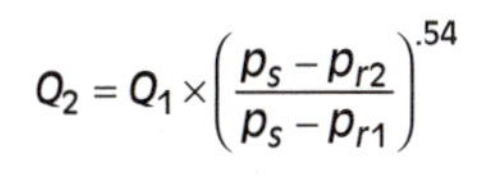

Calculating the capacity of our test hydrant is a matter of applying the findings of the above test to another version of the Hazen-Williams formula. That formula is $Q_2 = Q_1 \times (p_s - p_{r2}/p_s - p_{r1})^{.54}$, where Q_2 is the gpm we are calculating, Q_1 is the gpm flow we calculated as a result of opening the flow hydrants, p_s is the static pressure, p_{r2} is the pressure we are calculating the flow for, and p_{r1} is the residual pressure obtained from the test. Essentially, $p_s - p_{r2}$ is the pressure drop from static to the theoretical test pressure. The pressure drop $p_s - p_{r1}$ is the drop actually observed during the test.

Most fire department testing is designed to calculate the maximum capacity of the water main, that is, the gpm at 20 psi residual, so p_{r2} is 20. At times we may need to know the capacity of the water main at pressures not intended to produce the maximum capacity. To learn this, simply insert the pressure for which you need a corresponding capacity into the formula as p_{r2}. Q_2 will be the flow at pressure p_{r2}.

NFPA 291

Fire Flow Testing and Marking of Hydrants

The first step in calculating the capacity of the test hydrant is to determine how much water was flowing from the flow hydrant(s). This can be done in one of two ways: (1) look up the flow in a chart such as in NFPA 291, *Fire Flow Testing and Marking of Hydrants*, or (2) calculate the flow from each hydrant using the formula $gpm = 29.83 \times D^2 \times C \times \sqrt{P}$. This is the same formula used in Chapter 5 to calculate gpm, only the constant is slightly different and C is the hydrant coefficient.

Finding the flow in charts is simple and saves time doing the calculations. There is one caution however. If the chart used has not already taken the hydrant coefficient into account, you must. But that is simple, just multiply the gpm specified in the chart for the particular outlet size and pressure and multiply it by the coefficient.

Because this text is about hydraulics and how to use and apply various formulas, we will use the formula $gpm = 29.83 \times D^2 \times C \times \sqrt{P}$ to calculate the gpm from the flow hydrants. The formula is actually simple to use. Plug in the exact diameter for D, the pressure measured by the pitot gauge for P, and the hydrant coefficient for C. The answer will be the gpm flowing from a particular outlet during the test. If more than one outlet is used during a test, the calculation must be done for each outlet. The pressure obtained at any outlet is the same for all other outlets of the same hydrant. If more than one hydrant was used during the test the calculations must also be repeated for each hydrant. Caution: Each hydrant will not normally have the same pressure reading as taken by the pitot gauge, so each set of calculations must reflect this. (If several hydrants with different size outlets are used you quickly understand the value of charts to find gpm.) Once all the calculations are made, the gpm are added together and the result is Q_1.

Example 12-7 By testing a hydrant according to the foregoing procedures, you were only required to flow water from one outlet that has an actual diameter of $4\frac{9}{16}$ inches. The pitot reading gave a pressure of 15 psi and you determined the hydrant had a coefficient of .8. How much water was this hydrant flowing?

ANSWER $GPM = 29.83 \times D^2 \times C \times \sqrt{P}$

$= 29.83 \times 4\tfrac{9}{16}^2 \times .8 \times \sqrt{15}$

$= 29.83 \times 20.82 \times .8 \times 3.87$

$= 1{,}922.8$ or $1{,}923$

Now that we know how much water was flowing at the flow hydrant, we can determine the amount of water available in the test hydrant at any desired pressure.

Example 12-8 Calculate the maximum capacity of the test hydrant if the static pressure was 78 psi and the residual pressure was 52 psi.

ANSWER Use the Hazen-Williams formula for calculating the capacity of the hydrant,

$Q_2 = Q_1 \times (p_s - p_{r2}/p_s - p_{r1})^{.54}$. We have already calculated Q_1 in Example 12-7. In the formula, p_s is the static pressure read at the test hydrant. Because we want to find the maximum capacity of the hydrant, we are going to use 20 psi for p_{r2}. The residual pressure at the test hydrant when the flow hydrant(s) was flowing water is p_{r1}.

$Q_2 = Q_1 \times (p_s - p_{r2}/p_s - p_{r1})^{.54}$

$= 1{,}923 \times (78 - 20/78 - 52)^{.54}$

$= 1{,}923 \times (58/26)^{.54}$

$= 1{,}923 \times (2.23)^{.54}$

Now find the .54 power of 2.23 on your calculator

$Q_2 = 1{,}923 \times 1.54$

$= 2{,}961.42$ or $2{,}961$ gpm

Graphing Hydrant Flow

The process just described for calculating the flow from a hydrant at a specific pressure is fairly simple and very accurate. As previously mentioned, when we test a hydrant, we are usually calculating for maximum flow. There are times, however, when we need to know the amount of water available in the water main at a pressure other than 20 psi. We can either calculate the pressure at each and every pressure we need a flow for or we can make just one calculation and graph the capacity of the water supply.

Graphing the capacity of the water supply has the primary advantage of being fast. Its primary disadvantage is that it is not quite as accurate as calculating the flow. Before we can graph the capacity of the hydrant we must first do one flow test to determine the flow from our flow hydrant(s) and the static and residual pressure at the test hydrant. Once one flow test has been done, we can plot the curve of the hydrant capacity versus pressure.

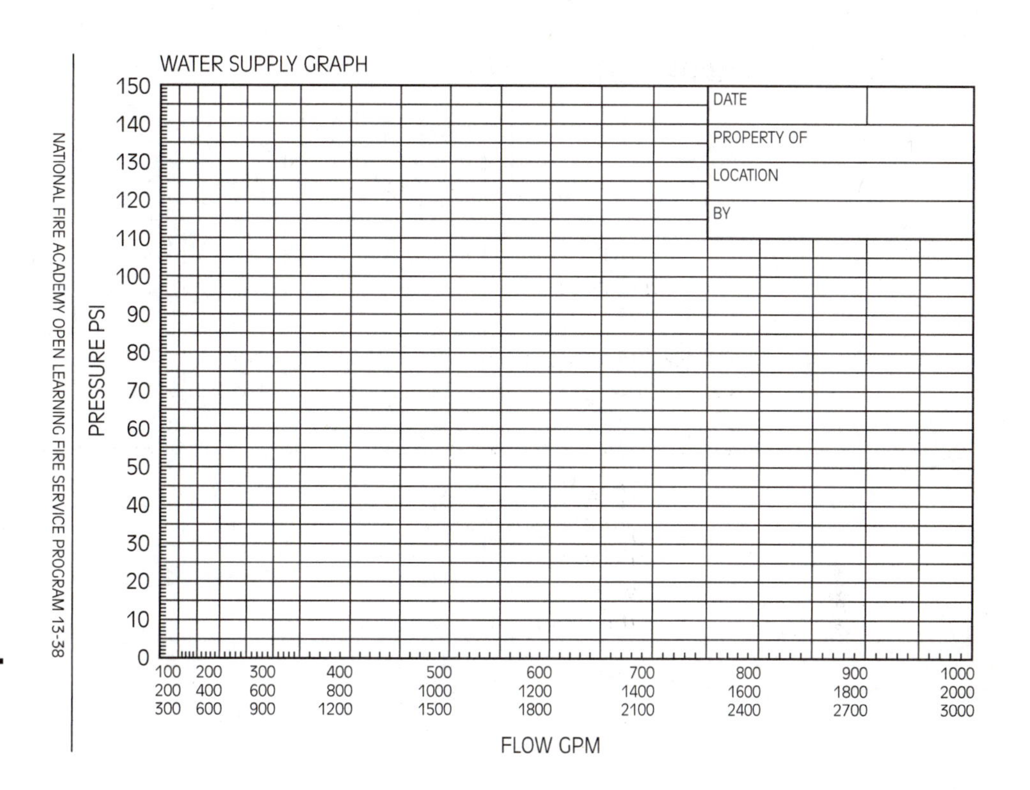

Figure 12-11 *Sample of 1.85 exponential paper.*

Graphing the flow of the hydrant requires special graph paper referred to as 1.85 exponential paper (see Figure 12-11). The Y-axis (vertical) of the paper is calibrated for hydrant pressure and the X-axis (horizontal) is calibrated for hydrant flow. There are usually three scales across the bottom in order to accommodate a wide range of pressure and flow. The vertical lines are not equal distance apart but represent the change in pressure versus flow at the 1.85 power.

Using the graph paper to illustrate the capacity of the hydrant over its pressure range is as easy as following the following steps:

1. Locate the static pressure of the test hydrant on the Y-axis at 0 gpm and make a mark.
2. Located the gpm flow from the flow hydrant(s) on the X-axis.
3. Determine which scale to use.
4. Extend a line up from the gpm of the flow hydrant, parallel with the reference lines, until you come to the residual pressure of the test hydrant. Mark the spot.
5. Take a straight edge and connect the two points with a line that extends all the way from the X-axis base line to the Y-axis base line.

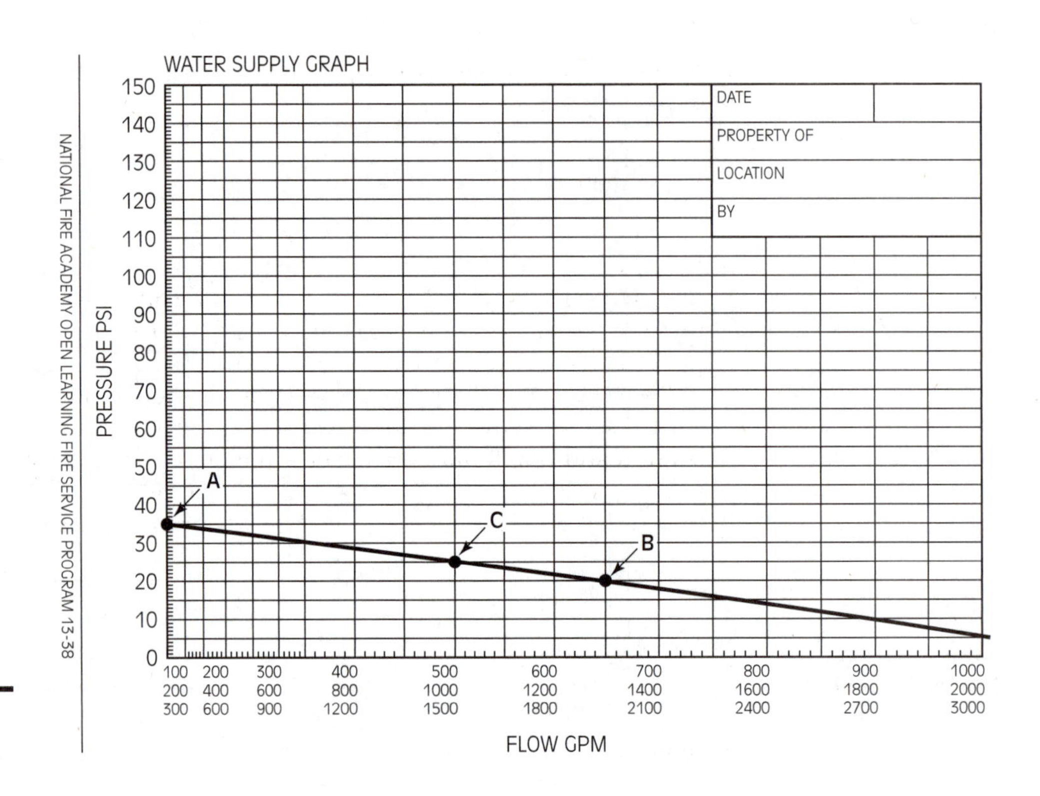

Figure 12-12 *Graph of hydrant test.*

Example 12-9 Graph the flow test of a hydrant that has a static pressure of 35 psi, a residual pressure of 20 psi, and had one flow hydrant with a flow of 1,300 gpm.

ANSWER Begin by marking the static pressure of the test hydrant, point A, on the graph (see Figure 12-12). Locate the hydrant flow along the bottom of the graph. Find the intersection of 20 psi residual (of the test hydrant) and 1,300 gpm and mark point B. Finally, draw a straight line connecting these two points and extending across the entire face of the graph.

Once we have the hydrant test graphed, it is easy to find a corresponding flow for any given pressure. For instance, what is the maximum flow from this hydrant at 25 psi? Point C on Figure 12-12 indicates a flow of approximately 1,000 gpm.

GETTING THE MOST OUT OF THE WATER SUPPLY

The single most important benefit of learning hydraulics is the ability to make maximum use of the local water supply. Getting the maximum amount of water from the source to the fire in the shortest time is the key to successful firefighting.

There are three general evolutions that are used to get water from the hydrant to the fire. Each is discussed in the following sections with attention to advantages and disadvantages.

Fire to Hydrant

Evolution 1 is where the pumper stops in front of the fire building and drops off a skid load of 1½-inch, 1¾-inch or 2½-inch hose, and a gated wye, then lays a line to the hydrant (see Figure 12-13). The only advantage to this evolution is that it allows the pumper to connect to the hydrant, which allows you to pump whatever pressure is needed.

There are several disadvantages to this evolution. First, the amount of water supplied is limited by the size of hose and attached nozzles dropped off at the fire. Second, if additional lines are needed, they must be laid from the pumper at the hydrant. Third, unless the hose lines attached to the wye are the same length, require the same nozzle pressure, and flow the same gpm, getting the correct pressure to any line is impossible.

Hydrant to Fire

Evolution 2 involves laying a line from the hydrant to the fire and connecting the supply line directly to the hydrant (see Figure 12-14). This procedure has the advantage of placing the pumper at the fire where multiple lines can be easily pumped, each with the correct pressure. All the tools and appliances carried on the pumper are readily available.

The primary disadvantage of this evolution is limited water supply. With the pumper relying on hydrant pressure to overcome friction loss and provide adequate intake pressure, the water supply available, even with LDH, is limited. To complicate things even more, as additional pumpers connect to other hydrants, the

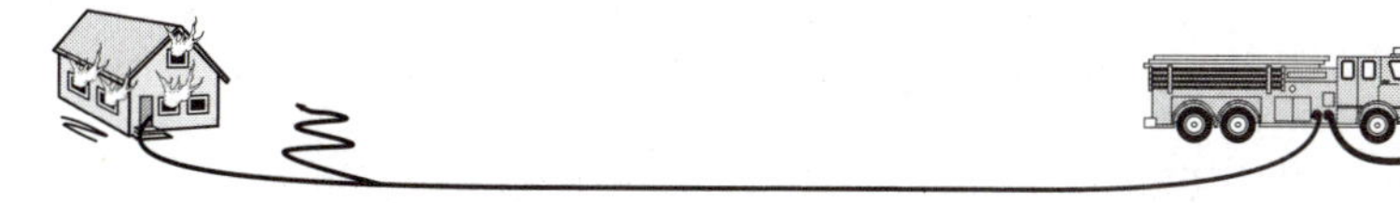

Figure 12-13 *Fire to hydrant.*

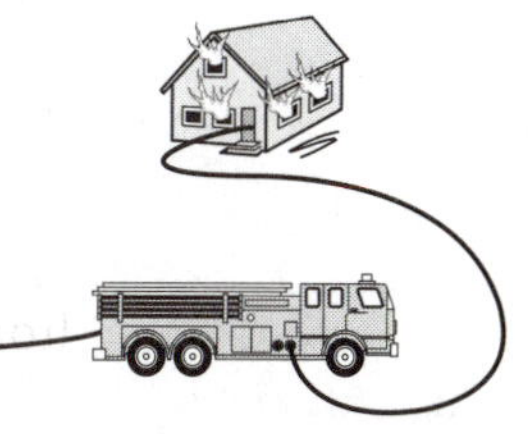

Figure 12-14 *Hydrant to fire.*

pressure in the system will fall and total pressure available to supply water will be reduced, reducing the amount of water available.

Example 12-10 How much water can a pumper expect to get from a hydrant if it has laid 400 feet of 4-inch hose and the hydrant has a static pressure of 75 psi?

ANSWER This is a maximum flow problem. First, subtract the needed intake pressure of 20 psi from the static pressure of 75 psi, leaving us with only 55 psi available to overcome friction loss. To find the maximum amount of water, first calculate the allowable friction loss per 100 feet of hose.

$$55 \div 4 = 13.75 \text{ psi} = FL\ 100$$

Now find the amount of water a 4-inch hose can flow with 13.75 psi of friction loss per 100 feet of hose. From Chapter 6 we use the formula $Q = \sqrt{FL\ 100/Cf \times 2}$.

$$\begin{aligned} Q &= \sqrt{FL\ 100/Cf \times 2} \\ &= \sqrt{13.75/.1 \times 2} \\ &= \sqrt{13.75/.2} \\ &= \sqrt{68.75} \\ &= 8.29 \text{ or } 829 \text{ gpm} \end{aligned}$$

While a flow of 829 gpm sounds good, remember the flow depends on the static pressure at the hydrant, which is usually less than 100 psi, and the length of the lay. LDH also has the disadvantage of forming a roadblock once it is charged. Finally, if the hydrant laid from were to have a capacity of 2,000 gpm, by relying on hydrant pressure to move the water there is no way you could ever get the maximum water from the hydrant. This means that if you had a needed fire flow of 1,500 gpm, even though the hydrant were capable of delivering the water there is no way you could get the water to the fire.

Two-Pump Operation

Evolution 3 involves the use of two pumpers and gets the most water from the hydrant (see Figure 12-15). The only disadvantage to this evolution is that it re-

Figure 12-15 *Two-pump operation.*

quires two pumpers. The first pumper lays hose from the hydrant to the fire, and the second connects to the hydrant and pumps to the first. In all other ways this is the most efficient way to get the maximum water out of the hydrant. This is a textbook example of a two-pump operation. Since it is technically a relay operation, engine pressure for the pumper at the hydrant is calculated as described in Chapter 10 in the section Relay Formula.

What gives this evolution the advantage over the second evolution is that you are no longer restricted to hydrant pressure. The pumper at the hydrant can pump as much water as is in the system as long as the intake pressure stays above 20 psi and the pumper at the hydrant has the capacity. Even as other pumpers hook up to the system and the residual pressure in the system is reduced, the pumper at the hydrant can compensate by adding more pressure with the pump. (The discharge pressure remains the same, but the pump has to contribute more of the pressure.)

This evolution does have some limiting factors that must be considered when calculating engine pressure. These factors are capacity of the hydrant, capacity of the pump, maximum capacity of the hose, and maximum pressure of the hose. In addition to straightforward relay calculations, the advantage of this evolution becomes very evident in maximum lay and maximum flow calculations.

Example 12-11 Using the hose lay in Example 12-10, calculate the maximum amount of water possible if a pumper were at the hydrant. For now, assume the hydrant and the pumper have the required capacity.

ANSWER LDH, unless it is special attack hose, has a maximum pressure of 185 psi. To find the maximum pressure available for friction loss, subtract 20 psi from the 185 psi, giving a maximum pressure allowable for friction loss of 165 psi. Now find the allowable friction loss per 100 feet.

$165 \div 4 = 41.25 \text{ psi} = FL\ 100$

Now find the amount of water 4-inch hose can flow with 41.25 psi of friction loss per 100 feet of hose. From Chapter 6 we use the formula $Q = \sqrt{FL\ 100/Cf \times 2}$.

$$Q = \sqrt{FL\ 100/Cf \times 2}.$$
$$= \sqrt{41.25/.1 \times 2}$$
$$= \sqrt{41.25/.2}$$
$$= \sqrt{206.25}$$
$$= 14.36 \text{ or } 1{,}436 \text{ gpm}$$

By placing a pumper on the hydrant in Example 12-10, we have increased the water supply from the hydrant by 73 percent. The pumper on the hydrant also allows us to compensate for residual pressure losses as other hydrants are used.

As mentioned previously, the only disadvantage of this system is that it requires two pumpers. This, however, is not a major disadvantage. Most structural fire responses mandate a response of at least two engine companies. By directing the second-in engine to "catch" the hydrant, the two-pump operation can be accomplished universally. Where it is desired to have the first two engines secure their own hydrants, hydrant valves should be used. Hydrant valves allow later responding engines to hook up to the steamer connection of the hydrant and boost pressure or water supply to the pumper at the fire. As additional pumpers respond, they can be directed to hydrants that are already in use to boost pressure, water supply, or both. Personnel from pumpers on hydrants can utilize hose and equipment from pumpers at the fire.

■ Note
Hydrant valves allow later responding engines to hook up to the steamer connection of the hydrant and boost pressure/water supply to the pumper at the fire.

Regardless of which of the foregoing evolutions is used to supply water to a working structure fire or major incident, you should never rely on a single source of water. If a single supply line is supplying the incident and it breaks, there will be no water for anyone.

APPLICATION ACTIVITIES

Use your knowledge of water supply to solve the following problems.

Problem 1 You are the fire chief of a small department in a town of 2,500 residents. What is the MDC?

ANSWER MDC = *Pop.* × 214.5/1440
= 2,500 × 214.5/1440
= 536,250/1440
= 372.4 gpm

This community would have a peak demand of 327.4 gpm.

Problem 2 If the community in Problem 1 had a single target hazard with an NFF of 1,500 gpm, what is the *MFr*?

ANSWER *MFr* = *MDC* + *NFF*
= 372.4 + 1,500
= 1,827.4 gpm

Problem 3 How much minimum storage capacity must this small town have in order to meet domestic needs and fire suppression requirements?

ANSWER Use the formula *SC* = (*MDC* + *NFF*) × *T*. Since the *NFF* is under 2,500 gpm we only need to store enough water for 2 hours. Using the answer to Problem 2, we have

SC = (*MDC* + *NFF*) × *T*
= 1,827.4 × 120
= 219,288 gallons

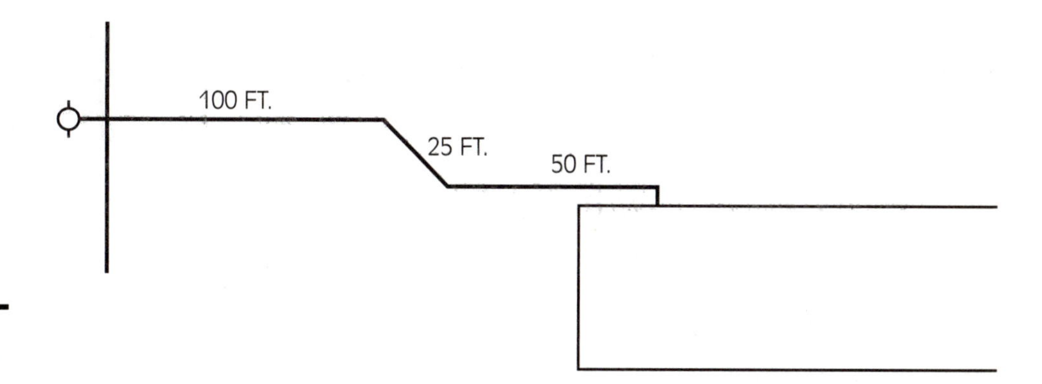

Figure 12-16 *How much friction loss?*

Problem 4 One commercial building in this small town has a sprinkler system. For the sprinkler system to work properly, it needs a minimum pressure of 35 psi at the entrance to the building. With the pipe layout shown in Figure 12-16, is it possible to deliver the required water if the sprinkler system is designed to flow a maximum of 300 gpm? The pipe is 8 inch Class 50 ductile iron pipe with a C of 100 and $D = 8.51$. Each 45-degree bend is equivalent to 9 feet of pipe and the 90-degree turn is equivalent to 18 feet of pipe. The hydrant on the public water system, at the point where the pipe enters the private system, has a static pressure of 65 psi.

ANSWER Use the Hazen-Williams formula to calculate pressure loss in the pipe.

$$Pt = 4.52 \times Q^{1.85} \times L/C^{1.85} \times D^{4.87}$$
$$= 4.52 \times 300^{1.85} \times 211/100^{1.85} \times 8.51^{4.87}$$
$$= 4.52 \times 38{,}253.75 \times 211/5{,}011.88 \times 33{,}787.52$$
$$= 36{,}483{,}366/169{,}338{,}990$$

= .22 psi pressure drop for the required 300 gpm flow. More than adequate pressure will reach the building.

Problem 5 As the result of a hydrant test, you have determined that the test hydrant has a static pressure of 63 psi, and a residual pressure of 53 psi. If the flow hydrant was flowing one outlet with an actual diameter of 2½ inches and had a flow pressure of 47 psi, how much water was it flowing? How much water will the test hydrant have available at 25 psi? The hydrant has a C of .9.

ANSWER Begin by finding out how much water was flowing from the hydrant.

$$gpm = 29.83 \times D^2 \times C \times \sqrt{P}$$
$$= 29.83 \times 2.5^2 \times .9 \times \sqrt{47}$$
$$= 29.83 \times 6.25 \times .9 \times 6.86$$
$$= 1151.07 \text{ or } 1151$$

Total gpm flow from this hydrant is 1,151 gpm.

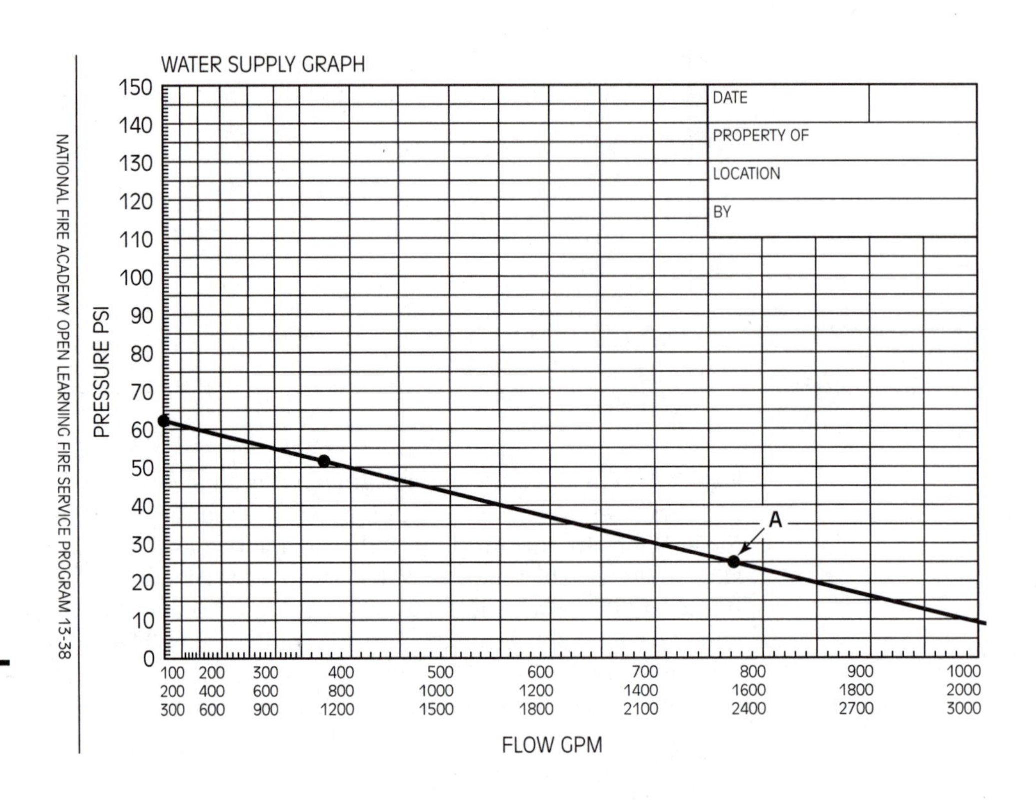

Figure 12-17 *Graph of water supply curve.*

Now chart the flow on a piece of 1.85 exponential paper (see Figure 12-17).

At 25 psi the test hydrant will flow 2,300 to 2,350 gpm (see point A on Figure 12-17).

Problem 6 Using the Hazen-Williams formula, verify the flow at 25 psi for the hydrant in Problem 5.

ANSWER $Q_2 = Q_1 \times (p_s - p_{r2}/p_s - p_{r1})^{.54}$

$= 1151 \times (63 - 25/63 - 53)^{.54}$

$= 1151 \times (38/10)^{.54}$

$= 1151 \times 3.8^{.54}$

$= 1151 \times 2.06$

$= 2371$ gpm

The actual flow at 25 psi residual pressure in Problem 5 would be 2,371 gpm. Plotting the curve on the water supply graph is off by a small amount. Where absolute accuracy is necessary, the Hazen-Williams formula should be used.

Problem 7 Go back to Example 12-10 and calculate the maximum lay possible for 829 gpm with 4-inch LDH with a two-pump operation.

ANSWER Remember, with 4-inch LDH the maximum pressure is 185 psi.

185 – 20 = 165 psi maximum discharge pressure

We already know the *FL* 100 for 829 gpm is 13.75 psi

165 ÷ 13.75 = 12 = 1,200 feet

The maximum lay possible with 4-inch LDH and flowing 829 gpm would be 1,200 feet. That is three times more than the distance possible if the hose is connected directly to the hydrant as in Example 12-10.

Summary

In this chapter you have learned the basic principles of water supply. Whether the water is from a municipal system with hydrants every 500 feet or brought in by tanker shuttle, having enough water at the right spot is the key to success in fire suppression. If you know how the water supply system works and how to calculate the pressure drop in the system by the Hazen-Williams formula, you can better understand system capabilities and limitations.

To assist you in determining if the water supply is adequate, you now know how to test hydrants to determine the capacity of the system. Being able to calculate the capacity of a hydrant and graph the curve of that hydrant's water supply capacity is a critical expertise learned by studying hydraulics. Even without graphing the capacity of the hydrant, being able to calculate the capacity of any hydrant at 20 psi will assist you in the preplan of your water supply needs.

What may be the single most important principle in this chapter is how to get the most water from the hydrant. It is never adequate to simply lay out from a hydrant and count on hydrant pressure alone to deliver the water, even with LDH. A pumper on the hydrant always allows you to get more water from the hydrant or allows you to use longer lays. With a pumper on the hydrant, as other pumpers take water from the system a steady discharge pressure can be maintained. These principles can be verified by maximum flow and maximum lay calculations.

Understanding water supply is the first step in becoming proactive. It represents the first step in truly preparing to meet the water supply needs of any given community, whether you are connected to a fire hydrant that is part of a municipal system with a few billion gallons of water in reserve, or all your water is trucked in. Unless you understand your strengths and weaknesses before the fire, you will never be able to solve problems during the fire.

Review Questions

1. You live in a community of 25,000 people, what is the MDC of your community?
2. If your community has a target hazard that requires a fire flow of 2,500 gpm, what gpm flow must the water system be able to supply?
3. What is the minimum amount of water that needs to be in storage for this community?
4. What are the two methods of creating pressure for a municipal water supply system?
5. What is the maximum pressure loss that will be realized in the piping from the town water tank to the water supply grid in your town (review questions 1–3) during a fire at the target hazard? The piping from the tank is 24-inch cast iron and has a straight run of 400 feet. $C = 100$, $D = 24.34$.
6. During a hydrant test, the test hydrant has a static pressure of 88 psi and a residual pressure of 75 psi. The flow hydrant had a resid-

ual pressure of 54 psi, a C of .9, and the opening was measured to be 2 9/16 inches.

A. How much water is flowing at the flow hydrant?

B. How much water will the test hydrant flow at a residual of 20 psi?

7. Now chart the flow curve on a piece of 1.85 exponential paper (Figure 12-11).
8. With a 5-inch supply line 700 feet long connected directly to the hydrant, what is the maximum gpm that can be expected with a hydrant pressure of 75 psi and the pumper 20 feet above the hydrant?
9. With 4-inch supply line, what is the maximum length of hose lay possible if you want to deliver 1,000 gpm and you have a 1,250-gpm-rated pumper connected to the hydrant?
10. During a hydrant test you have a residual pressure at the flow hydrant of 38 psi while flowing water from a single 2½-inch discharge that has a C of .9. The test hydrant has a static pressure of 65 psi and a residual pressure of 40 psi.

A. How much water is flowing from the flow hydrant?

B. How much water will the test hydrant flow at 20 psi?

C. Graph the flow.

List of Formulas

Calculating a community's maximum daily consumption:

$$MDC = \frac{Pop. \times 214.5}{1440}$$

For calculating a community's needed storage capacity:

$$SC = (MDC + NFF) \times T$$

For calculating the minimum flow rate:

$$MFr = MDC + NFF$$

Hazen-Williams formula:

$$P_f = \frac{4.52 \times Q^{1.85}}{C^{1.85} \times D^{4.87}}$$

Total pressure loss for system:

$$P_t = P_f \times L$$

Direct calculation and total pressure loss:

$$P_t = \frac{4.52 \times Q^{1.85} \times L}{C^{1.85} \times D^{4.87}}$$

Hydrant capacity formula:

$$Q_2 = Q_1 \times \left(\frac{P_s - P_{r2}}{P_s - P_{r1}}\right)^{.54}$$

References

Cote, Arthur E., P.E., Editor-in-Chief, *Fire Protection Handbook*, 18th ed. Quincy, MA: National Fire Protection Association, 1997.

Hickey, Harry E. *Fire Suppression Rating Schedule Handbook.* U.S.A.: Professional Loss Control Foundation, 1993.

NFPA 13, *Installation of Sprinkler Systems.* Quincy, MA: National Fire Protection Association, 1996.

NFPA 24, *Private Service Mains and Their Appurtenances.* Quincy, MA: National Fire Protection Association, 1995.

NFPA 291, *Fire Flow Testing and Marking of Hydrants.* Quincy, MA: National Fire Protection Association, 1995.

NFPA 1231, *Water Supplies for Suburban and Rural Fire Fighting.* Quincy, MA: National Fire Protection Association, 1993.

NFPA 1901, *Automotive Fire Apparatus.* Quincy, MA: National Fire Protection Association, 1996.

Standpipes, Sprinklers, and Fireground Formulas

Learning Objectives

Upon completion of this chapter, you should be able to:

- Better understand the principles of installation and operation of a standpipe system.
- Understand the basic concepts of the design and installation of sprinkler systems.
- Test a sprinkler system for proper performance.
- Verify water supply needs for a sprinkler system.
- Know how to calculate hydrant capacity at a fire.
- Understand the use and weaknesses of fireground formulas.

INTRODUCTION

Fire suppression systems such as standpipes and sprinkler systems are so common in fire suppression today that we often take them for granted and we often misuse them. In this chapter you will learn about basic design requirements and testing of both standpipe and sprinkler systems, not to the point of qualifying you to design such systems, but to give you a better understanding of the systems so you can better use them at fires and for testing them.

In addition to standpipe and sprinkler systems, this chapter also explores the subject of fireground formulas. Some firefighters use fireground formulas exclusively instead of learning hydraulics. This practice may be adequate for someone who only operates the pumps at drills, but for actual fireground operation we must be as exact as possible. Some fireground formulas can actually be useful, but most just substitute for learning the correct way.

As you study this chapter, a word of caution is in order. The specifications for both standpipe and sprinkler systems mentioned in this chapter come from their respective NFPA standards. These standards do not always agree with model or local codes. In fact, in many places NFPA standards are flexible and allow the authority having jurisdiction (AHJ) to modify the standard to meet local preferences. Where the AHJ has modified the standard, the modifications must be followed.

STANDPIPE SYSTEMS

Standpipe systems are traditionally thought of as vertical piping in the stairways of buildings that firefighters can attach hose to for firefighting. Standpipe systems, however, do not need to be either vertical or in buildings. In buildings that cover large expanses of property, standpipe systems can be helpful even when the building is only one story. Instead of dragging several hundred feet of hose in from the street, a horizontal standpipe system allows us to hook up our standpipe hose to the system at the outlet closest to the fire.

Systems can be installed to gain access to difficult to reach portions of any structure. Here are several unique applications of standpipe systems.

- A town house community built on the roof of another building; there is no vehicle access to the town homes, so a standpipe system was required with risers spaced throughout the courtyard
- Bridges with standpipe systems intended either to get water up to the bridge or down to the highway below
- A most ingenious system that runs through some woods and up a hill to provide water to an elementary school on top of the hill
- Standpipe systems installed on long piers

standpipe system
a water distribution system designed to get water for firefighting purposes from point A to point B

Installation of Standpipe and Hose Systems

Each of the systems mentioned in the list, as well as more traditional standpipe systems, is designed to solve one problem. They allow firefighters to get water from a siamese (point A) to a riser closer to the fire (point B), which leads us to the definition of standpipe system. A **standpipe system** is a water distribution system designed to get water for firefighting purposes, from point A to point B.

The following information about standpipe system design and testing is general information only. NFPA 14, *Installation of Standpipe and Hose Systems* contains specific criteria for the installation and maintenance of standpipe systems and should be consulted before doing design review or testing.

Classes of Systems

Three classes of standpipe systems are the following:

- Class 1 standpipe systems are intended solely for use by fire department personnel. Each outlet is gated and is fitted with a 2½-inch hose connection.
- Class 2 standpipe systems are intended for use by building occupants and are fitted only with 1½-inch gated outlets and are usually fitted with 100 feet of 1½-inch hose and nozzle.
- Class 3 standpipe systems are also intended for use by building occupants, but are fitted with 2½-inch outlets reduced to 1½-inch with 1½-inch hose and nozzle, which allows fire department personnel to use the system as they would a Class 1 system as well as allowing the occupants to use the system.

The primary difference between these systems is the water supply requirements. Class 1 and 3 systems require a minimum of 500 gpm for the first riser and 250 gpm for each additional riser, to a maximum of 1,250 gpm, with 100 psi residual. A Class 2 system only requires a flow of 100 gpm with 65 psi residual. Where a water supply is required for any class system, it must be for a minimum of 30-minute duration. Table 13-1 summarizes the classes of standpipe systems by intended use, water supply, size outlet, and hose requirement.

Table 13-1 *Summary of standpipe classes.*

Class	Intended Use	Water Supply	Size Outlet	Hose
1	FD personnel	30 min	2½-inch	none
2	Occupants	30 min	1½-inch	100 feet 1½-inch with nozzle
3	Both	30 min	2½-inch reduced to 1½-inch	100 feet 1½-inch with nozzle

Types of Systems Standpipe systems can be subdivided into the following five types of systems.

1. Automatic-Wet. This system is connected to a water source and is capable of automatically supplying the system demand.
2. Manual-Wet. This system has water in it, but is unable to meet system demand. The water is there to indicate if a leak develops and eliminate the need to discharge large amounts of air before water begins to flow. Water is supplied by the fire department.
3. Automatic-Dry. This system is dry, under air pressure, causing a dry pipe valve to activate when an outlet is opened. The water supply for this system must meet system demand.
4. Semiautomatic-Dry. This system is similar to Automatic-Dry, but there is no pressurized air in the system. Instead, this system has a remote device at the hose connection that activates a deluge valve. The water supply has to meet system demand.
5. Manual-Dry. This system is always dry and needs water from the fire department to meet demand.

Standpipe Design

actual length concept
hose connections must be located so that 100 feet of hose and a 30-foot stream are able to reach any area of the building

exit location concept
allows hose connections to be located in exit stairwells, horizontal exits, and exit passageways

pipe schedule
the size of the pipe is determined by a flow versus pipe size schedule in NFPA 14

hydraulic design
pipe sizes are calculated to provide the minimum design flow and pressure

Standpipe systems are designed with one of two overall design concepts: actual length or exit location.

The actual length concept is used only for Class 2 systems. **Actual length concept** stipulates that hose connections be located so that 100 feet of hose and a 30-foot stream are able to reach any area of the building. This 100 feet is not a straight-line measurement but an actual travel distance measurement. This means distance must be measured around obstructions so that every square foot of the building can be reached with 100 feet of hose and a 30-foot stream.

The **exit location concept** allows hose connections to be located in exit stairwells, horizontal exits, and exit passageways. This method is used for Class 1 and 3 systems. The idea is that in a properly designed building, exit structures will be adequately spaced to provide sufficient spacing and number of 2½-inch outlets. NFPA 14, however, allows for additional 2½-inch hose connections under specific situations.

Determining the size of pipe in a standpipe system can be done by either of two methods: pipe schedule or hydraulic design. With the **pipe schedule** method the size of the pipe is determined by a flow versus pipe size schedule in NFPA 14. The size of pipe depends on the type of system and height of the standpipe. Under the pipe schedule method, the minimum size pipe for a Class 1 or 3 system is 4 inch, and the minimum size pipe for a Class 2 system is 2 inch. With **hydraulic design** the system pipe sizes are calculated to provide the minimum design flow and pressure. To hydraulically design a system, the designer must know the minimum pressure and flow at the hydraulically most remote outlet. From there, the design is calculated back to the water supply, using hydraulic calculations (Hazen-Williams formula) to determine the size of the pipe.

Water Supply

Water supply for standpipes comes from one of two sources: The fire department pumps into the system, or the system has its own supply. On systems of type 2 and type 5, the fire department siamese is the only source of water for the system. Therefore before the system can be built, the local water supply must be tested to make sure the fire department can supply the system requirements.

Systems types 1, 3, and 4 must be provided with an independent source of water. This source can be as simple as a connection to the municipal, or private, system, or more complicated such as water tank and pump. Pressure tanks and elevated water tanks also serve as water supplies for standpipes.

Regardless of the source of water for a standpipe system, each system must also be fitted with siameses. These siameses should be conveniently located. Ideally there should be one located only a few feet away from the main entrance to the building for use by the first-in engine company. Additional siameses should be located for convenient use by additional responding engine companies. At no time should contractors ever be allowed to put siameses in the rear of buildings, they *must* be located where it is *convenient for the fire department* to use them.

Testing Standpipe Systems

Testing standpipe systems is a simple process. The testing is divided into three phases: (1) a pretest, (2) a hydrostatic test, and (3) a flow test. Additional specifics of design and testing of standpipe systems are found in NFPA 14.

Pretest The pretest is nothing more than a physical check of the system. You need to make sure every outlet is properly capped and closed off. Each outlet must have a working valve and a handle to operate the valve. Check the threads on the system; it is not unheard of for contractors to install connections with the wrong threads. Look at the method used to tie the pipe to the building; it needs to be secure to prevent the pipe from moving. Finally, while conducting the pretest, check the overall physical condition of the pipe and all components. The system should give the impression that it was installed by professionals with neat welds (where required) and joints that line up.

Hydrostatic Test Once the pretest is completed, the system is filled with water and pressurized for 2 hours. The minimum test pressure is 200 psi, or 50 psi above the minimum design pressure if the minimum pressure is over 150 psi. While the system is under pressure, walk the system searching for leaks—not just at valves and obvious points, but at joints and along long runs of pipe.

The contractor should be able to provide a pump for the hydrostatic test. Once the system is pressurized, the pump is shut off and should be disconnected to prevent the contractor from adding pressure if the system is leaking. After 2 hours, if the system has not lost any pressure, it passes.

Before beginning timing of the hydrostatic test make sure *all* air is exhausted from the system. To do this it is necessary to flow water from the topmost outlet of each riser. Once you are certain the air is exhausted, the test time begins.

Flow Test The flow test is conducted after the system has passed the hydrostatic test. The flow test determines if the system is capable of flowing the required gpm. Remember, the most remote outlet must flow 500 gpm and each additional riser must flow 250 gpm. There is no time period on the flow test. Once you can verify the flow, the test is over.

Verifying the flow is a relatively simple process. Use a smooth-bore tip and take a reading with a pitot gauge. By now you should be adept at calculating the flow when the nozzle pressure is known. To make things easier at the test site, calculate in advance the minimum pressure needed to get the required flow with the size tip used. As long as you meet or exceed the calculated nozzle pressure, you will have the required minimum flow. This procedure must be done at each riser while all risers are flowing water.

Example 13-1 Calculate the nozzle pressure necessary to get a flow of 500 gpm from a 1⅜-inch tip. Assume a C of .97.

ANSWER Use the formula $P = (gpm/29.77 \times D^2 \times C)^2$

$$P = (gpm/29.77 \times D^2 \times C)^2$$
$$= (500/29.77 \times 1\tfrac{3}{8}^2 \times .97)^2$$
$$= (500/29.77 \times 1.89 \times .97)^2$$
$$= (500/54.58)^2$$
$$= 9.16^2$$
$$= 83.9 \text{ or } 84 \text{ psi}$$

Verifying the residual pressure is a bit more complicated. One method is to put an in-line gauge on the outlet with an in-line valve after the gauge. With water flowing, if the in-line gauge does not read 100 psi, the valve can be closed down until a pressure of 100 psi is read on the gauge. With a minimum of 100 psi on the gauge there must be sufficient pressure at the nozzle, measured with a pitot gauge, to verify a 500-gpm flow.

Another way to ensure the minimum residual pressure is to simply use enough hose and properly sized smooth-bore tip to ensure a minimum pressure of 100 psi at the outlet. For example: 100 feet of 2½-inch hose has a FL 100 of 50 psi when flowing 500 gpm. Take half the 50 psi, using only 50 feet of hose, and add it to the 84 psi required nozzle pressure from Example 13-1 and you now have a required pressure of 109 psi. With this setup as long as you have a pitot reading of 84 psi at the tip, you are assured of having both the minimum residual pressure and the minimum flow.

If there is more than one riser, you only have to verify the residual pressure at the most remote outlet. Remember that pressure throughout should be equal in a properly designed system. Caution: Make sure additional outlets are flowing at least 250 gpm, but no more than 250 gpm. The valve on the outlet can be closed off a bit to restrict the flow if necessary. If additional outlets are allowed to flow more water than the minimum required, they will reduce pressure available to the system by creating excessive friction loss.

Special notes on testing standpipe systems: When flow testing automatic and semiautomatic systems, the flow test should include a test of the entire system, including the dry pipe or deluge valves. When testing a Class 2 system, the flow and residual pressure requirements should be reduced appropriately.

Finally, a safety note: When doing flow tests from the roofs of buildings, nozzles should either be anchored or some sort of diffusion device used. Opening and closing nozzles creates a reaction that has the potential to knock a firefighter off balance and off a building if too close to the edge.

SPRINKLER SYSTEMS

Installation of Sprinkler Systems

Installation of Sprinkler Systems in Residential Occupancies up to and Including Four Stories in Height

Installation of Sprinkler Systems in One and Two Family Dwellings and Manufactured Homes

In the entire history of mankind, the single most important fire protection invention has been the sprinkler system. When used in a residential application, it has the ability to cut loss of life to approximately one-third the rate of unprotected homes. And as an added advantage, it reduces property damage and can significantly reduce insurance rates.

Useful applications for sprinkler systems are virtually unlimited. Systems can be designed to protect the largest buildings imaginable or a single apartment. Transformer vaults and piers can also be protected by sprinkler systems.

Installation and testing of sprinkler systems is governed by one of the following three NFPA standards:

NFPA 13, *Installation of Sprinkler Systems*

NFPA 13 R, *Installation of Sprinkler Systems in Residential Occupancies up to and Including Four Stories in Height*

NFPA 13 D, *Installation of Sprinkler Systems in One and Two Family Dwellings and Manufactured Homes*

Types of Systems

There are four types of sprinkler systems. Specific elements of each type system can be altered to some extent, but they always fall into one of the following types.

Wet-Pipe System A wet pipe sprinkler system is designed to have water, under pressure, in it at all times, which allows water to flow immediately when the sprinkler head opens. It is the simplest type system to design and maintain. The wet-pipe system is the most common system and should be used whenever possible. The only disadvantage of the wet-pipe system is that it must be kept at a temperature of at least 40° F to prevent freezing.

The wet-pipe system is designed with a wet pipe valve between the incoming water supply and the system. The primary purpose of the valve is to prevent back flow when the system operates. In some instances the valve also activates the flow alarm.

Dry-Pipe System The dry-pipe system is found in environments subject to freezing conditions. Normally the dry-pipe system has compressed air in the piping. When a sprinkler head goes off, the air escapes and water enters the system. The dry-pipe valve is designed to cause the air in the sprinkler system to hold back the water in the supply system. The dry pipe valve is often referred to as a differential valve, because it is designed so a small amount of air pressure can hold back a larger amount of water pressure. Thus, a differential in pressure exists. A dry pipe valve can require as little as 1 psi of air pressure on the system side for every 5 or 6 psi of water pressure on the supply side.

The dry-pipe system has several disadvantages. Because the system is used where it can freeze, the sprinkler valve must be placed in a heated room. Additionally, because compressed air is required in the system, a means to maintain the air pressure is needed. Finally, because the system is full of air, water will not flow immediately. However, NFPA 13 requires that a dry system be designed so that water will flow from the most remote outlet within 60 seconds.

Preaction System The preaction sprinkler system is not a single system, but three variations of the same system. In general, the preaction system is very similar to the dry-pipe system. The primary difference is that a preaction valve replaces the dry-pipe valve. A preaction valve is activated by a remote fire sensor and is independent of the operation of the sprinkler heads. Another difference is that the air in the preaction system may or may not be pressurized.

In the first type of preaction system, when the fire detection system detects a fire, it automatically trips the preaction valve, allowing water to flow into the system. When the sprinkler head(s) closest to the fire activate, water is already well on the way, reducing the time it takes to get water to the fire.

The second mutation of the preaction system maintains pressurized air in the system. This variation requires that the fire detection system trip the preaction valve, *and* the sprinkler head(s) must operate before water is admitted to the system. For example, if the preaction valve is tripped by the fire detection system before the sprinkler head(s) activate, water will not flow.

The final mutation of the preaction system is most similar to the second. There is pressurized air in the system, but in this system water begins to flow when *either* the preaction valve is tripped or a sprinkler head operates.

Deluge System A deluge system operates most similarly to the first variation of the preaction system. A fire detection system senses the fire and activates the deluge valve, allowing water to flow. All sprinkler heads on a deluge system are open allowing water to flow from every head in the system.

Sprinkler System Design

Just as in the standpipe system, sprinkler systems can be either pipe schedule or hydraulically calculated. Pipe schedule systems have limited use today. Most of the systems designed today are hydraulically calculated systems. In fact, most systems today are designed, at least in part, with the use of computers. A word of caution is in order here: Do not assume a computer-generated sprinkler system design is correct. The computer only does what its operator tells it to do. Qualified reviewers should review all sprinkler plans before any system is built.

Sprinkler systems are not intended to extinguish a fire if the building is fully involved. They are only intended to hold a fire in check if they are unable to extinguish the fire in its incipient stage. In large buildings, enough water cannot be delivered to supply every sprinkler head if they all activate, so the sprinkler system design is separated into design areas. A **design area** is a portion of a larger building, wherein all sprinklers in that area are expected to operate. This fact is primarily important in determining the required gpm. You only need enough water to supply the sprinkler demand in the largest design area. Design areas are also referred to as zones.

design area
portion of a larger building wherein all sprinklers in that area are expected to operate

Figure 13-1 is a photograph of a typical wet standpipe valve. Note the three system specification plates (A, B, and C) hanging from the system. Each plate specifies the water flow requirement and residual pressure for each of three zones supplied by this one valve. Also, notice the OS&Y (outside stem and yoke) valve. When the stem is protruding, as it is in this picture, the valve is open. If the stem is not showing the valve is closed. This is a good visual reference that the system is open or closed. Finally, note the fire department connection (FDC or siamese) and how it feeds into the system.

Figure 13-2 is a photograph of a combination wet and dry system. The dry pipe valve only feeds one zone of the system that protects unheated portions of the building. The remainder of the system, to the left of the dry pipe valve, serves three zones and is a wet system. Note the air compressor to maintain the air pressure on the dry side of the system.

The basic philosophy in calculating the number of heads for any given design area is to achieve a specified water delivery density. The **density** is the amount of

density
amount of water delivered per square foot of surface or floor area

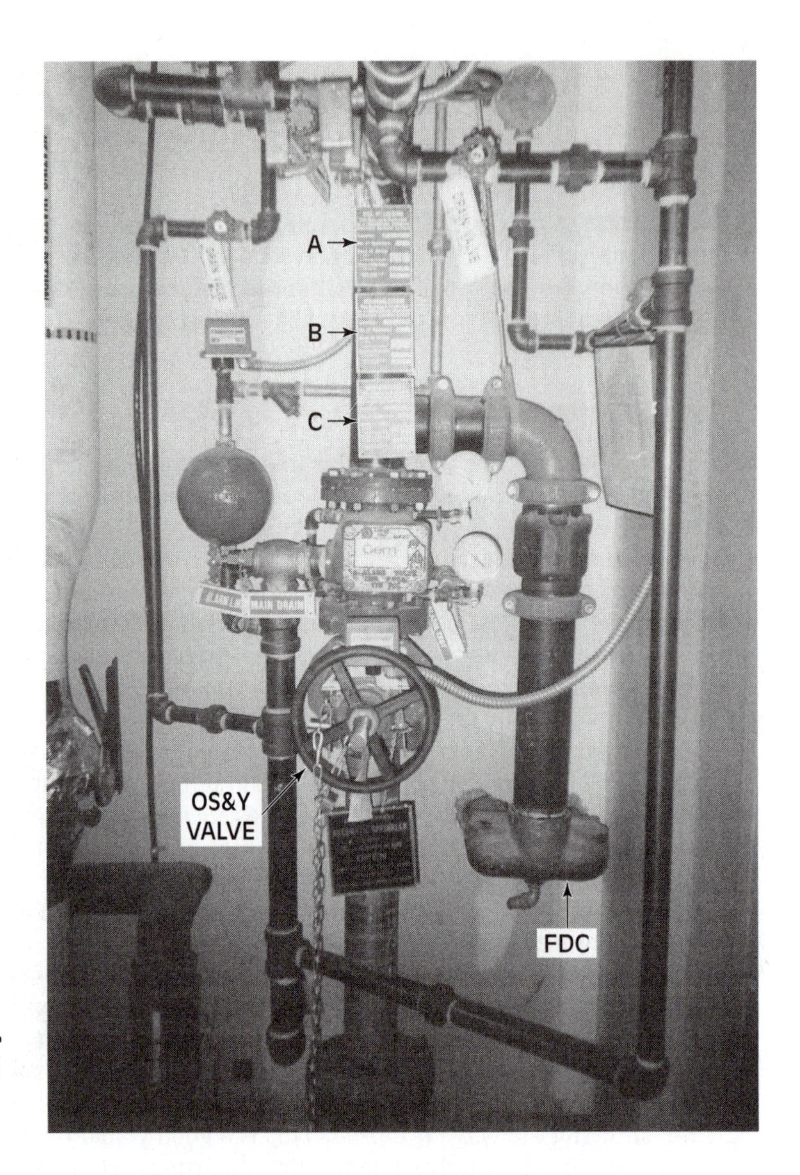

Figure 13-1 *A typical wet sprinkler system.*

water delivered per square foot of surface or floor area. The required density is dependent on the type of fuel, fuel geometry, distance the sprinkler head is from the fuel, fire load, and heat release of the fuel. These factors are summarized into three basic **hazard levels:** light, ordinary, and extra hazard.

hazard levels
light, ordinary, and extra hazard

To achieve a specified density depends on the heads used and the spacing between heads. The same sprinkler heads can be used to achieve different densities simply by varying the spacing between them. When heads are spaced closer

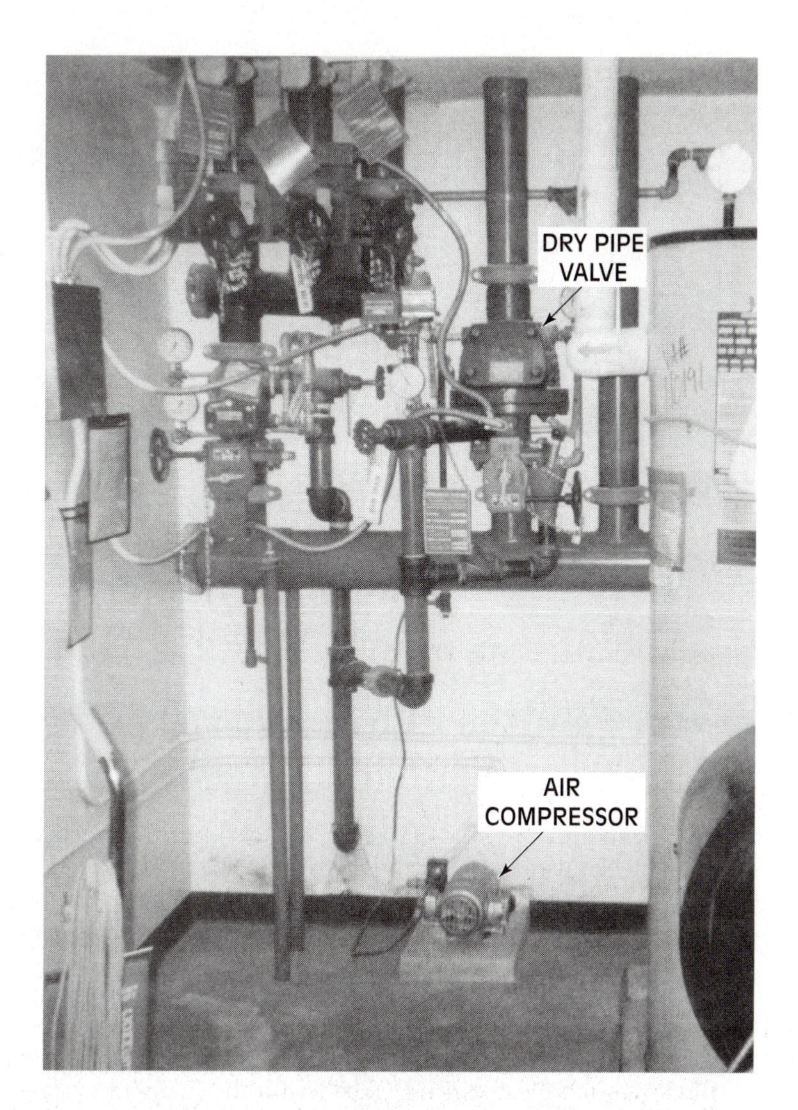

Figure 13-2 *A split sprinkler system; left side is wet and the right side is dry.*

together, a higher density is achieved, and when heads are spaced further apart, a lower density is obtained.

Sprinkler Heads

The heart of any sprinkler system is the sprinkler head itself. Sprinkler heads come in one of three versions (see Figure 13-3): upright, pendant, and sidewall. There are sprinkler heads intended for special application, but they are just special applications of one of these three versions. Also notice in Figure 13-3 that each

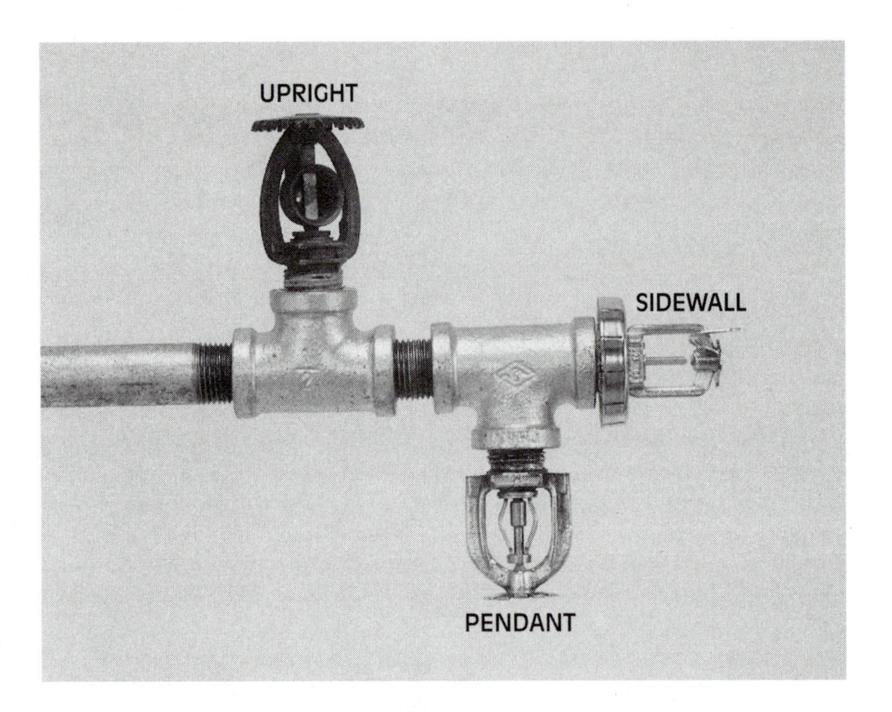

Figure 13-3 *Types of sprinkler heads.*

head has a different kind of fusible link, or operating mechanism. The upright head has a mechanism held together by a eutectic metal, or solder sensitive to a specific temperature. The pendant head has a mechanism that has a temperature-sensitive chemical pellet in the mechanism. Finally, the sidewall head has a glass bulb with a temperature-sensitive liquid.

The operation of a sprinkler head is very basic. It is simply a frame that holds a deflector, (upright, pendant, or sidewall) in place. To prevent water from flowing when not needed the fusible link holds a cap or plunger in place. Figure 13-4 illustrates a typical sprinkler head with the fusible link removed. The plunger is held in place by the fusible link. The gasket prevents water from leaking prior to operation. When the heat of a fire operates the fusible link, water pressure forces the cap or plunger out and allows water to flow.

Testing Sprinkler Systems

Testing of sprinkler systems is divided into three different parts, just as with standpipe systems: pretest, pressure test, and flow test. Since details of testing of sprinkler systems can be specific, it is recommended that the appropriate NFPA standard be consulted.

During the following tests it is not necessary to have sprinkler heads in place. The outlets for the heads can be capped if the heads will prevent proper

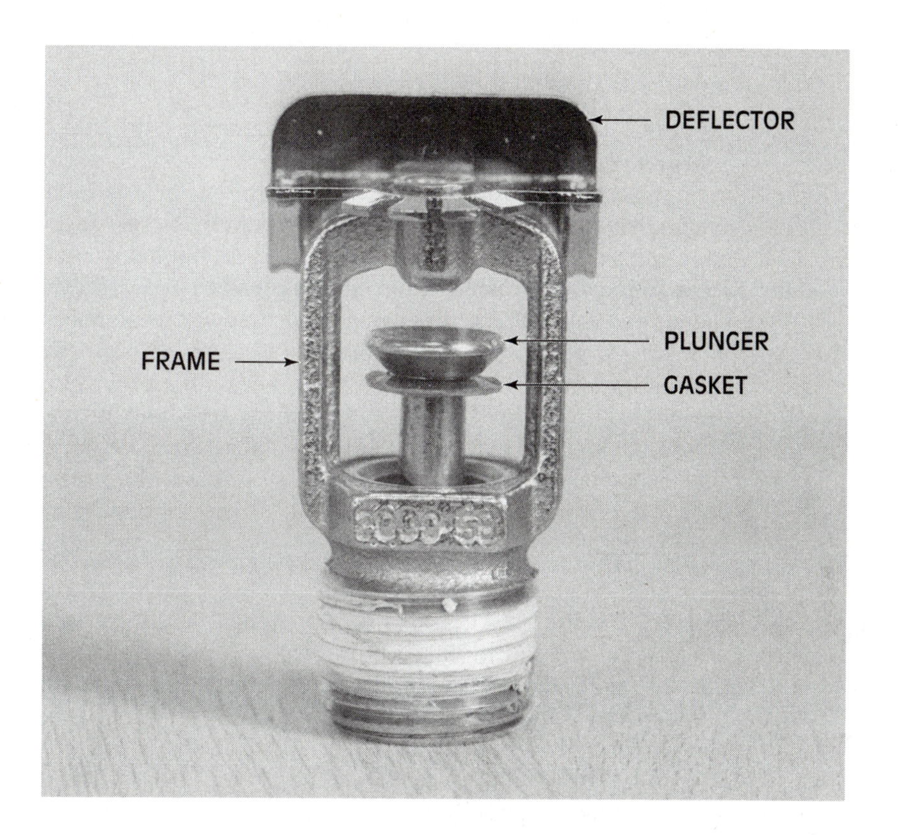

Figure 13-4 *Parts of a sprinkler head.*

installation of a ceiling or wall later. For obvious reasons, deluge systems cannot have heads in place to conduct a hydrostatic test. However, when the system is made of plastic pipe, all heads *must* be in place before conducting any test, because the plastic pipe may not be able to take the stress of putting on caps, removing the caps, and then putting on the sprinkler heads.

Pretest Before testing a sprinkler system, a walk-through inspection should be made of the entire system. During this visual part of the inspection the inspector should be looking for obvious defects, making sure the proper hangers were used and verifying that *all* piping is visible. It is absolutely necessary that the inspector can see every inch of pipe to make sure it is not leaking during the hydrostatic test.

Pressure Test Once the pretest has been completed the entire sprinkler system must be put under a hydrostatic test for two hours. After the pressure has been achieved, the pump is disconnected as in the standpipe test. The pressure during the test is 200 psi or 50 psi over the maximum operating pressure of the system if the maximum pressure is over 150 psi. A word of caution is in order here. Sprinkler systems that are put together using steel pipe and joined using rubber gaskets and split compression fittings at joints can cause a problem. These rubber gaskets can actually seal leaks at high pressure. If the test pressure in these systems is too high, low-pressure leaks will not be evident.

During the 2 hours of the test, the inspector must inspect every inch of the system, checking and rechecking for leaks. At the end of 2 hours if no leaks have been found, the system can be certified as having passed the hydrostatic test. Sprinkler systems installed in accordance with NFPA 13 D are only required to be tested for 15 minutes at the pressure provided by the local water supply.

Dry sprinkler systems require an additional pressure test. Since these systems are required to hold back the flow of water with air pressure, it is necessary to test the system for air leaks. To do so requires a 24-hour pressure test at 40 psi of air pressure. Once the test is begun, the system cannot be pumped up for the remainder of the test. At the end of the 24-hour period, the pressure in the system cannot have dropped more than 1.5 psi. If a drop of more than 1.5 psi is observed, the system must be repaired and retested before being placed in service.

inspector's test connection simulates water flowing from the most remote head in the system

Note

When conducting water flow tests on a dry system, water must flow from the most remote outlet within 60 seconds.

Flow Test The flow test of a sprinkler system actually consists of two tests. The first test is conducted at the inspector's test connection. The **inspector's test connection** simulates water flowing from the most remote head in the system. When the inspector's test connection is opened you are looking for two functions. First, you want to make sure water flows. At this point there is no need to measure the flow, you just want to verify that water will flow when the test connection is opened. When conducting water flow tests on a dry system, water must flow from the most remote outlet within 60 seconds. Second, you want to determine that the water flow alarm is working. The water flow alarm should take no less than 30 seconds (to prevent water surges from causing alarms), or longer than 60 seconds to activate.

Main Drain Test The final test of the hydraulics of the sprinkler system involves verifying that the system has adequate water. Verification is done with a main drain test which is part of the flow test. The main drain is located at the sprinkler system valve on the supply side. The proper procedure for testing is to first record the static pressure, then open the main drain valve fully and record the residual pressure. The pressure drop should be in line with the anticipated pressure drop as specified on the system specification plate, located at the sprinkler valve riser.

Water Supply

In general, water supply sources are the same as for standpipe systems. Again, it is worth repeating that siameses should be located for ease of use by the fire department, not the convenience of the contractor.

The hazard level of the occupancy governs the amount of water required by a sprinkler system. The desired density as determined by the hazard level and number of sprinkler heads in the design area determines the gpm required. The sprinkler design engineer should supply this information.

Determining the amount of water flowing from a sprinkler head is not a difficult task for anyone who has gotten this far in this book. The basic formula for calculating the gpm of a sprinkler head was first given in Chapter 5. Recall from Chapter 5 the formula for calculating gpm is gpm = $29.77 \times D^2 \times C \times \sqrt{P}$. You can calculate the gpm from a sprinkler head simply by inserting the diameter of the opening and the *C* factor for the head. Since this might have to be done over and over again with heads of the exact same type for a particular design area, it makes sense to shorten the formula. Instead of having to multiply $29.77 \times D^2 \times C$ for all the heads in a design area due to minor pressure variations, these factors are multiplied together to create a *K* factor for a particular sprinkler head. The formula for calculating gpm from a sprinkler head then becomes $gpm = K \times \sqrt{P}$.

$gpm = K \times \sqrt{P}$

Example 13-2 What is the flow from a sprinkler head at 7 psi if it has a *K* factor of 5.3?

ANSWER

$$\begin{aligned} GPM &= K \times \sqrt{P} \\ &= 5.3 \times \sqrt{7} \\ &= 5.3 \times 2.646 \\ &= 14.02 \text{ or } 14 \text{ gpm} \end{aligned}$$

With the formula, gpm = $K \times \sqrt{P}$, calculating the gpm for each individual head requires minimal effort.

INSPECTION OF WATER-BASED SUPPRESSION SYSTEMS

Inspections and testing of water-based suppression systems should be conducted on a regular basis. These systems should be inspected and tested in conjunction with their appropriate NFPA standard, that is NFPA 13, NFPA 14, and with NFPA 25, *Inspection, Testing, and Maintenance of Water-Based Fire Protection Systems.*

NFPA 25

Inspection, Testing, and Maintenance of Water-Based Fire Protection Systems

When inspecting sprinkler systems, special attention must be given to the condition of sprinkler heads. For a sprinkler head to operate, it must be in the same condition it was on the day it was installed. Figure 13-5 shows several defective heads. A has been painted and does not work as intended. Even if the chemical pellet is melted at the correct temperature, the paint will probably prevent the mechanism from releasing. B is a glass bulb with some of the liquid missing. If this head operates at all, it will be at the wrong temperature. Finally, C is also a glass bulb and all the liquid has leaked out, rendering it as effective as a plug on the end of a pipe.

FIREGROUND FORMULAS

By now you understand the principles of hydraulics as they pertain to the fire service. You can use any number of formulas in order to calculated gpm, friction loss,

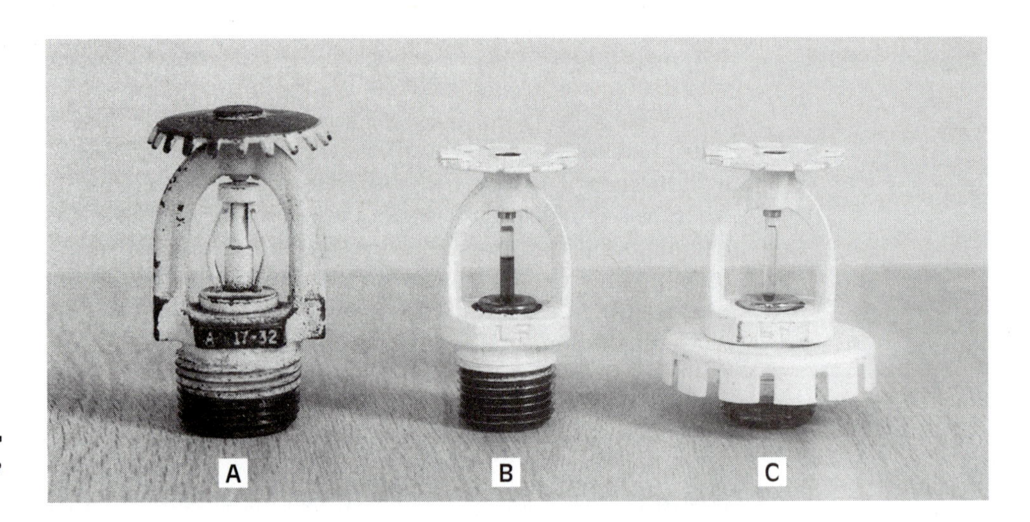

Figure 13-5 *Defective sprinkler heads.*

nozzle reaction, and so forth. But even armed with all these formulas, using them on the fireground is impossible. For this reason it is necessary to develop a means to apply your knowledge at the scene of a fire without getting bogged down with calculations.

There are several methods of remembering the correct pressure and friction loss for various hose lines and nozzles. One such method relies on knowing how many fingers you have on a hand and assigning numbers to them that represent gpm. These rules of thumb or hand method of calculating friction loss should be avoided. They are not sufficiently accurate, and by now accuracy should be ingrained in your understanding of hydraulics.

The only acceptable method of simplifying fireground calculations should involve a two-step approach. First, personnel who drive and operate pumping apparatus should be required to learn the gpm each nozzle will deliver. Along with that they should also know the *FL* 100 and nozzle pressure associated with the correct gpm for all appliances carried. On the fireground it should only take a moment to mentally recalculate the correct engine pressure if a tip is changed or if the length of a line is altered.

The second step is to provide a friction loss chart, accessible to the operator, for quick reference. The friction loss chart can be custom designed to reflect your department's hose size, flow requirements for different nozzles, and special appliance recommendations. The biggest advantage of the friction loss chart is that friction loss in supply lines can be calculated at any flow. It takes a little time to calculate the *FL* 100s for a friction loss chart such as in Figure 13-6, but in the long run it is far preferable than relying on some rule of thumb to find the friction loss. Calculations for the sample friction loss chart in Figure 13-6 took less than 1 hour.

APPLIANCE PRESSURE	GPM	HOSE SIZE AND NUMBER 2 1/2 IN.	3 IN.	2–3 IN.
	100	2	1	
	150	5	2	
	200	8	4	
1 IN. TIP @ 50 PSI	205	8	4	
	225	10	4	
CVFSS @ 100 PSI	240	12	5	
1 1/8 IN. TIP @ 50 PSI	260	14	6	
	300	18	8	
1 1/4 IN. TIP @ 50 PSI	330	22	9	
	350	25	10	
	375	29	12	
	400	32	13	4
1 1/4 IN. TIP @ 80 PSI	405	32	13	4
	425	37	15	4
	450	41	17	4
1 3/8 IN. TIP @ 80 PSI	490	48	20	5
	500	50	20	5
	525		22	6
	550		25	6
	575		27	7
1 1/2 IN. TIP @ 80 PSI	585		28	7
	600		29	8
	625		32	8
	650		34	9
	675		37	10
1 5/8 IN. TIP @ 80 PSI	685		38	10
	700		40	10
	725		42	11
	750		45	12
	775		48	12
1 3/4 IN. TIP @ 80 PSI	790		50	13
	800		52	13
	850			15
	900			17
	950			18
	1000			20
	1050			22
	1100			25
	1150			27
	1200			29
	1250			32
	1300			34
	1350			37
	1400			40
	1450			42
	1500			45
	1550			48
	1600			52

SPECIAL APPLIANCES	
STANDPIPE	25
MONITOR NOZZLE	20
SPRINKLER SYSTEM	125
FOAM EDUCTOR	200 PSI AT THE EDUCTOR

1 3/4 IN. NOZZLE	GPM	N.P.	FL 100
CVFSS	150	100 PSI	35 PSI
15/16 TIP	163	36 PSI	42 PSI

Figure 13-6 *Sample friction loss chart.*

In Figure 13-6, the first column is the appliance and its required nozzle pressure. Column two is the gpm for each appliance in column 1, and at various increments of flow. The next three columns are the *FL* 100 for a single 2½-inch hose, a single 3-inch hose, and two three-inch hose lines. Allowances for special appliances are also included as well as gpm, nozzle pressure, and *FL* 100 for various 1¾-inch appliances.

This friction loss chart can be protected in a piece of plastic and taped inside a cabinet convenient to the pump panel, or even silk screened onto a piece of aluminum and attached to the apparatus.

Elevation

In Chapters 10 and 11, whenever elevation was calculated we used the formula $P = .433 \times H$. On the fireground, calculating elevation to this extent is impossible. A good standard practice is to calculate elevation on the fireground at ½ pound per foot of elevation. If pumping up hill, or down hill, to another pumper or a nozzle, just estimate the elevation and use half of it as the elevation allowance. In Problem 4 of the Application Activities in Chapter 10, the practice of allowing ½ pound of pressure for every foot of extension of an aerial device was already established as an acceptable practice.

Calculating Available Water on the Fireground

One of the most valuable tools a fire chief can have on the fireground is knowledge of how much water is available at the time of the fire. If the department has been well organized, it has already made flow tests and has labeled each fire hydrant with its expected flow. While this is very useful, it is not a perfect solution. The primary weakness with flow testing hydrants is that you only test one hydrant at a time. Subsequently, the water that is calculated as being available in that hydrant is dependent on other hydrants not being used. At a major fire, using only one hydrant is a virtual impossibility. Now we need to develop a means of determining how much water is available while the fire is burning.

In order to determine the amount of water available at a hydrant during a fire requires that a pumper be attached to the hydrant. (Here is another argument in favor of a two-pump operation.) The pump operator should first connect his soft sleeve to the hydrant and then charge the hydrant. After the hydrant has been charged, the operator must note the static pressure. Knowing the static pressure is critical in determining the amount of water left in the hydrant.

When it becomes necessary to estimate the amount of water left in the hydrant, the current residual pressure must be noted. The amount of water left in the hydrant is determined by the difference between the static pressure observed when the hydrant was first charged and the current residual pressure. Additionally, the gpm flow of the pumper must be known. The percentage drop in pressure, from static to residual, corresponds to the amount of water available at that hydrant. A drop of up to 10 percent indicates that the hydrant can supply three times more water than it is already supplying. A pressure drop of more than 10 percent but not more than 15 percent means the hydrant can supply two times more water than it is already flowing. Finally, if the pressure drop is more than 15 percent but not more than 25 percent, the hydrant is capable of supplying one time the amount of water it is already supplying. Table 13-2 summarizes these figures.

Example 13-3 You are operating a pumper that has hooked up to a hydrant near the fire. Your static pressure was 70 psi and your residual pressure is 55 psi. You are supplying a ladder pipe that is flowing 500 gpm. Assuming the capacity of the pumper is not an issue, how much more water can you supply?

ANSWER $70 - 55 = 15$

Since 7 is 10% of 70 and you have a 15 psi drop, your total drop is just over 20 percent. This puts you in the over 15 percent but not over 25 percent category; you can supply another 500 gpm.

Two final points about this method of determining hydrant capacity. First, this method is only an approximation. Unfortunately, there is no way to determine precisely how much water is available on the fireground in real time. Second, to be even close, the residual pressure must be taken at the time you want to know the available water remaining. As additional pumpers hook up to the water main, they take water and pressure away from you. For example, when you first hook up to the hydrant and are the only pumper flowing water, you may have only a 10 percent drop while supplying 500 gpm. After two or three other pumpers hook up to hydrants and begin flowing water, you may discover you now have a 25 percent drop in pressure. If the water supply officer were to ask you how much water you had left and you based your calculation on your initial residual pressure, you would have mistakenly told him you had 1,500 gpm more. In reality you only have approximately another 500 gpm left.

Table 13-2 *Calculating hydrant capacity.*

Pressure drop	Hydrant Capacity
Up to 10%	3 × current flow
Over 10% but not over 15%	2 × current flow
Over 15% but not over 25%	1 × current flow

High-Rise Operations

High-rise firefighting can be the most challenging of all firefighting operations. The high-rise fire can challenge the endurance of both firefighters and equipment. Our concern is how it will challenge the ability to pump to hose lines on upper floors.

Typically high-rise buildings can require excessively high pressures. The maximum output of a typical pumper is 300 psi. In high-rise operations it is easy to require pump pressures this high. For instance, a 150-foot, 2½-inch line flowing 322 gpm from a 1¼-inch tip and operating on the thirtieth floor requires an engine pressure of 261 psi. Two points to remember here: First, the test pressure of the standpipe system will be reached or exceeded, and second, the test pressure of the hose used to supply the standpipe will be exceeded. In short, regardless of your best efforts, it may be necessary to exceed maximum pressures and test your equipment, as well as the firefighters, to the limit.

APPLICATION ACTIVITIES

Problem 1 Your fire chief is concerned that the allowance of 25 psi for standpipes in your city is excessive. He points out that the tallest building is only 6 stories and has only two risers. You respond that you have just finished reading the book, *Hydraulics for Firefighting*, and can calculate the actual friction loss for a minimum flow in the most demanding system in your city. You can then determine from that if an allowance of 25 psi is excessive. The system you calculate is made of schedule 10 iron pipe with a *C* of 100 and *D* for 4-inch pipe is 4.26 and *D* of 6.357 for 6-inch pipe. The system has 4-inch 90-degree elbows that are equivalent to 10 feet; 6-inch 90-degree elbows worth 14 feet; and a 6-inch tee worth 30 feet (see Figure 13-7).

ANSWER Be sure to account for the correct flow in various portions of the system. Include allowances for fittings that change the direction of the water flow.

From the siamese to point A, the system is flowing 750 gpm. The tee has been included in this part of the calculation.

$$Pt = (4.52) \times (Q^{1.85}) \times (L)/(C^{1.85}) \times (D^{4.87})$$
$$= (4.52) \times (750^{1.85}) \times (141^*)/(100^{1.85}) \times (6.357^{4.87})$$
$$= (4.52) \times (208{,}383.5) \times (141)/(5{,}011.9) \times (8{,}162.78)$$
$$= 132{,}806{,}972.2/40{,}911{,}037.1$$
$$= 3.25 \text{ psi}$$

*(83 feet of pipe, 2 90-degree elbows equivalent to 28 feet, 1 tee equivalent to 30 feet.)

From point A to point B the system is flowing 500 gpm. By using the pipe schedule chart in NFPA 14 you will find that no more than 100 feet of

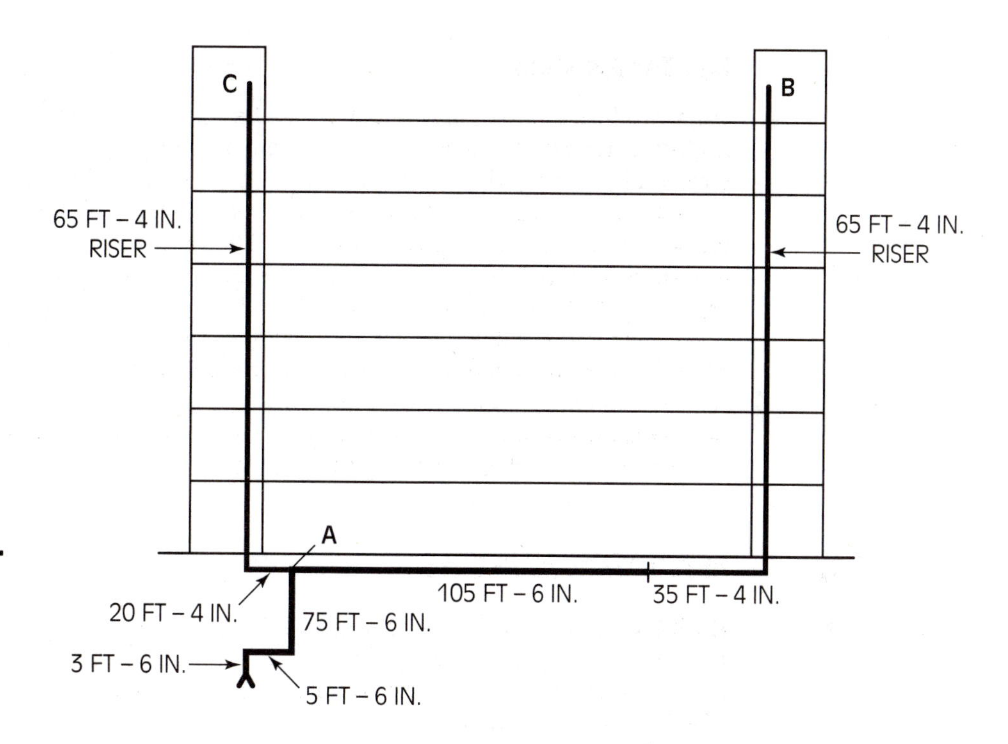

Figure 13-7
Calculate the special appliance allowance for this system.

4-inch pipe can be used when the flow is 500 gpm. Beyond that, 6-inch pipe must be used. This means that of the 205 feet of pipe from point A to point B, 100 feet is 4-inch and 105 feet is 6-inch. First, calculate for 500 gpm through 105 feet of 6 inch pipe.

$$Pt = (4.52) \times (Q^{1.85}) \times (L)/(C^{1.85}) \times (D^{4.87})$$
$$= (4.52) \times (500^{1.85}) \times (105)/(100^{1.85}) \times (6.357^{4.87})$$
$$= (4.52) \times (98{,}422.5) \times (105)/(5{,}011.9) \times (8{,}162.78)$$
$$= 46{,}711{,}318.5/40{,}911{,}037.08$$
$$= 1.14 \text{ psi}$$

Now calculate friction loss for the remainder of the pipe from point A to point B.

$$Pt = (4.52) \times (Q^{1.85}) \times (L)/(C^{1.85}) \times (D^{4.87})$$
$$= (4.52) \times (500^{1.85}) \times (110^{*})/(100^{1.85}) \times (4.26^{4.87})$$
$$= (4.52) \times (98{,}422.5) \times (110)/(5{,}011.9) \times (1{,}162.05)$$
$$= 48{,}935{,}667/5{,}824{,}078.4$$
$$= 8.4 \text{ psi}$$

*(Includes 10 feet for the elbow.)

We have a total friction loss in the system of 3.25, plus 1.14, plus 8.4 or 12.79 psi. There is no need to calculate the friction loss in the system from point A to point C because it is flowing half the water as the same size pipe to the most distant outlet. Based on this calculation you recommend to the fire chief that the department adopt an allowance of 15 psi for standpipe systems.

Problem 2 What pressure would be required on a 1-inch tip to obtain a flow of 250 gpm?

ANSWER Use the formula $P = (gpm/29.77 \times D^2 \times C)^2$ and assume a C of .97.

$$P = (gpm/29.77 \times D^2 \times C)^2$$
$$= (250/29.77 \times 1^2 \times .97)^2$$
$$= (250/28.88)^2$$
$$= 8.66^2$$
$$= 75 \text{ psi}$$

Problem 3 How many gpm will flow from a sprinkler head with a K factor of 5.6 and operating at 10.33 psi?

ANSWER Use the formula $gpm = K \times \sqrt{P}$.

$$gpm = K \times \sqrt{P}$$
$$= 5.6 \times \sqrt{10.33}$$
$$= 5.6 \times 3.21$$
$$= 17.98 \text{ or } 18$$

Summary

Standpipe and sprinkler systems are two of the most useful tools in the arsenal of the fire service today. It should be the goal of all fire officers to know as much about these systems as possible. Only when we fully understand something can we use it to its greatest potential.

Now that you have studied this book and know the basic principles of hydraulics, the principles of pumps and drafting, and several pages worth of formulas, you should also appreciate the need for accuracy. This means that rules of thumb and hand methods should be discouraged. Instead a well thought out and properly calculated friction loss chart is just as easy to use and much more accurate.

Review Questions

1. Define a standpipe system.
2. What NFPA standard specifically addresses standpipe systems?
3. How many classes of standpipe system are there?
4. Is it possible for a standpipe system to be maintained without water but operate automatically when needed?
5. What is the difference between a pipe schedule and a hydraulically designed system, either standpipe or sprinkler?
6. What is the maximum flow a standpipe system is designed for?
7. How many types of sprinkler systems are there?
8. What is the design area when referring to a sprinkler system?
9. Are all sprinkler systems required to have a 24-hour air pressure test?
10. What is the purpose of flowing water from the inspector's test connection?
11. What NFPA standard addresses inspection, testing, and maintenance of water-based fire protection systems?
12. Why should rules of thumb be avoided?
13. If you are operating a pumper connected to a hydrant, what do you need to know in order to estimate how much more water the hydrant can supply?
14. You are the pump operator pumping through 100 feet of 3-inch hose into a standpipe system of a high-rise building. Your company is operating with 150 feet of 2½-inch hose and an automatic nozzle flowing 225 gpm on the thirtieth floor. After operating with 225 gpm for a short time, the officer notifies you that a second company has arrived and is operating with the same hose and nozzle configuration. Since the company that has just joined up with your officer is from the same engine that is your water supply company, you will have to supply both lines.
    A. How can you do this without changing your engine pressure?
    B. What is the minimum capacity pump needed to do this?

List of Formulas

gpm from a sprinkler head: $gpm = K \times \sqrt{P}$

References

NFPA 13, *Installation of Sprinkler Systems.* Quincy, MA: National Fire Protection Association, 1996.

NFPA 13D, *Installation of Sprinkler Systems in One and Two Family Dwellings and Manufactured Homes.* Quincy, MA: National Fire Protection Association, 1996.

NFPA 13R, *Installation of Sprinkler Systems in Residential Occupancies up to and Including Four Stories in Height.* Quincy, MA: National Fire Protection Association, 1996.

NFPA 14, *Installation of Standpipe and Hose Systems.* Quincy, MA: National Fire Protection Association, 1996.

NFPA 25, *Inspection, Testing, and Maintenance of Water-Based Fire Protection Systems.* Quincy, MA: National Fire Protection Association, 1998.

Formulas for Calculating Area and Volume for Selected Shapes

Circle

Area = $.7854 \times D^2$

Cylinder

Vol. = *Area of end* $\times H$

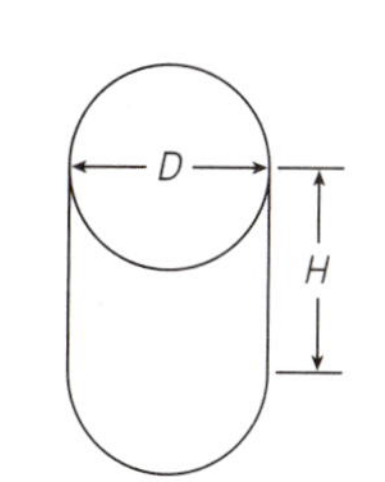

Ellipse

Area = $3.14 \times A \times B$

Elliptical Tank

Vol. = *Area of end* $\times H$

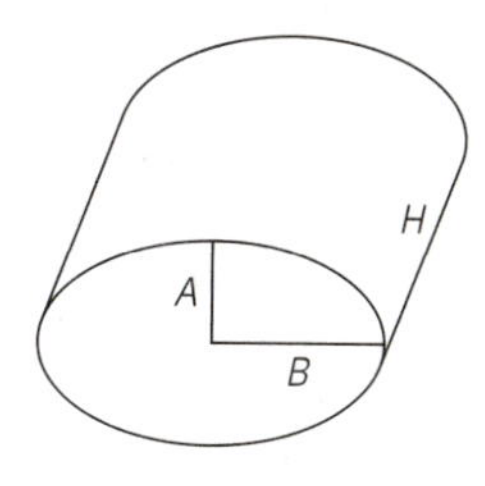

Parallelogram

Area = $W \times L$

Vol. = *Area of end* $\times H$

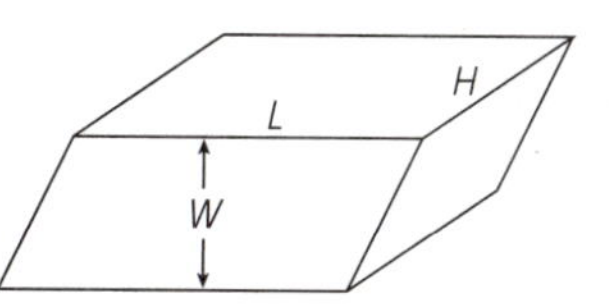

Square or Rectangle

Area = $L \times W$

Vol. = *Area of end* $\times H$

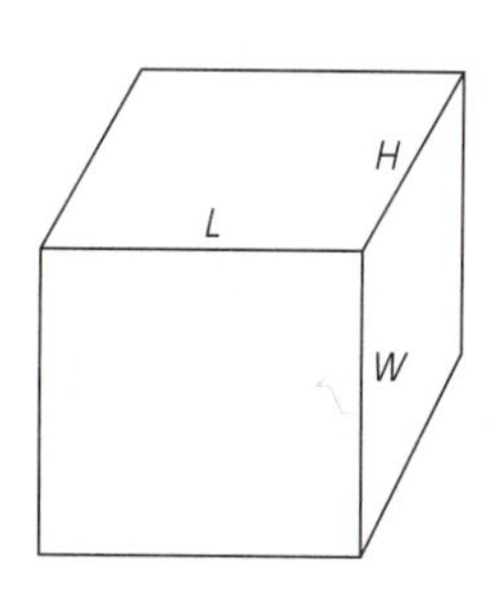

Trapezoid

$Area = W \frac{L_1 + L_2}{2}$

$Vol. = Area\ of\ end \times H$

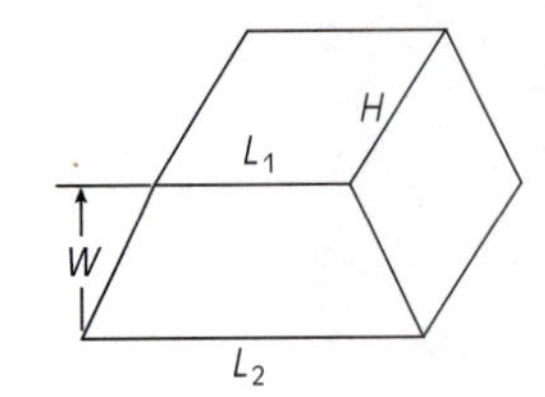

Sphere

$Area = 12.56 \times r^2$

$Vol. = 4.188 \times r^3$

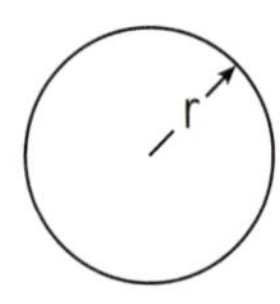

Triangle

$Area = W \times \frac{H}{2}$

$Vol. = Area\ of\ end \times L$

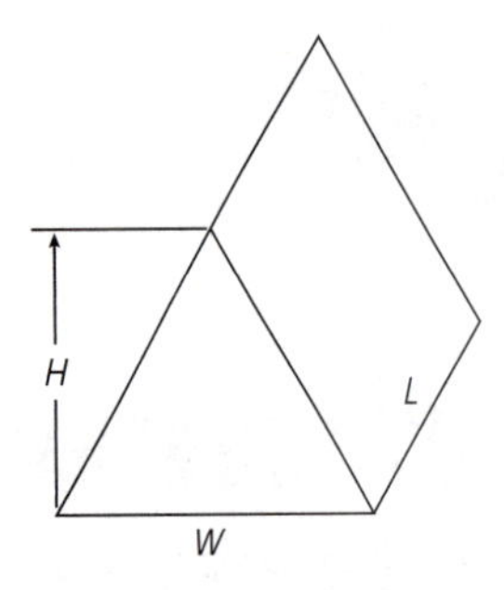

Capacity in gallons = *Vol.*/231 when *Vol.* is in cubic inches

Capacity in gallons = 7.48 x *Vol.* when *Vol.* is in cubic feet

Formulas

CHAPTER 1 INTRODUCTION TO HYDRAULICS

Finding specific heat:

$$\text{Specific heat} = \text{weight of water} \times \text{change in temperature}$$

Calculating latent heat of a given weight of water:

$$\text{Latent heat} = \text{weight of water} \times 970.3$$

Finding the amount of water needed to absorb the BTUs being given off:

$$\text{Gallons of water} = \text{BTUs given off} \div 9{,}346$$

Fire load formula:

$$FLD = A \times fld$$

Finding how much water is needed to fill a room with steam:

$$\text{Cubic feet of water} = V \div 1{,}600$$

Calculating how much water is in a container:

$$\text{Gallons} = V \times 7.48$$

Finding how many BTUs will be absorbed:

$$\text{BTUs absorbed} = 9{,}346 \times \text{gallons}$$

Volume of all rectangular boxes:

$$V = L \times W \times H$$

Finding gallons when volume is in cubic inches:

$$\text{Gallons} = V \div 231$$

Volume of a cylinder:

$$V = .7854 \times D^2 \times H \text{ (or } L\text{)}$$

Volume of a cylinder in gallons:

$$V = 7.48 \times .7854 \times D^2 \times H \text{ (or } L\text{)}$$

Direct calculation of gallons of water in a cylinder:

$$V = 5.87 \times D^2 \times H \text{ (or } L\text{)}$$

Weight of water in a container:

$$W = V \times 8.34$$

Direct calculation of weight of water in a cylinder:

$$W = 49 \times D^2 \times H \text{ (or } L\text{)}$$

Gallons of water needed to fill a room with steam:

$$\text{Gallons} = V \div 214$$

CHAPTER 2 FORCE AND PRESSURE

Finding pressure when force and area are known:

$$P = \frac{F}{A}$$

Finding pressure when the height (elevation) of water is known:

$$P = .433 \times H$$

Finding the height of water when the pressure it creates is known:

$$H = 2.31 \times P$$

Finding force when the height of the water and the area it acts on are known:

$$F = .433 \times H \times A$$

Finding force when the pressure and area are known:

$$F = P \times A$$

Finding area from diameter:

$$A = .7854 \times D^2$$

(This formula is more convenient than the conventional $A = \Pi \times r^2$ because we universally use diameter in the fire service. Either one will work, so use what you feel comfortable with.)

Finding absolute pressure:

$$\text{Absolute pressure} = \text{relative pressure} + \text{atmospheric pressure}$$

CHAPTER 3 BERNOULLI'S THEOREM

Conservation of energy:

$$\text{Total } PE = \text{Total } KE$$

Bernoulli's equation:

$$PH_1 + EH_1 + VH_1 = PH_2 + EH_2 + VH_2$$

Torricelli's equation:

$$Ve = \sqrt{2g(y_2 - y_1)}$$

Finding velocity when the height is known; Torricelli's equation simplified:

$$Ve = 8\sqrt{H}$$

Revised Bernoulli's equation:

$$\frac{P_1}{.433} + Z_1 + \frac{Ve_1^2}{64} = \frac{P_2}{.433} + Z_2 + \frac{Ve_2^2}{64}$$

CHAPTER 4 VELOCITY AND FLOW

Finding velocity when pressure is known:

$$Ve = \sqrt{2 \times g \times 2.31 \times P}$$

Finding velocity when pressure is known, simplified:

$$Ve = 12.16\sqrt{P}$$

Finding height when velocity is known:

$$H = \left(\frac{Ve}{8}\right)^2$$

Finding pressure when velocity is known:

$$P = \left(\frac{Ve}{12.16}\right)^2$$

Comparison of area and velocity in two sizes of hose for same flow:

$$A_1 \times Ve_1 = A_2 \times Ve_2$$

Finding the diameter of a circle when area is known:

$$D = \sqrt{\frac{A}{.7854}}$$

Basic quantity flow formula:

$$Q = A \times Ve$$

Finding flow from circular opening:

$$Q = .7854 \times D^2 \times Ve$$

Finding flow from circular opening when diameter is in inches:

$$Q = .7854 \times D^2 \times \frac{1}{144} \times Ve$$

Finding area of opening when quantity of flow and velocity are known:

$$A = \frac{Q}{Ve}$$

Finding diameter in inches when area is in square feet:

$$D = \sqrt{\frac{A}{.7854} \times 144}$$

Finding diameter in inches when quantity and velocity are known:

$$D = \sqrt{\frac{\frac{Q}{Ve}}{.7854} \times 144}$$

Finding velocity when quantity and area are known:

$$Ve = \frac{Q}{A}$$

Finding velocity when quantity and diameter in inches are known:

$$Ve = \frac{Q}{\left(.7854 \times D^2 \times \frac{1}{144}\right)}$$

Calculating flow in 1 minute:

$$Q = .7854 \times D^2 \times \frac{1}{144} \times Ve \times 60$$

CHAPTER 5 GPM

Calculating gpm when diameter and velocity are known:

$$gpm = 2.448 \times D^2 \times Ve$$

Calculating gpm when diameter and pressure are known:

$$gpm = 2.448 \times D^2 \times 12.16 \times \sqrt{P}$$

Basic gpm formula:

$$gpm = .7854 \times D^2 \times \frac{1}{144} \times 12.16 \times \sqrt{P} \times 60 \times 7.48$$

Freeman's formula:

$$gpm = 29.77 \times D^2 \times \sqrt{P}$$

AIA formula:

$$gpm = 29.83 \times D^2 \times C \times \sqrt{P}$$

Freeman's formula with coefficient of discharge:

$$gpm = 29.77 \times D^2 \times C \times \sqrt{P}$$

Finding diameter when gpm and pressure are known:

$$D = \sqrt{\frac{gpm}{29.77 \times C \times \sqrt{P}}}$$

Finding pressure when gpm and diameter are known:

$$P = \left(\frac{gpm}{29.77 \times D^2 \times C}\right)^2$$

Finding velocity when gpm and diameter are known:

$$Ve = \frac{gpm}{2.448 \times D^2}$$

CHAPTER 6 FRICTION LOSS

Finding the friction loss multiplier:

$$Fm = \left(\frac{Ve_2}{Ve_1}\right)^2$$

For finding the conversion factor:

$$Cf = \frac{D_1^5}{D_2^5}$$

Friction loss per 100 feet of 2½-inch hose:

$$FL\ 100 = 2Q^2$$

Friction loss per 100 feet of hose other than 2½-inch:

$$FL\ 100 = Cf \times 2Q^2$$

Finding conversion factor when friction loss and quantity of flow are known:

$$Cf = \frac{FL\ 100}{2Q^2}$$

Equivalent length formula:

$$L_2 = L_1 \times \left(\frac{Cf_1}{Cf_2}\right)$$

Equivalent friction loss formula:

$$FL_2\ 100 = FL_1\ 100 \times \left(\frac{Cf_2}{Cf_1}\right)$$

Finding quantity flow when friction loss and conversion factor are known:

$$Q = \sqrt{\frac{FL\ 100}{(Cf \times 2)}}$$

CHAPTER 7 PUMP THEORY

Finding net pump pressure:

Discharge pressure – intake pressure = pressure generated by the pump

CHAPTER 8 THEORY OF DRAFTING

Finding pressure from inches of mercury:

$$P = .489 \times Hg$$

Finding height of water from inches of mercury:

$$H = 1.13 \times Hg$$

Reverse lift calculation:

$$Hg = .882 \times H$$

Finding remaining pressure when height is known:

Remaining pressure = atmospheric pressure – (.433 × H)

Or

Finding remaining pressure when inches of mercury are known:

Remaining pressure = atmospheric pressure – (.489 × Hg)

Finding friction loss on intake side of pump at draft:

FL = (dynamic reading – static reading) × .489

CHAPTER 9 FIRE STREAMS

Horizontal range of streams:

$$HR = \frac{1}{2}\,NP + 26^*$$

Vertical range of streams:

$$VR = \frac{5}{8}\,NP + 26^*$$

Nozzle reaction:

$$NR = 1.5 \times D^2 \times P$$

Nozzle reaction when gpm and pressure are known:

$$NR = .0504 \times gpm \times \sqrt{P}$$

---

*Add 5 for each ⅛ inch of nozzle diameter more than ¾ inch.

CHAPTER 10 CALCULATING ENGINE PRESSURE

Finding total friction loss:

$$FL = FL\,100 \times L$$

Engine pressure formula:

$$EP = NP + FL\,1 + FL\,2 \pm E + SA$$

Relay formula:

$$EP = 20 + FL \pm E$$

CHAPTER 11 ADVANCED PROBLEMS IN HYDRAULICS

For finding the actual nozzle pressure (INP) when an incorrect engine pressure is used:

$$\frac{CNP}{CEP} = \frac{INP}{IEP}$$

CHAPTER 12 WATER SUPPLY

Calculating a community's maximum daily consumption:

$$MDC = \frac{Pop. \times 214.5}{1440}$$

For calculating a community's needed storage capacity:

$$SC = (MDC + NFF) \times T$$

For calculating the minimum flow rate:

$$MFr = MDC + NFF$$

Hazen-Williams formula:

$$P_f = \frac{4.52 \times Q^{1.85}}{C^{1.85} \times D^{4.87}}$$

Total pressure loss for system:

$$P_t = P_f \times L$$

Direct calculation and total pressure loss:

$$P_t = \frac{4.52 \times Q^{1.85} \times L}{C^{1.85} \times D^{4.87}}$$

Hydrant capacity formula:

$$Q_2 = Q_1 \times \left(\frac{p_s - p_{r2}}{p_s - p_{r1}} \right)^{.54}$$

CHAPTER 13 STANDPIPES, SPRINKLERS, AND FIREGROUND FORMULAS

gpm from a sprinkler head:

$$gpm = K \times \sqrt{P}$$

Acronyms

ADC	average daily consumption
AHJ	authority having jurisdiction
AIA	American Insurance Association
AWWA	American Water Works Association
BTU	British thermal unit
CAFS	compressed air foam systems
CVFSS	combination variable fog and straight stream
FDC	fire department connection
ISO	Insurance Services Office
LDH	large diameter hose
MDC	maximum daily consumption
NBFU	National Board of Fire Underwriters
NFF	needed fire flow
NFPA	National Fire Protection Association
OS&Y	outside stem and yoke

Glossary

Absolute pressure The pressure indicated when 14.7 psi = atmospheric pressure.

Actual length concept Stipulates that hose connections be located so that 100 feet of hose and a 30-foot stream are able to reach any area of the building.

AIA formula Formula for determining flow from hydrants.

Automatic nozzle Delivers water at any gpm within its design range and maintains a nozzle pressure of 100 psi.

Axial flow pump The flow of liquid is in line with the axis of the impeller.

Back pressure The pressure exerted by a column of water against the discharge of a pump.

Barometric pressure The current atmospheric pressure.

Bell mouthing Excess wear on the clearance rings due to lack of hydraulic balance.

Bernoulli, Daniel (1700–1782) Swiss mathematician and physicist.

Bernoulli's theory Where the velocity of a fluid is high, the pressure is low, and where the velocity is low, the pressure is high.

British thermal unit (BTU) The amount of heat needed to raise the temperature of 1 pound of water by 1°F at 60°F.

BTU *See* British thermal unit.

Cavitation The formation of vapor (steam) bubbles in the impeller of the pump.

Changing over The process of shutting off water from the on-board tank, opening the intake, and admitting water from the charged supply line.

Conversion factor The factor for calculating friction loss for hose other than 2½-inch.

Coriolis effect Natural tendency of fluids to rotate due to the rotation of the Earth.

CVFSS nozzle Nozzle capable of producing any pattern from a straight stream to 100-degree fog.

Density The amount of water delivered per square foot of surface and floor area.

Design area A portion of a larger building, wherein all sprinklers in that area are expected to operate.

Discharge coefficient *C* factor.

Distribution main The part of the supply system that has the hookup to each individual home, apartment, school, business, or factory.

Drafting Removing water from a static source by the influence of the atmosphere.

Dynamic intake reading Reading on the intake gauge when water is flowing.

Eccentric Off center.

Eddying Water running contrary to the direction of the main flow.

Elevation The need to compensate for pressure, either added or subtracted, due to the vertical position of the nozzle in reference to the pump.

Elevation head The elevation or height of water, in feet.

End thrust Produced when the direction of flow of liquid is abruptly changed.

Energy The ability to do work.

Engine pressure formula
$EP = NP + FL\,1 + FL\,2 \pm E + SA$.

Eutectic metal Solder sensitive to a specific temperature.

Exit location concept Allows hose connections to be located in exit stairwells, horizontal exits, and exit passageways.

Feeder mains Carry water from the point of storage to the secondary feeders.

Fire load The amount of combustibles per room or structure.

Flow pressure The pressure measured by means of a pitot gauge as the water leaves the nozzle or other opening.

Force Pressure multiplied by the area it is exerted against.

Friction The resistance to movement of two surfaces in contact.

Friction loss The conversion of useful energy into nonuseful energy due to friction.

Gate back Partially closing off the discharge until the correct pressure is obtained.

Gate up Partially opening the discharge until the correct pressure is obtained.

Gross pump pressure Total pressure discharged by the pump.

Hazard levels Light, ordinary, and extra hazard.

Hazen-Williams formula Used to determine pressure (friction) loss in pipe.

Hydrant valve Allows later responding engines to hook up to the steamer connection of the hydrant and boost pressure or water supply to the pumper at the fire.

Hydraulic design System pipe sizes are calculated to provide the minimum design flow and pressure.

Hydraulics The science of water (or other fluids), at rest and in motion.

Hydrodynamics The study of water in motion.

Hydrostatics The study of water at rest.

Impeller The part of the nonpositive displacement pump that imparts energy to the liquid.

Inspector's test connection Simulates water flowing from the most remote sprinkler head in the system.

Kinetic energy Energy of motion.

Laminar flow The flow of the water is smooth and orderly, with layers, or cores, of water effortlessly gliding over the next layer of water.

Latent heat The amount of heat absorbed by a substance when it changes state.

Latent heat of fusion The amount of BTUs absorbed when a pound of water goes from its solid state to its liquid state.

Latent heat of vaporization The amount of BTUs absorbed when a pound of water goes from its liquid state to its vapor state.

MDC Maximum daily consumption, or the amount of water the system demands, per minute, at its peak demand time.

Net force A partial force.

Net pump pressure Pressure generated by the pump.

Newton's first law of motion Every body continues in its state of rest or uniform speed in a straight line unless acted upon by a nonzero force.

Newton's third law of motion Whenever one object exerts a force on a second object, the second exerts an equal and opposite force on the first.

NFF An estimation of the amount of water, in gpm, needed to extinguish a fire in a specific building.

Nonpositive displacement pump The volume of liquid discharged is dependent on the resistance offered to the movement of the liquid.

Nozzle pressure The pressure required at the nozzle to allow the nozzle to deliver the designed gpm and pattern.

Nozzle reaction The opposing force created by water exiting the nozzle.

Pascal, Blaise (1623–1662) French philosopher and scientist.

Pascal's principle If pressure is applied to a confined liquid, that pressure is transmitted to every point within the liquid without reduction in intensity.

Piezometric plane The equivalent elevation head (H) of water at the point the force is measured.

Pipe schedule Size of the pipe is determined by a flow versus pipe size schedule.

Positive displacement pump Discharges a *fixed quantity* of fluid with each cycle of the pump.

Potential energy Energy of position or stored energy.

Pressure The force per unit of area.

Priming the pump Reducing the pressure inside the pump, allowing atmospheric pressure to fill the pump with water.

Radial flow pump Liquid is discharged out from the center of the impeller to its circumference.

Radial hydraulic balance The equal discharge of liquid around the circumference of the impeller.

Rated capacity of a pump The maximum amount of water a pump can deliver at 150 psi discharge and 10 feet of lift.

Relative pressure The pressure indicated with 0 psi = atmospheric pressure.

Relay Any situation in which one pumper supplies water to another pumper.

Relay formula $EP = 20 + FL \pm E$.

Residual pressure The pressure remaining when water is flowing.

Secondary feeder mains Supply water from the feeder main to the distribution mains.

Sine A trigonometric function of an angle that when multiplied by the length of the hypotenuse of a right triangle will find the length of the side opposite the angle.

Slippage The tendency of water on the discharge side of the pump to slip back to the intake side.

Smooth-bore nozzle Only capable of producing a solid stream of water.

Specific heat The amount of heat, in BTUs, needed to raise the temperature of 1 pound of a substance 1°F.

Standpipe system A water distribution system designed to get water for firefighting purposes, from point A to point B.

Static intake reading The reading on the intake gauge before water is flowing.

Static pressure The pressure of a fluid at rest.

Steamer connection Large discharge of hydrants.

Tangent A line that touches but does not intersect a circle.

Torricelli, Evangelista (1608–1647) Italian physicist and mathematician.

Torricelli's theorem The velocity of water escaping from an opening below the surface of a container of water will have the same velocity, minus exit losses, as if it were to fall the same distance.

Transfer valve The mechanism that transfers the pump from parallel configuration to series configuration.

Turbulent flow Water flows in a disorganized, random manner.

Vacuum A relative pressure below the surrounding atmospheric pressure.

Vectors The measure of strength and direction of forces.

Velocity pressure The measure of the kinetic energy of flowing water.

Vena contracta The point of maximum stream contraction.

Venturi tube A restriction in a conduit intended to increase velocity with a corresponding reduction in pressure.

Viscosity The resistance to flow of a liquid.

Water A chemical compound made up of two parts of elemental hydrogen and one part of elemental oxygen.

Work done by the pump A combination of net pump pressure and how much liquid is being discharged.

Bibliography

Brady, James E., and Gerald E. Humiston. *General Chemistry: Principles and Structure.* New York: John Wiley & Sons, 1975.

Brock, Pat D. *Fire Protection Hydraulics and Water Supply Analysis.* Stillwater, OK: Fire Protection Publications, Oklahoma State University, 1990.

Bryan, John L. *Automatic Sprinkler and Standpipe Systems*, 3rd ed. Quincy, MA: National Fire Protection Association, 1997.

Burrell, Brian. *Merriam-Webster's Guide to Everyday Math.* Springfield, MA: Merriam-Webster, 1998.

Cote, Arthur E., ed. *Fire Protection Handbook*, 18th ed. Quincy, MA: National Fire Protection Association, 1997.

Fire Suppression Rating Schedule. New York: Insurance Services Offices, 1980.

Gagnon, Robert M. *Design of Water-Based Fire Protection Systems.* Albany, NY: Delmar Publishers, 1997.

Giancoli, Douglas C. *Physics: Principles with Applications*, 5th ed. Englewood Cliffs, NJ: Prentice Hall, 1998.

Hickey, Harry E. *Fire Suppression Rating Schedule Handbook.* U.S.A.: Professional Loss Control Foundation, 1993.

Hickey, Harry E. *Hydraulics for Fire Protection.* Boston: National Fire Protection Association, 1980.

International Fire Service Training Association. *Water Supplies for Fire Protection.* Stillwater, OK: Fire Protection Publications, Oklahoma State University, 1988.

Purington, Robert G. *Fire-Fighting Hydraulics.* U.S.A.: McGraw-Hill, 1974.

Rouse, Hunter. *Elementary Mechanics of Fluid.* New York: Dover Publications, 1946.

Shepherd, Fred. *Fire Service Hydraulics.* New York: Case-Shepherd-Mann Publishing Corporation, 1941.

Sturtevant, Thomas B. *Introduction to Fire Pump Operations.* Albany, NY: Delmar Publishers, 1997.

Index